TECHNOLOGY
AT WORK

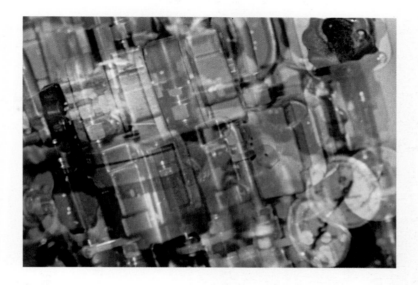

Above: a robot produces reed
relays. Machinery increasingly
replaces human labor.

Endpapers: an electrostatic
generator built by the self-
educated British scientist
Francis Hauksbee in 1706.

Facts On File New York

TECHNOLOGY AT WORK

ANTHONY FELDMAN
AND BILL GUNSTON

Editorial Coordinator: John Mason
Art Editor: John Fitzmaurice
Designer: Gill Mouqué
Editors: Paul Hutchinson, Damian Grint
Research: Frances Vargo

Printed and bound in Hong Kong by
Leefung-Asco.

Library of Congress Cataloging in Publication Data

Feldman, Anthony.
Technology at Work.
Includes index.
1. Technology — popular works
2. Materials — popular works
I. Gunston, Bill, joint author
II. Title.   600   79-20201
ISBN 0-87196-413-9

Above: this was the first
helicopter to fly. It was built
by Paul Cornu in 1907.

Title pages: an oil refinery at
night. Oil has growing political
and economic significance.

INTRODUCTION

Technology today is a highly complicated business, largely taken for granted, and frequently ignored until it becomes controversial. The aim of this book is to provide a clear but thorough survey of the numerous ways in which that technology has shaped and is reshaping the world we live in. Chapters describe the recent advances and possible future developments in world transportation on land, at sea, and in the air. Other chapters show how the natural phenomena of light and sound have provided the basis for astounding technical advances, from laser power to ultrasonics. A complete chapter is devoted to the search for alternatives to the Earth's dwindling resources of energy, with accounts of the latest developments in tapping the power of the sun, the wind, the sea, and even of space. New man-made materials, military and space technology, advances in the field of communication and control – all these are clearly and fully described. Technological advance often changes long established ways of life, and this means that such progress is often opposed by well-meaning people. For this reason the moral questions of the uses and abuses of technology are fully discussed in the final chapter.

CONTENTS

CHAPTER 1

RESHAPING OUR WORLD

Unlike most other animals, which are unable to control their environments and must fit themselves into the natural conditions as they find them, man is largely able to dominate his surroundings. The tool that provides him with this mastery is what is known as technology. From the moment the first man began chipping at a flint the story of man's use of technology has been a constant search for ways to improve the quality of his life. In fact, in cosmic terms, it is only a few moments since advanced human technology meant chipping at a flint to fashion an arrowhead. The complex achievements of modern technology provide the subject matter for this book. What of the future? The very speed of recent developments makes it impossible to predict the technology of a century to come. All we can say is that fears about machines taking over the vital human functions of management, control, and decision-making, or acquiring intelligence, are likely to be as unfounded as they are now. For ultimately, technology is only as good or bad as the use to which it is put or the amount of thought given to its applications.

Opposite: the world's largest solar power station at Odeillo in the French Pyrenees. As reserves of fossil fuels dwindle, the effective exploitation of alternative energy sources becomes more urgent. Solar power is one of these.

A Brief History

When, some 2,000,000 years ago, men began to fashion hand tools for specific purposes, they were starting on a technological road that was to lead to an ever greater mastery of their environment. The next big advance, the control and use of fire, came much later, about 50,000 years ago. Agriculture is even more recent. The sowing of crops and the domestication of animals was not widespread until a mere 10,000 years ago, though sometime before this tools had been invented for cutting grasses and grain, and for grinding corn. Technologies based on metals appear to date from about 8000 years ago, and by 5000 BC civilized societies were advancing rapidly.

By this time, too, skilled smiths and artisans were working with copper, silver, gold, ivory, and glass, and two of the most basic of man's tools, writing and arithmetic, were in common use. In Sumeria the wheel was

Below: the invention of the printing machine revolutionized people's ability to communicate. From primitive shops like this grew a massive industry.

soon in use, though it remained unknown in ancient Egypt during most of the pyramid-building period. By 2500 BC large ships were able to navigate out of sight of land, and animals were being used to pull plows or carry loads and people. Basic inventions such as the horseshoe, efficient harness, and heavy plow did not mature until after 1000 AD, together with a rudimentary understanding of metallurgy (so that iron could

ABRAHAM VON WERDT.

Above: this detail from the Standard of Ur shows the wheel being put to warlike use in Sumeria. Most land travel would still be seriously handicapped without wheels, and the smooth roads they need to run on.

Above: a print showing Newcomen's steam engine raising water from a tin mine in the mid-18th century. Later refinements of the steam engine provided the motive power that drove the industrial revolution in Britain.

be turned into flawless steel), and the generation of power by waterwheels and windmills, as well as the treadmill.

After the fall of the Roman Empire the center of progress shifted to the East where the Arabs, Chinese, and many other races made advances in chemistry, mathematics, astronomy, and engineering. Printing with moveable type had been in use in China since the 14th century, and by the middle of the 15th century the first printing presses were appearing in Europe. The invention of printing encouraged literacy and brought about widespread discussion that challenged centuries-old ideas. It was the Polish astronomer Copernicus who realized that the Earth was not the hub of the Universe but actually orbited the Sun. He delayed publishing his theories until 1543, when he was on his death-bed, for fear of offending religious and political leaders of the time. In a way Copernicus symbolized man's awakening into a new era of discovery.

By the mid 1700s Europe was on the threshold of the first Industrial Revolution, with the mechanization of spinning and weaving, and the invention of new techniques and new tools that revolutionized agriculture. Long before 1800 the steam engine had, to quote James Watt's monument, "multiplied the power of men." It dramatically changed society, polarizing it around the factory, coal mine, iron foundry, and steam-hauled railroad. This Industrial Revolution spurred unprecedented increase in property and population, a move from the countryside to the swelling towns, and a life based on goods bought in shops for money, and increasingly "mass produced." By 1910 the American engineer Henry Ford and a handful of other

visionary engineers had perfected a mass-production system in which not only did the price per component fall sharply but increased accuracy ensured that all components were interchangeable. This meant that an automobile, for instance, could swiftly be assembled without having to hand-pick parts to fit together, and also that a defective part could later be replaced by a "spare" in the knowledge that that would fit equally well.

By 1930 most of industry was mechanized, and mechanical aids were common in the home. By 1950 automation in its simplest forms was beginning to eliminate the need for humans to feed repetitive commands into machines, and today we are on the brink of an age that will be dominated by the silicon integrated microcircuit, which on a microscopic scale of size can offer a myriad of new possibilities. Modern man also enjoys revolutions in materials, the conquest of disease, and countless other assets, but there is another side to technological progress: the refinements of weapons that can kill tens of thousands and destroy vast areas of our planet, or, even without war, the pollution of air, earth, and water with the waste products of the same technological skill that can give future generations peace, prosperity, and material satisfaction.

Technology for a Better Environment

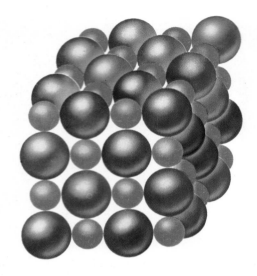

From the gradual perfection of new materials, tools, and processes to the development of machines that range from the very simple to those of amazing complexity, the story of the achievements of technology has been one of constant improvement in the material benefits and the prolonging of human life.

In almost every field of endeavor, increasing knowledge has merely taught man how little he really knows. In the science of materials, early man was delighted with his stone and refined metals, which served his purposes well. It was only the incessant demands of modern engineers that forced the development of ever-stronger materials. Now, by focusing attention on crystalline lattices and atomic structures, the modern researcher has discovered how woefully imperfect are most of our present materials – especially metals, which consist of immensely strong crystals loosely thrown together; when loaded, they simply split apart at the weak joints. But through

Below: the astonishing pace of technological advance has forced mankind to search for new materials to suit his purposes. This rocket motor case is made of epoxy resin reinforced with 40 miles of boron filament.

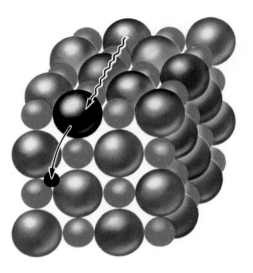

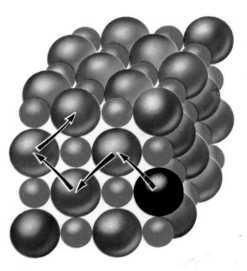

Above: how light affects cubic crystal lattice of silver bromide. Positively charged silver ions (shown red). Negative bromide ions are shown blue. **Center:** when a quantum of light is absorbed by a bromide ion, an electron (black) is ejected, leaving an uncharged bromide atom (gray) in lattice. **Top:** this "positive hole" has moved through crystal lattice by a series of *electron jumps* between adjacent bromide ions.

Above: the workhorse of the United States' Moon missions, the rocket Saturn V. The object of the space rocket is to provide a big enough thrust to escape the Earth's atmosphere. Weight has to be kept to a minimum and lightweight alloys were developed to achieve this.

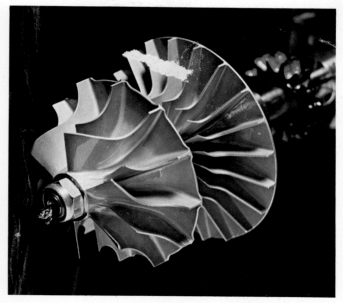

Above: a gas turbine compressor from the Rolls-Royce Dart turboprop used in some modern airliners. Improved cooling systems allow the turbines to run in hotter gas without themselves becoming as hot as earlier engines. Thus the operating efficiency of the airliner is considerably increased.

knowledge of this fact lies the route to achievement. Today we are in possession of single crystals of various materials, such as sapphire, dozens of times stronger yet more elastic than the finest steel. We have gone into mass-production with silky fibers of graphite and boron that can be bonded with resins and other adhesives into composite materials much stronger than steel yet weighing about one third as much. Hundreds of laboratories are trying to make useful engineering structures out of perfect single crystals. Every day the results are better. Soon it will be possible to slash the weight of engineering structures, or span great distances, by, for instance, building a graceful bridge from England to France.

Of course, the technology of single crystals, with all their atoms lined up in absolutely perfect rows to make a three-dimensional lattice, also lies at the heart of the new technology of semiconductors and microelectronics. Here the objective is not to build engineering structures but to make small (often submicro-

scopic) pieces of material within which we can play with individual atoms. Without straining the perfect crystal lattice we can replace some atoms by others of different elements, changing the electrical properties of the material so that electrons (individual negative charges) can flow in a controlled manner. Our ability to operate at this level of what is an incredibly small size is quite recent – almost all of it dates from much later than the invention of the transistor in 1948 – and for the first time it gives man, so often presumptuous and condescending toward nature, at least a partial capability of rivaling nature in the precision of our manipulations. While better materials lead to previously impossible large-scale engineering products – such as the million-ton tanker, with its ability to reduce bulk transport costs – better ability to tinker at the atomic level is opening up wholly new vistas not only in electronics but also with living organisms.

There are almost limitless areas in which today's materials and techniques are superior to those of even the most recent past. Gas turbines and rockets are two among many spurs to the development of metal alloys and other materials that retain their strength even when white hot. New superalloys refined in a vacuum or inert-gas atmosphere far outperform their predecessors. They can be made into finished parts by new techniques, such as directional solidification in which the crystalline structure is aligned along the direction of most severe stress, which again multiplies the performance. And some of the most critical components, such as the turbine rotor blades of gas turbines, can today be made with such intricate internal cooling by high-velocity airflows that, although they run in much hotter gas flows, the actual tempera-

13

ture of the metal is cooler than in older engines, and thus the blade can be subjected to higher mechanical loads.

The list is endless. Today we have whole new families of harder cutting tools and abrasives, and new techniques for shaping hard material rapidly and accurately that our fathers could never have fashioned at all. The diversity of our lubricants, at one time confined to no more than four animal, vegetable, and mineral "oils," has proliferated into totally new regions of chemistry and physics to meet ever more severe demands, so that today it is possible to lubricate surfaces rubbing together with almost any conceivable combination of material, contact pressure, relative velocity, and temperature. Sometimes the choice seems bizarre – who would guess that glass can lubricate white-hot steel being squeezed like toothpaste through a die? – but each means a new capability that we did not possess previously. One extremely important family of special materials are those compatible with our own bodies, able to spend a lifetime in contact with bone, tissue, or blood without mechanical, chemical, or electrochemical incompatibility. Early in this century not one of these materials even existed. Searching for materials compatible with our bodies has often shown us how little we knew about the amazing complexity of living things; a mere century ago we thought of blood as a mere red fluid containing microscopic corpuscles, whereas today we recognize the limitations of our knowledge of not only this amazingly complex fluid but also living cells in general. It is no exaggeration to

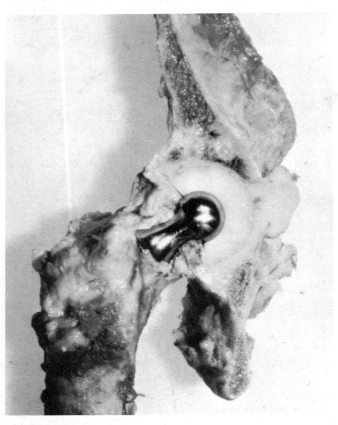

Above: the Starr Edwards heart valve being placed in position. It is made to cause minimum tissue reaction. **Left:** artificial hip joints bring renewed mobility.

claim that in today's world we learn more each year about the cell and how advanced life-forms work than the sum total of our knowledge at the start of this century. This enables us to apply our man-made technology to problems as diverse as repairing the body after disease or injury, helping damaged or otherwise subnormal infants grow into minimally handicapped adults, and doubling our output of crops, meat, or dairy products.

Technology is absolutely crucial to improving the quality of all our lives. Most of us might think the doctor's involvement with technology ceased with his car and stethoscope, and the farmer's with his tractor. In fact technology plays the central role in every route to preventing starvation, fighting disease, and prolonging our active life. In many parts of the world crop

Above: this cabbage field in Lesotho has been sprayed against pests. Fertilisers and insecticides are doing their part in providing food for the world's people.

production from an unchanged area under cultivation has been multiplied 25–28 times by the elimination of harmful practices (some founded on superstition), introduction of improved plants, fertilizers, and other chemicals, and many other changes including proper irrigation and satellite surveillance. The vast increase in output feeds those kept alive by the elimination of dreaded diseases and the provision of pharmaceuticals and medical care where previously none existed. There is no more exciting branch of technology than pharmaceuticals, though nine tenths of it consists of tireless plugging away and testing. It is an area of constant challenge and discovery that is gradually replacing superstition, just as did scientific enquiry itself in the so-called advanced nations in medieval times.

In general, technology is easier when it is concerned with man's artefacts rather than living things, though the engineer can never escape from the need to minimize cost. In countless areas new engineering has not only brought other directly sought improvements but has also dramatically reduced cost. Transportation and telecommunications are among the most visible forms of technology, and in general among the most successful. In the case of the family car the success has been only partial; cars are today found in almost every town on earth, but they still burn petroleum and have been unable to make much headway in carrying greater loads or traveling faster because the owner seldom wants to carry more and the speed is limited by the environment. Railroads, on the other hand, have progressively carried greater loads at higher speeds, and because they have succeeded in this objective with existing track, they have remained viable despite the emergence of alternative tracked systems which would demand immense capital investment. Ships and aircraft, too, have been able to make dramatic progress, and in their case they have been able to increase in size to many times that of their predecessors, and thus reap the great benefits of enormously increased traffic at reduced prices. A fortuitous by-product of the new generation of giant transport aircraft is a new family of gas-turbine engines that at one stroke have slashed the fuel bills (per unit of payload moved) and virtually eliminated noise as a problem associated with aircraft. Retirement of older, noisier aircraft is being spurred by increasingly severe noise legislation.

In parallel with the unprecedented advances in speed and carrying power of transportation – especially by sea and air – safety and reliability have been dramatically improved by the ability of various forms of electronics to pierce fog and darkness, give instant exact read-out of position, warn of other traffic, and enable control and communications to function at all times. Man's ability to navigate has reached the point where he can travel with precision through the depths of the oceans, or at supersonic speed skim the surfaces of jagged mountains in darkness, or voyage to an exact landfall on the Moon or a planet. Tele-

Above: the Post Office Tower stands over 600 feet above Central London. Part of the world's communication system, the Tower is equipped to handle thousands of telephone lines and hundreds of television channels.

communications, another integral part in man's conquest of time and space, is the biggest success story of all in reducing costs by increasing the ability to handle traffic. Communications satellites relay telephone messages, television programs, and radio broadcasts between continents to link up even the most remote places in the world.

15

Problems of Technology

In the industrialized nations, and especially among the more affluent people, it has become fashionable to make technology the whipping-boy for nearly all human ills. Broadly, it is held up as something menacing that pollutes the Earth, dehumanizes people, and reduces them to the status of mindless ants in a jungle of concrete and noise. The truth is, of course, that the machines and instruments of technology are as good or bad as the people who create and manipulate them. The purpose of technology is to offer us fresh options

Above: stress probably affects people today more severely than ever in the past. The incidence of nervous breakdowns as well as mental illness has never been higher. Technology may impose a hostile environment on millions of people.

Above: the all too familiar scene of a congested city street. The automobile is a wasteful vehicle – using a great amount of space and consuming huge quantities of oil to move only four or five people at most.

that we did not previously possess. If something is judged, on balance, to leave us worse off than before then we can simply stop using it. So many people in modern society are awake to the dangers – real or imagined – that on the whole most of the things that survive their scrutiny probably benefit humanity.

But problems there are, in plenty. The most intractable are those associated with man himself, because while technology is based chiefly upon such aspects of the material Universe as atomic structures, electromagnetic radiation, and superficially understood chemical reactions, we still know very little about such things as life, thought, emotion, and stress. Our own behavior is often destructive rather than constructive, and highly changeable. As soon as the basic pressure for survival is removed we become oddly critical. The modern urban environment is said to impose on us such pressures and such dehumanizing influences as to excuse any act of assault on either the environment or its inhabitants. Technology is ill equipped to answer this kind of problem, or the damage we inflict on our lives by drugs, smoking, and alcohol. They are problems that are strangely resistant to our otherwise impressive research and development system.

Indeed, the very success of the modern research and development process has itself caused problems that are likely to become more acute. For millions of years our world changed very slowly, and even man evolved at a pace in keeping with that of the rest of the Earth's creatures. But his ability to solve problems, discover how the Universe works, and invent things to perform useful tasks have accelerated the pace of change many millionfold. Today we change our capabilities and environment at such a rate that we do not even know

put of waste material, the result is frightening indeed. In the single year 1990 we will litter the Earth with several times more waste than in the whole of the 19th century, and most of it will be virtually indestructible. This is part of the familiar problem we lump under the general heading of pollution. Radioactive wastes, for instance, generate violent emotions in spite of assurances by the experts. Sometimes our emotions are hard to explain and appear not to be based upon any reasoned thought. Thus, throughout the world nuclear power stations have been the scene of angry and often violent protest. Yet the safety record of nuclear power is certainly better than that of stations burning fossil fuel. Yet the protests are directed against the use of stations that offer an alternative to fossil fuels – which, incidentally, cause millions of times more pollution than uranium. This is one of many cases where technology has solved a problem but, apparently, the original problem is preferred. Almost certainly, the objection to nuclear power rests upon a lack of sufficient evidence to the contrary as well as

Above: Pitsea Tip, Essex, England. There is great local concern over the fact that chemical companies are dumping up to 50 million tons of toxic waste here annually. Liquid and solid waste comes from all over Britain.

Above: oil spillages have wreaked havoc on wildlife and coastlines around the world. At least 10 million tons of oil is spilt into the seas every year. This crab was a victim of spillage many miles out to sea.

the kind of world in which our children will live in the next century. In New York City a 20-year-old dwelling is judged ready for demolition, but replacement of totally unsuitable cars by vehicles tailored to city use runs contrary to the dictates of fashion.

On the face of it, it is foolish for one human to drive in crowded traffic in a vehicle four times longer and ten times more powerful than necessary. It is a misuse of precious raw material and energy supplies, just as are many of the products of the world packaging industry, which are designed to serve the whims of fashion rather than the needs of the product. The problem becomes even more serious as traditional materials are replaced by plastics, whose solid, liquid, and gaseous waste, in the opinion of a number of experts, pose a threat to man's environment to which we do not yet have answers. If the growing number of humans is multiplied by the growing individual out-

ignorance, just as in the case of fluoridation of water supplies. There is something in the human character, it seems, that welcomes controversy and prefers passion to understanding. Often, it appears to be simpler to join a crusade than to study the alternatives.

The generation of electricity and reduction of dental caries are precisely the kind of problems technology is geared to solve. It would be simple to put the cause of public outcry in this and many other aspects of technology down to a failure in communications; if people were told the full facts, perhaps they would have better understanding? But this overlooks the fact that it is improved mass communications that spreads all manner of superficially plausible doctrines, nearly all of which contain the odd germ of truth, that in fact present a wholly distorted picture. As well as man's social structure, religion, politics, and employment, the media strongly influence our views on technology, and the wish to create "a story" invariably results in a one-sided picture that probably does more harm than leaving the public in ignorance. Sometimes the view is unjustly negative, presenting a new development as an unmitigated menace; but this is no more common than the equally unthinking approach that hails each new development as an unalloyed marvel. Feeding a public with a diet of distorted reporting is itself one of the major problems of modern technology and the cause of immense aggravation and worry. The widespread belief of 1830 that steam railroads would stop cows from giving

to buy a new one, further increasing the rate of consumption of raw materials and adding to our mountain of wastes.

There is just a chance that the power and versatility of microelectronics may to a large extent overcome the problems caused by the soaring price of labor, so that many tasks can be done automatically. Typical of the kind of work that in 50 years has gone from simple to difficult is repairing a hole in the road. There is nothing intrinsically difficult about doing this, but today the price of such a job has become so high that it is increasingly left undone. New York City, and an

Above: demonstrators protest against the use of the airport at Washington DC by the supersonic airliner *Concorde*. They were concerned about the aircraft's noise levels, although it is not much noisier than subsonic airliners. *Concorde's* enormous consumption of fuel would have been a better ground for protest.

milk, and do a thousand other dreadful things, has countless modern equivalents.

This fixation of the media on controversy is part of our difficulty in accepting change, especially rapid change. Those of us who want our way of life to avoid too much disruption should note that a rate of inflation of even five percent per year is, by any possible yardstick, horrific. Inflation determines human capability at least as much as do breakthroughs in technology. Already we have so changed industry that, while it is possible to set up a manufacturing system that will produce astronomic quantities at a very low price, even simple repairs are extremely costly. Increasingly we are driven by inflation into a situation where the cheapest way to repair a faulty product is

increasing number of other centers of today's affluent world, can find no technology able to keep its streets in proper repair, whereas 50 years ago it had an army of laborers earning a few cents an hour. All other labor-intensive tasks, such as collecting refuse or delivering mail, have now become so expensive that the service declines while costs soar. In many countries rural bus and train services have gradually disappeared, while the myriad of small shops have been replaced by central supermarkets. All these things, which are happening in pace with inflation, not only change our society but leave an increasing proportion of the adult population out of work. We have yet to find answers to this particularly deep problem, which – if one extrapolates a few decades ahead – appears to lead to a society with a vast network of computerized machines, a handful of highly paid managers, and an unemployment rate of 95 percent.

In a world of rapid change trends are more important than existing situations. Often they are hard to discern. Workers in the cotton mills of the early days of

Britain's industrial revolution had no reason to doubt that the same mills would hum for many hundreds of years, supplying cloth and finished products for most of the world. Who could predict that the price of labor would at first improve the lot of those workers but then put the mills out of business? Who could predict that a later generation would regard those mills not with pride but distaste? Who 40 years ago could predict that we now regard as equally undesirable the sprawling growth of suburbs with "ribbon development" of endless tentacles of small houses? Who a mere 10 years ago could predict the sudden antipathy toward city office buildings and tall apartment buildings? In a world dominated by costly labor, scarce energy and raw material, and unprecedented inflation we cannot afford to get any major decision wrong. Mills, single-story houses, and skyscrapers are meant to be solid structures that last, not mistaken blots on the landscape.

This focuses on one of the most intractable problems. We cannot at the same time have increasing costs and also rapid change. Yet both costs and the pace of change show signs of being a permanent feature of later-20th century life.

Left: a United States cotton mill photographed in the early 19th century. Textile production is declining in all the developed countries, while the "Third World" fills the demand for cheap clothes.

Below left: a mountain of crushed automobiles in an American car-breaker's yard. In the West it is often cheaper to throw away and buy another, so natural resources are wasted on a huge scale.

Below: in Alabama the world's largest atomic power station caught fire in 1975. Its fail-safe systems were put out of action. There is growing concern about the safety of atomic energy; opposition is growing.

CHAPTER 2

OVERLAND TRAVEL

Transportation is the most visible of man's technologies. It is also among the most controversial. In the "developing" countries people grow up in the self-sufficient village community where they were born, while in the "developed" countries people are continually on the move with wheels and wings, and live in communities which depend upon transportation for their survival. Modern speedy, comfortable, reliable transportation is one of the boons of technological advance. But recent years have witnessed many protests. Unlike those of the previous century who merely resisted change, today's opponents of transportation draw attention to its often catastrophic effect on the environment, on human society, and on the resources of the Earth.

Opposite: a flyover at night on the British highway *M4*. Private ownership of automobiles has vastly increased since World War II, and enormous sums have been spent on roads – both cross-country and within cities.

A Brief History

It is not known when the first wheeled vehicle made its appearance, nor whether it was preceded by the first ice sledge or boat. The first vehicles, it is thought, were various types of boat. The Earth's rivers provided naturally level and usually navigable routes. But on land there was little incentive to build vehicles when there were no proper roads.

In the absence of roads human backs and heads, and pack-animals transported loads. The American Indian devised the travois, or horse litter, constructed from two tent poles and a few strips of leather. It is hard to imagine anything better for carrying heavy loads across trackless country. In steep mountains, as in parts of Bolivia and Peru, beasts of burden like llamas are more useful than the most sophisticated automobile. Should a road be built, the situation is at once reversed. But a road changes the economy of the region. Distances are increased and a new set of equations are introduced into the economy which revolves around petroleum instead of animal feed. Petroleum has become one of the largest transported commodities and it is temporarily distorting the world economy by dividing it into the petroleum exporting and the petroleum importing countries.

Below: a chariot pictured on the Tomb of Nebamun, Thebes 1400 BC. The horse-drawn chariot was effective over flat ground and could reach considerable speeds. It emerged as a fighting vehicle around 2000 BC, and seems to have disappeared about 500 AD.

Above: Sioux Indians moving camp. The travois (a piece of hide stretched between two poles) was towed by a horse and could carry up to 200 pounds. It was a simple and efficient means of transportation on the endless flat plains the Sioux inhabited.

Along with the construction of shelter, transportation is man's oldest technology. Recognizable sledges pulled by dogs or reindeer are believed to have been used in Scandinavia almost 10,000 years ago. But land vehicles proper had to await two further developments before they reached any kind of maturity – the road and the harness. Egyptian paintings of about 1600 BC depict light chariots with modern-looking spoked wheels. This was shortly after the introduction of the horse, and such chariots doubtless could have broken the speed limit of most modern cities. But this overlooks three important factors: they could not carry a useful load; the horse harness was not only inefficient but positively harmed the tractive power and health of the horse; and the vehicle worked only because of the flat, firm sand on which it ran.

Over many centuries, the road and the harness evolved. The Chinese led with the harness. They devised a collar ahead of the animal's shoulders. This enabled it to exert its full tractive force without constriction on its breathing or circulation. The Romans failed to design such harnesses but built marvelous roads. These were noted for their robust all-weather construction, and for their straightness. Throughout what had been the Roman Empire, the roads were generally allowed to fall into disrepair. By late Medieval times the major routes of Europe were deeply rutted rivers of mud. In many places, carts or carriages had to be taken to pieces. For a fee, they were carried by hand over the worst sections. Most countries enacted laws to improve matters. Toll gates were set up to collect money on a scale which varied from peddlers on foot to large wagons. The money was supposed to finance the upkeep of the road.

Wagons were forced to use broad wheels – in Britain, a minimum of 16 inches after 1773 – to reduce rutting. But even in quite small towns the streets were chaotic.

In open country the highways were blocked by droves of sheep, flocks of turkeys, and carriages with broken springs.

In the 18th century the decrepit roads of Europe had to cope with increasing industrialism. Burgeoning demand for coal and heavy raw materials encouraged the development of canals. A water-borne craft like a canal barge, towed by a horse, can move as much payload in a week as a road wagon needing a team of eight horses. Britain led the industrial revolution, and was the first and most intensive builder of canals. Many other countries, including North America, had better roads. These were often originally built for military purposes, and in the most hilly regions canals were uneconomic to build. The canal era was short-lived anyway. In the early 19th century the combination of the steel wheel running on the steel rail (one or the other having a flange), and the steam engine, gave birth to the railroads. These revolutionized not only transportation but society itself. In 1838 Dr Arnold of Rugby said the train meant "the end of feudality." With hindsight it is easy to see what he meant. Today there are no widely traveled landlords and helpless village serfs. Nearly everybody travels, and so do goods. Many people in the prosperous West travel to different countries. Television brings the world to those who stay at home. As "transportation" and "communication" become intermingled, so to some degree does the need for human travel diminish. When people all over the world can instantly not only talk to each other but also see each other, the need to conserve energy may mean that transportation concentrates even more on commodities rather than passengers.

Left: this early 19th century print shows the Rolle Canal as it crosses the aqueduct near Torrington, in Devon. The British canal system was essential to the success of the industrial revolution. The mileage of navigable waterways increased from 1400 miles in 1760 to 4025 miles in 1850.

Below: Britain's economy was transformed by the building of the railroads. People could, for the first time, travel miles from their homes and be back within a day. Goods and services could easily reach their markets within a matter of hours. Though many rural lines have closed the railroad is still vital.

Automobiles

Above: Goldsworthy Gurney's 1828 steam carriage. It was driven from London to Bath and back in 1829. The 84-mile return journey to Hounslow Barracks took 9 hours 20 minutes.

The first car was the cumbersome steam-driven three-wheeler, built by Nicolas Cugnot in 1769. All the early mechanically propelled road vehicles had steam engines. They were heavy, unwieldy, slow, needed stoking with coal (rarely wood) throughout a journey, and had to be started an hour or two before they were needed. Because of petroleum availability and price, pollution, and several other factors, the steam car enjoyed a brief rebirth in the 1960s, but soon disappeared again. Even the attractive and economic electric car has failed to attract buyers. Overwhelmingly, gasoline has been the fuel for the car, though young people alive today will see alternative energy sources used.

Early cars really were – as they were called at the time – horseless carriages. It was natural to base them upon the carriage's large narrow wheels, crude elliptic or semielliptic steel leaf springs, high wooden body, and giant oil or acetylene brass lamps. Many denounced them as inventions of the devil. Each of the noisy, bellowing, smoking monsters invariably attracted a mob of jeering onlookers.

There was some substance behind the protests. From the very beginning, in 1885, the motor car had tires of rubber. Whether these were solid or of the new pneumatic variety, they pulled apart the granite chippings with which the roads were surfaced, and made the roads break up. By the beginning of the 20th century the roads of Europe and the United

Below: after about 1910 automobiles began to become more sophisticated. Pneumatic rubber tires tended to be standard, as well as an inclined steering column with a wheel to make driving safer as well as more comfortable. This 1916 Ford exhibits these features.

States were once more disintegrating into white dust or mud. The response was to repress the driver. Almost everywhere, state or local taxes were imposed.

Originally, cars were for the rich. It was the accepted thing to hire a chauffeur to look after and drive it. But after 1910 a few bold manufacturers saw that there might be a much larger market among ordinary people who could drive their own cars. Manufacturers proliferated. More than 200 established themselves in Britain alone, and the lack of uniformity, even among the simplest spare parts, was almost total. But in the United States Henry Ford pioneered the new concept of mass-production on an assembly line basis. Standardized parts were used, made to such close tolerances that any part would fit any car. Ford's famous Model T was a masterpiece of simple, robust engineering that transformed not only the car industry but the lifestyle of millions of people. Over 15,000,000 were made between 1908 and 1927. Even today a few are still at work, especially in remote regions of developing countries where the Model T withstands the elements better than a modern car.

After 1910 the steam and electric car were sharply in decline, and design became virtually standardized. The engine was placed in front, driving through a clutch (so that the drive could be disconnected when changing gear) and gearbox to the back axle. Initially, the final drive was usually by a belt or chain, but by 1910 the propeller shaft had become almost universal. This arrangement left the front wheels free for steering. Early cars had an upright steering column with a large wheel or tiller on top, connected to a pivoted front axle. The standard layout after 1910 was an inclined column driving either a worm gear or a rack and pinion, to swing a crank arm. This arm was connected to a cross shaft linked to the separately pivoted left and right wheels in such a way that the inner wheel always rotated to a sharper angle than the outer wheel on a bend to avoid scrubbing of the tires. By the 1960s hydraulically assisted steering had become common on all larger or more expensive cars. So had collapsible steering columns which in a head-on impact absorbed energy by plastic deformation, instead of piercing the driver's chest. Improvements in suspension, brakes, and tires have been central to the ever-increasing safety and efficiency of cars. Most developed countries today have smooth surfaced roads linking almost every major town and industrial development. But even on the best roads it is necessary to suspend the bodies of the car's occupants over the wheels in such a way that they have a smooth ride and the vehicle behaves correctly on bends. A soft suspension for high-speed highway cruising is not ideal for fast cornering, and immense effort has been devoted to perfecting compromises.

Most cars are fitted with independent front suspension, the left and right wheels each having their own vertical coil springs. Leaf springs in various arrangements are used at the rear, while all four wheels have shock-absorbers. These store the energy of the

Above: a Ford assembly line in Belgium. Mass production has made automobiles available to many.

Below: a Citroën-Maserati engine. Citroën pioneered the front-wheel-drive car.

25

compressed springs and release it in such a way that the passenger compartment is kept more or less stable over the bumpiest surfaces. Many cars have patented fluid suspension systems using trapped air and/or oil. The French Citroen company has also perfected a self-leveling quality that keeps the car body horizontal at its most efficient height.

Citroen also pioneered the front-wheel-drive car (FWD). Theoretically, FWD offers advantages in safety and maneuverability. But there are problems in driving and steering on the same wheels. By 1930, Citroen had made FWD a practical proposition. After 1950 this French maker had been joined by other manufacturers in the production of FWD cars.

Cars differed enormously 50 years ago – though many makers followed Ford's dictum that customers could have "any color they like so long as it's black." Almost half the cars produced did not have a rigid roof. Some had no roof at all, while many had a canvas or rubberized-fabric roof. This could, with effort, be unfolded, fastened to the top of the windshield and the sides filled in with plug-in celluloid windows. A popular feature was the rumble seat, a folding lid at the rear (where the trunk is today) concealing two extra seats. Wealthier customers favored the Sedanca de Ville in which the chauffeur sat in the open, separated by glass partitions from the well-appointed compartment reserved for his employers. Bodies were made of steel, wood, and even steel tube covered with fabric like early airplanes. The only uniformity was the basic engineering of the chassis.

Today the reverse is true. Apart from specialized sport cars, embracing such curious beasts as dragsters

Above: hundreds of the successful Volkswagen Beetle awaiting export. World War II crippled the German motor industry, but it now runs second behind the United States. Still made under licence in a number of countries, nearly 20 million Beetles have been sold.

and dune buggies, there are really only two body styles. Both these are usually mass produced from steel, and both are utilitarian. There is the standard two- or four-door hardtop. Those with more to carry buy a station wagon. But underneath this superficial similarity is a wealth of technical innovation and change.

Today's car can have its engine either at the front or rear. The Volkswagen "beetle," the best-selling car of all time, had its engine mounted at the rear, driving the back wheels. This engine was also unusual in being air-cooled, which was particularly useful in countries with severe winters. Another of the best-sellers, the British Austin or Morris Mini introduced the engine mounted transversely at the front, driving the front wheels. Some cars have a mid-engined layout, though this restricts passenger space.

At one time, nearly all cars employed four or six

Below: an Aston Martin Lagonda undergoing crash tests. Passenger and driver safety are given top priority in automobile design. The interior is usually protected by a rigid steel frame, while the front is designed to collapse to absorb impact.

Above: DAF automatic transmission fitted to a Volvo 66. Introduced in 1939 by Oldsmobile as Hydramatic Drive, automatic transmission became popular in automobiles in the 1950s and 60s. Most automobile models now offer an automatic version.

cylinders. Now, Saab cars from Sweden offer three cylinders, and the high-quality German Audi five. Saab has again broken new ground in adding a turbocharger, a supercharger driven by an exhaust-gas turbine. This automatically boosts the output of a basically small engine. Most of the advantages of both small and large engines are achieved. A few cars employ the smooth power of the Wankel RC (rotating-combustion) engine, and an even smaller number use gas turbines.

Automatic transmission, almost universal in North America, is less common in Europe where the need to change gear is still accepted. But one of the smallest and cheapest European cars, the Dutch Daf, does have a simple form of automatic transmission. Drive belts are gripped between two pairs of conical pulleys which can be moved nearer to or further apart from each other so as to vary the ratio between engine and wheels.

Cars that look outwardly similar may be fundamentally different in design and construction. Before the early 1950s almost every car was made in the form of a chassis, comprising a strong frame, wheels, suspension, engine, and drive system. Any desired style of body could be mounted on this. Today, the chassis has almost ceased to exist, particularly in Europe and Japan. The body itself is the basis of the car, and the engine, suspension, and other parts are mounted on it. This can reduce weight, but the structure must be designed to give adequate rigidity. Since 1965, much greater emphasis has been placed on "crashworthiness," and car designers pay more attention to rigidity.

A few special safety cars have been built and have proved marketing failures because people did not like their appearance. But even standard production cars are designed with a crushable front and rear, and a very strong passenger compartment in between. A car has to be able to withstand impacts from front, rear, or sides. There has to be considerable strength in the door columns and doors, and they have to be fitted with burst-proof locks. Seat belts for the front seats are required by law in most countries. Tempered multi-laminate glass and a collapsible steering column are other common safety features. The effect of cars on those around it has also been taken into account. Gone are the jagged monsters of the 1950s. Today's cars are invariably smooth and rounded, and free from unnecessary projections.

Car designers have made impressive progress in minimizing the pollution caused by their products. California – the worst sufferer from automobile pollution – introduced tough state laws to curb it. The United States government soon followed suit. Car manufacturers are engaged in finding means dramatically to reduce harmful emissions from cars without seriously affecting their performance. Crankcase breathers (vapors escaping from the crankcase) cause considerable air pollution. In many new cars such escapes are piped straight into the engine inlet manifold so the emissions are mostly burned in the cylinders. Likewise, the fuel tank breather no longer escapes straight into the atmosphere, but is piped to the engine.

The major problem of what to do with the exhaust gases remains. Even in a well-tuned car these fumes contain deadly carbon monoxide, harmful oxides of nitrogen, and traces of lead from antiknock gasoline. Engines today pour out much less of these products than their more primitive predecessors. Fumes are further reduced by injecting fresh air into the exhaust to burn the hydrocarbon residues and convert monoxides to dioxides. But the seemingly vital catalytic converter, which alone can chemically convert the remaining residues to harmless products efficiently, is rapidly choked and inhibited by lead from the fuel. Research continues constantly to defeat this problem. It is likely that within a few years the lead content of gasoline will have been reduced in many countries.

Below: exhaust pollution has become so serious in Tokyo that policemen testing its exhaust fumes wear breathing apparatus. They are also relieved every hour as an extra safeguard.

27

Record-breaking and Racing

The earliest racing cars appeared in the late 19th century. They looked nothing like horseless carriages, but were big-engined streamliners unlike anything seen before. The four-stroke internal combustion engine invented by Nikolaus Otto had not yet become the dominant means of powering a car. In the United States the gasoline engine trailed in third place behind steam and electric power. The earliest speed records for cars were set by electric vehicles. Camille Jenatzy's

Above: Camille Jenatzy in his electric car *"la Jamais Contente"* (Never Satisfied). Most early speed vehicles were powered by gasoline or steam, but Jenatzy's electric car achieved a record 66 mph in 1899.

Above: a motor racing meeting in New York in 1937. By the late 1930s motor racing was a major sport, and like many other fields of human endeavor was the subject of intense national rivalry.

1899 electric car was especially futuristic in appearance and attained a speed of 66 mph. Then in 1902 Leon Serpollet built a steam car of more conventional appearance. This reached 75 mph. Four years later a wooden-bodied Stanley Steamer weighing only 1624 pounds – the lightest speed-record car in history – set a figure of 121.57 mph, and subsequently crashed at over 150 mph. The gasoline engine racers found it difficult to beat this kind of performance, and they did it by having bigger engines, or more engines. Ray Keech's Triplex of 1928 had three of the largest aircraft engines mounted on one overloaded chassis.

Aviation technology made the building of most racing and record-attempting cars possible. The part that caused most headaches to the designers was the tires. A lot of attention was given to designing tires that would not blow out at high speeds, and would maintain their grip on the road.

In the 1930s highly rated aircraft piston engines were used, giving 2500 hp each. After World War II aircraft gas turbines were installed. Initially, these were of the converted turboprop type, driving the

wheels. Today, they are exclusively of the turbojet or rocket variety. The 1978 land speed record holder used a single rocket chamber burning hydrogen peroxide and LNG (liquefied natural gas). This is a possible combination for ordinary cars of the next century, though not burned in a rocket.

Today's speed-record cars and projects look more like spaceships or missiles than cars. Their only parallel on wheels is found in the more exotic members of the dragster fraternity. Originating in southern California in 1949, dragsters pose problems as challenging as any other kind of sport on wheels. The object of the sport is to accelerate from a standing start and cross a finish line a quarter-mile ahead in the shortest possible time. This calls for the most powerful engine, lightest overall weight of vehicle, and nearly all of what weight there is at the back over the largest possible pair of "slicks" (driving tires). Before the timed run, the slicks are given a covering of molten rubber by a few vicious kicks on the accelerator to spin the wheels while the car remains stationary. The front end of a dragster tails off to next to nothing, with a flimsy looking structure and wheels like a bicycle. Engines are fueled by special blends of nitromethane, costing $14 to $20 per gallon, and giving 30 times the power of an ordinary family car in an engine only slightly larger.

Dragsters teach ordinary car designers as much as the longer-established Grand Prix racing, despite the fact that in the latter the cars have to weave around on an ordinary track, possibly in the rain, and negotiate unbanked bends. While backroom workers have never ceased to wring additional power from engines of great reliability, the greatest effort has been devoted to such mundane matters as the tires, clutch, gearbox, and generally keeping the car under control. By 1938 the most powerful racing cars, especially the German Auto Union with a 16-cylinder engine behind the driver, were very difficult to control. Today "Formula" racing puts limits on size and power. It takes ingenuity and skill to build faster cars. In the mid-1960s a racer in the United States, Jim Hall, fitted a small upside-down wing above his car to give downthrust and keep the tires firmly on the track. Then came variable-incidence wings adjusted from the cockpit, as well as front and rear wings. By 1969 wings were not only multiplying but causing dangerous crashes. Now, only fixed wings of limited size are allowed, and they must be fixed to the body and not the suspension. Jim Hall's team then turned one car into a hovercraft in reverse, sucking itself to the track with fans and skirt. This worked well, but it sucked up debris and threw it at drivers following behind.

Even more recent than dragsters is ORR (Off-Road Racing). This sport is having a most beneficial effect on the design of cars and trucks. For excitement and challenge, ORR offers as much as most people, participant or spectator, can wish. By racing over hard rough tracks, pockmarked with crevasses, large

bumps, and every other hazard, at speeds nudging 100 mph, the whole technology of the motor vehicle is given a most severe test. Competitors choose anything from a VW beetle to a truck, and rebuild it so it can withstand treatment that would cause other racers, or standard cars, to fall to pieces. Equipment includes completely redesigned suspension, large air filters mounted as high as possible, armor to protect the underside of the engine, and a strong steel superstructure framing to withstand any amount of overturning.

Above: Carlos Reutemann in a Brabham at the Monaco Grand Prix of 1975. The car is equipped with an aerofoil. Formula 1 rules have restricted the use of aerofoils since.

Below: drag racing contestants are timed over a set distance from a standing start. The modern dragster has very large rear wheels with a lightweight small-wheeled front end. Many dragsters can clear 400 yards in about six seconds.

Vehicles for Industry

In almost every country private owners such as farmers, engineers, and self-employed specialists use station wagons, vans, and pickup trucks which serve both to carry people and goods. But true commercial vehicles are invariably designed to carry either the one or the other, when they are not built to perform a myriad other tasks. Fire engines, ambulances, tractors, bulldozers, mechanical shovels, refuse collectors, and concrete mixers are just some of the vehicles at work to keep modern society running.

Long distance passenger vehicles are almost always diesel engine single-deckers, though the quiet and efficient gas turbine is beginning to capture a little of the "top end of the market." Here vehicles are lavishly appointed for transcontinental service with toilets, air-conditioning, and a snack bar, and can cruise at 80 mph. For urban service there has been a wider

Above: a modern DMS 10 London Transport double-decker bus. This replacement for the previous generation double-deckers is equipped with two automatic doors and a middle staircase.

choice, with buses having one or two decks, trams, and trolleybuses all offering their own advantages and disadvantages. Surprisingly, in view of man's wish to reduce dependence on petroleum and minimize pollution, the electrically propelled tram and trolley have been almost eliminated by the noisier, less efficient, and oil-burning bus, perhaps because the latter can use any assigned route and needs no "infrastructure"

of substations and supply cables.

Except in the smallest sizes, where engines are the same as those of cars, freight vehicles invariably use diesel engines. Nearly all have a six-cylinder in-line engine rated at about 180 hp, though the big articulated trucks have an eight-cylinder engine of about 250 hp. Sometimes the engine is fitted with an exhaust-driven turbocharger to increase power and efficiency. An increasing number of the very largest trucks use gas turbines with a heat exchanger (regenerator), which transfers waste exhaust heat to the incoming air to increase efficiency. In North America less constricted highways have allowed very large vehicles. But in Europe, vehicles of similar size did not become common until the "container revolution" put all kinds of general goods into International Standards Organization (ISO) containers. These containers were designed to be as large as road, rail, sea, and air transportation allows, and they have a cross-section eight feet square, and a length of up to 40 feet. It is rare for

Above: heavy goods vehicles (HGVs) on a highway near Toronto. More and more goods are being carried by trucks like this. Some people argue that more use should be made of the railroads to move heavy goods.

Below: a road train heads north from Alice Springs to pick up a load of cattle. These diesel-powered vehicles are common in many agricultural regions, but many industrial countries impose weight limits on trucks.

Above: a multi-axle truck equipped with a hydraulic tipping body. Trucks like this are used the world over to help in such tasks as road mending and work on building sites. Tipping trucks carry building materials like cement, sand, or gravel. The biggest vehicles, like the one shown here, can carry more than 40 tons.

any container truck to tow a trailer, but in Australia cattle "on the hoof" is transported in "beef trains." As many as six payload carriers are all pulled by one prime mover equipped with exceptionally low gearbox ratios to pull the heavy train up hills.

Significant progress has been made in the design of bulk payload vehicles. Liquids are carried in modern tankers with their own pumping and measurement systems. To make better use of the available space, box-shaped tanks are replacing the drum-shaped variety so that the largest look rather like a trailer-home without windows. For corrosive or inflammable cargoes, specially designed tanks must be used. They are sometimes filled with a shatterproof glass liner and in a few cases are divided into numerous sections with self-sealing laminates, and built-in equipment to neutralize spilled fluid or prevent fire. Most countries demand that a prominent notice be displayed on the outside giving details of the contents, and a telephone number where advice may be obtained in case of a serious accident. Handling of powders, such as flour and cement, also poses safety problems. Modern powder trucks usually have a self-contained pressurization system for pumping the contents out at the destination. Contamination of the contents must be scrupulously avoided. The consequences of heavy rain getting into a cement tank can be imagined.

Much has been done since World War II to improve accessibility and the comfort of the driver. Modern cabs, often made of glass-reinforced plastics (GRP), often hinge forward as a unit to expose the engine and ancillaries. In nearly all passenger vehicles the engine is mounted at the rear or in the middle, with its cylinders arranged horizontally so that it can be accommodated under the floor. Vehicles arranged entirely on one frame, called "rigid" or "straight" vehicles, invariably drive on the rear axle(s) which in all but the lightest sizes have two wheels on each side to reduce "footprint" pressure and allow the use of normal mass produced tires. The tractor unit of an articulated truck likewise drives on the rear axles. They are arranged to bear about half the weight of the load. Earth movers and other off the road vehicles usually have four-wheel drive, and a few have caterpillar tracks. The largest off the road vehicles, such as monster dump trucks weighing over 100 tons, have individual electric drive to each wheel. The carriage of outsize or super-heavy loads calls for special low-slung trailers with from 48 to over 100 tires, propelled by a tractor at front and rear. Several of these transporters have an air-cushion system, like a hovercraft, to spread the load on bridges and soft road surfaces.

31

Roads and Road Traffic Control

Modern roads are built with modern road vehicles in mind. These, almost without exception, use rubber-treaded pneumatic tires. When rubber tires first came into use at the end of the 19th century they did not grind, press down, and consolidate the crushed-stone surface of the roads. Instead, they plucked stones out, gradually ruining the roads, and causing each car to create a pall of dust. The first answer was tarmacadam. Tar was poured on to act as a binder. More than half the made roads of many major countries still have a "blacktop pavement," composed of asphalt or tar binding together a layer of gravel. This has to be renewed every few years, but it is relatively cheap and gives a good all-weather surface.

The name tarmacadam came from "tar" and "McAdam." John McAdam was a Scot, who 200 years ago demonstrated how simple roads could best be built. He showed that the subsoil itself could, if it was properly drained and if the road had an arched surface (camber) to assist good drainage, support any traffic without subsiding. In recent years soil mechanics has vastly increased man's understanding of how roads can be supported indefinitely without appreciable movement of the subsoil. There are numerous techniques for stabilizing or strengthening the soil. Bentonite (a kind of clay), cement, bituminous mixtures, and special chemicals can all be pumped in. When roads are needed in a hurry these methods alone sometimes suffice. But the long-distance highway is today made chiefly of concrete. The basic technology was learned in building airfields.

The modern highway is almost always supported directly by the subsoil, compacted by a vibrating roller after removal of vegetation and topsoil. If the subsoil proves not to be strong enough to bear the load of the road, a layer of crushed stone or other granular material is put down. The main pavement is a layer of concrete at least 12 inches thick, reinforced by a mesh of heavy steel wires about three inches

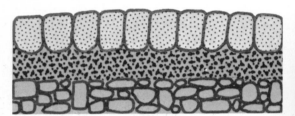

Above: a Roman road. A foundation of stone slabs and blocks was set in mortar. Next came a layer of small stones bonded with cement. The surface was made of stone blocks.

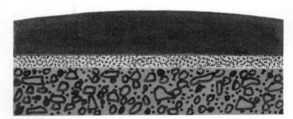

Above: modern roads have a foundation of broken concrete, bricks, and stones. Ashes or clinker is then rolled solid. Tarmacadam (black-top) is rolled smooth on top.

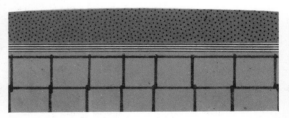

Above: a modern highway made to bear heavy loads has a foundation of reinforced concrete. Next comes a layer of waterproof paper. The surface is a thick layer of concrete.

Above: Wilkinson's Garage, Uxbridge, during tests by the Challenge Tyre Co's in August 1912. The new tires needed better road surfaces.

Below: a truck preparing a freeway roadbed for black-top paving. The picture shows the truck spreading cement in windrowed natural sand.

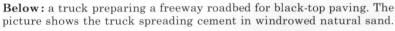

below the surface. Concrete pavements are called "rigid" to distinguish them from the more flexible pavements based on layers of asphalt. They have to be laid a section at a time, by a train of large machines running on rails along each side of the pavement and able to lay the entire width in one operation. A narrow gap, filled with bitumen, is left between each section to allow for slight relative movement, expansion and contraction, and any other disturbances which might otherwise result in a crack in the pavement. A typical length for sections is 120 feet. A few of the newest highways have been laid in continuous strips several miles long. A novel kind of reinforcement has been used that assures a crack-free membrane impervious to water.

In urban areas, highways have been built by various methods. Because of the much greater pressure on land use they sometimes have to run under or over the surface. Elevated roads are also necessary at major junctions, even far from cities, in order to avoid conflicts between opposing traffic flows of the kind unavoidable when the converging roads are all at the same level. A basic two-level crossing is achieved by an overpass or underpass. But in the most complex interchanges there are sometimes three or even four

highway levels with associated link roads or slip roads, with radius of curvature matched to the speed of traffic. In any underpass or tunnel it is essential not only to provide adequate "soft" (shadow-free) lighting, but also to add extra lighting at the extremities of the underground section to minimize any sudden change in light intensity as drivers go in or out.

Great attention has been paid to the road surface. This is as important as the tire in maintaining good vehicle adhesion and thus preventing accidents caused by skids and overturns. The basic concrete or asphalt surface provides adequate grip to avoid any skid under normal driving conditions, and the danger arises when something is interposed between the tire and road. Ice and snow pose a major hazard and are tackled as an emergency by vehicles equipped with dozer blades or plows, for spreading grit and/or coarse salt. In urban areas on the newest and most important highways, electric heater mats incorporated in the road surface are switched on to prevent ice or snow forming. Standing water must be avoided, especially on fast roads, because above quite modest speeds vehicles ride up on the water and "aquaplane" in an uncontrollable manner. While tire designers produce ever-better tires able to pump water out

Above: the technology of road laying is becoming ever-more sophisticated, with the greatest advances taking place in the last 50 years. Spurred on by the relentless expansion in road traffic, the pace of technological advance is unlikely to slow. Here, a giant concrete mixer pours concrete over steel mesh reinforcement in a specially prepared road base. A levelling machine follows, and a smoothing machine next.

Left: the Gravelly Hill interchange near Birmingham, England – better known as "Spaghetti Junction." This is the largest highway interchange in Britain. It is part of the Midland Links Motorway scheme. The highest road in the interchange is more than 80 feet from the ground. Lead pollution in the air is causing concern.

Above: a traffic roundabout on the British highway *M4*. Roundabouts are built to help the free flow of traffic coming from a number of different directions. An alternative method is the multi-level interchange. The cardinal rule that allows the roundabout to work at all is that drivers must always remember to give way to traffic already on the roundabout.

Left: a typical pedestrian crossing. Visual symbols are safer than words as they are readily understandable by everyone. Some crossings are also equipped with bleepers to signal when it is safe to cross.

through lateral jets in the tread surface, road engineers avoid any absolutely level surface and even use diamond-tipped saws to cut grooves along which water can escape.

Until the early 20th century road traffic control was almost nonexistent. Urban roads were often clogged with "traffic" which included flocks of turkeys and herds of sheep. One of the first steps to produce order from chaos was to legislate a unilateral rule of the road. In most countries this required users to occupy the right-hand half of the road. Much later many urban streets were made one-way only. Today all the most important intercity highways have four lanes which effectively make each two-lane half into a one-way street. Traffic on such routes is generally prohibited from making U-turns or crossing from one side to another.

The next situation on which attention was focused from about 1920 was the crossroad or intersection. Always a source of congestion, danger, and frequent accidents, they are sometimes blanked off, converted into a traffic circle, staggered, or provided with traffic lights. There have been many forms of circle. In towns and cities the lack of space often means that the best that can be done is to open out the intersection by perhaps 20 feet into a near-round shape large enough for traffic to proceed in a circular fashion. This avoids the former conflict, and replaces it with a gradual confluence of two streams at each of the four entrances where it is clearly understood which stream has priority. Often the rule is "traffic gives way to that approaching from the right" or "traffic yields to that already on the circle." On large highways, intersections are avoided, and when circles have to be built on four-lane highways they are big enough for traffic to get into the correct lane to turn off at the right point, sometimes with little need to reduce speed.

Traffic signals were introduced in 1927, and are found in nearly all the world's cities. The use of red and green for "stop" and "go" is universal, and in an increasing number of signals amber is added as the result of inability of drivers to comply accurately with a two-color system. Early systems operated on a fixed time cycle, often biassed in favor of a supposed major road. This often held up traffic to no useful purpose, which is the opposite of the objective of traffic control – to facilitate its easy movement. Strip detectors were then recessed into the road surface so that traffic approaching a light could send electrical signals to speed up the changeover. Often a "filter" green light is added to allow traffic turning off that does not interfere with other traffic streams to proceed, while that proceeding straight ahead is stopped. In some cases the green light is presented as an arrowhead to indicate that crossing straight over is either prohibited, or the only course permitted. Some signals are on permanent "go" aspect except when a pedestrian wishing to cross the highway presses a button. There are several types of pedestrian crossing signals, often combined with distinctive markings in the road surface and road signs. One type common in Britain combines green, amber, red, flashing amber, and green sequences with an audible warbling to urge pedestrians to hurry when the lights are about to release the traffic flow.

Wherever space permits, roads are divided into lanes, each wide enough for all normal vehicles, and marked by various white lines or light-reflecting studs. A few urban expressways have 16 lanes, while major highways usually have six or eight. All modern intercity routes try to separate conflicting lanes as far as possible, and often where an old road is used as one lane the modern route in the opposite direction does not run alongside but may be 100 to 500 feet (30–150 meters) distant. Where both lanes have to be close together there is a central reservation along which is a barrier designed to deflect an out-of-control vehicle back onto its own side with minimum damage and danger to others. When turning off across the opposing traffic is permitted, the road has to be constructed in such a way that there is room for a line of cars to wait without encroaching on either fast lane. If possible, the much more expensive answer to avoiding all conflicts is adopted by building a cloverleaf, which automatically dispenses with the need for control other than the provision of signs which can be unambiguously read by fast traffic.

Modern highways may have a flood of vehicles proceeding at 60 mph along four or more lanes simultaneously. Experience has shown that drivers drive too close behind the vehicle in front. Especially in conditions of fog or ice everything possible must be done to avoid disrupting the steady flow. Traffic entering the highway has to have a long "ramp" along which it can accelerate to match that on the main highway. Where there is no central reservation and several busy lanes run alongside each other it is often helpful to fit traffic lights above the center of each lane. To increase traffic capacity to match the need at different times of day, such as morning and evening rush hours, lanes can be shut off for a while until it can be confirmed (by TV surveillance, for example) that it is completely clear, and then opened to traffic in the reverse direction. This is called tidal flow control.

Traffic control is becoming an exact science. Prolonged traffic censuses, surveys, and computer study have led to costly but sophisticated computer-controlled systems that enable each section of road to carry the maximum traffic flow, with a minimum of holdups. There are plans for much more advanced systems still, with control positively exercised by cables in the road surface interacting with receiver loops in the vehicles and the whole system monitored by central controllers. But funds have not yet been available for their introduction.

Above: a fog warning traffic control installation on the British highway *M1*. The lights round the main signal are warning of the fog hazard. A maximum speed of 40 mph is being advised. Research is still going on to discover why many drivers ignore such warnings and drive recklessly fast – causing pile-ups.

Below: advanced technology is taking an increasing hand in the management of traffic, especially in the congested towns and cities. This Programmable Logic Control (PLC) controls traffic. Linked to traffic lights and other signals, it regulates the flow of traffic to minimize accidents and ease congestion.

Travel off the Highway

At one time all land transportation was off-highway because paved roads did not exist. In "undeveloped" countries there is still a great need for off-highway transport. Nonmilitary users of off-highway transportation include geologists, oil and other prospectors, construction engineers building pipelines, electric supply networks, emergency services at places like airports, farmers, and dwellers in remote areas where roads peter out into unpaved tracks.

There exists today a wide range of choice of commercially marketed vehicles ranging from near-conventional cars and trucks through a diversity of sporting and exploration vehicles to highly specialized working vehicles. In a few cases these are amphibious and can go almost anywhere.

Best known of the off-highway vehicles that are often seen on ordinary roads are the Jeep and Land-Rover. Both are simple to maintain and drive, and offer four-wheel traction to cross difficult surfaces with loads of four or five adults or about 650 pounds of goods. Immediately above these in size are an equally wide range of "safari" vehicles seating six to ten. These combine all-wheel drive with higher speed and greater comfort. One of their jobs is conveying tourists, armed only with cameras, through wildlife parks.

Polar regions call for specially designed vehicles. Early in this century the traditional dog sledge was joined by various derived versions including the sporting one-person toboggan, the steerable two to four-person bobsled, and various powered devices including the dangerous Russian sled with air propeller and the Arctic-Canadian skidoo or snowmobile with powered rear track and steerable front skis. In the Antarctic much exploration has been done with vehicles designed for crossing the polar icecap which, though often offering an excellent surface, is covered with deadly deep crevasses. The most common species have three or four sets of snowtracks with an overall wheelbase long enough to bridge any likely crevasse before the center of gravity crosses the brink and the vehicle overbalances.

Another specialized family of vehicles developed since the mid-1930s is the marsh buggy whose tires are so large that they make the whole vehicle buoyant. The original buggies had larger versions of conventional tractor tires, but today the usual "high flotation" tire has immense width and looks rather like the broad "slicks" used on the rear wheels of dragsters and Formula racing cars. On a much smaller

Above: an Amphicar under test. This German product was the first amphibious automobile to be sold in significant numbers. Made from 1961 to 1968, the Amphicar carried four passengers at a land speed of 68 mph and a water speed of 6½ knots.

Below: some types of terrain are impassible to conventional wheeled vehicles, and the development of the caterpillar track has proved invaluable. Such tracks were fitted to the Antarctic Snowcat. It crossed hundreds of miles of icy wilderness. Vital to its survival were the crevasse detectors at the front.

Left: with the aid of an outboard motor this British amphibious vehicle can achieve a water speed of about nine knots. It has a land speed of 22 mph. Useful on marshy sites, its nine-gallon fuel tank provides an adequate supply for an eight-hour working day.

scale these high flotation tires serve a profusion of "fun" vehicles ranging from golf karts, through beach-buggies and snow-climbers for hauling teams of skiers, into a large family of true amphibians. In general these vehicles have GRP hulls, a small gasoline engine and four or six high-flotation tires with drive to all axles. Some can reach 50 mph and even when carrying a load equal to their own weight can climb a slope of 45 degrees. Others, widely used for exploration, can carry twice their own weight, and are so light two men can pick them up. One in Wales is used by its owner to drive to the summit of a mountain, after being lifted *en route* over a garden fence.

Below right: the Apollo 15 Moon mission was launched in July 1971. The Lunar Rover was developed to carry Astronauts David Scott and James Irwin as they collected the samples of the Moon's surface. In the end they brought back the richest scientific haul of the whole lunar program.

All-terrain vehicles have proved a rich field for inventors. Many of the amphibians drive across water propelled only by their high-flotation tires. Some have outboard motors, swung down after they enter the water. One drives on left and right revolving pontoons arranged axially, with spiral metal blades that thrust it along over water, ice, snow, or thick mud. A few have peculiar wheel arrangements, an example being a military proposal that found civilian use. It had wheels in the form of cloverleaf units with a small wheel at each "leaf." For ordinary running on roads the units are locked with two of the small wheels on each in contact with the ground. For off-the-road operation the driver moves a changeover lever which locks the small wheels and transfers the drive to the cloverleaf units. These rotate to haul the vehicle over all obstacles onto which the "leaves" can climb.

Perhaps the most exotic class of off-highway vehicles are those built to explore the lunar surface. These have unique features tailored to the low gravity and absence of an atmosphere. Tires are woven from springy steel wire and strip which spread the load over the soft almost ashlike surface. The American Apollo astronauts used "buggies", called Lunar Rover vehicles, with four wheels, while the remotely controlled Russian Lunokhod had eight. In both cases propulsion was provided by battery powered electric motors.

The Railroads

The railroads are man's biggest fixed transportation infrastructure, in the sense that, unlike the roads, they exert a commanding influence on the vehicles that run on them. This immense worldwide system has far greater potential than was suspected even 15 years ago. The importance, and possible disadvantage, of this is that a system so large is almost impossible to replace. Its very existence acts as a deterrent to any change, as several national rail systems have demonstrated, when they have refused to support alternative forms of high-speed guided transportation (HSGT) system because they are not compatible with existing track.

The concept of the railroad is much older even than the steam locomotive, and crude systems were in industrial use in various European countries 500 years ago. Its advantage is chiefly that, as a wheel running on a smooth hard surface sinks in only a microscopic amount, it not only gives a smoother ride

but also suffers much less friction than one rolling over unprepared earth, or even a rough unpaved road. This was clearly demonstrated in New York in 1832 when rails were laid along the streets and the first horse-drawn trams began operation. Not only could a horse pull more than double the number of passengers previously possible but the ride was smooth and comfortable – and that was in comparison with the best existing city streets. Compared with the rough country roads of the day the commercial railroads in the early 19th century offered immense advantages. Much heavier loads could be carried, hills avoided, and much higher speeds attained.

The basis of the system has always been the track. From medieval times it was recognized that there had to be some means of keeping the wheels running on the rails. Though there were many alternative plans the best answer was considered to be the two-rail system with flanges either on the track or on the wheel. At first, the flange was put on the track, which in the earliest systems was made of wood. After making each rail in the form of a three-sided channel for many years it was appreciated that two-sided angles are sufficient, because the left and right sides work together. By 1800 the rails were always of iron, and by 1810 it was generally agreed that it was preferable to put the flange on the wheel, making the rail in the form of a bull-head support. This greatly increased the rails' carrying power. Especially when the cumbersome early steam locomotives replaced horses, the flat-

Left: one of the earliest illustration of a railroad in use comes from George Bauer's "Agricola" *De Re Metallica*. A track made of parallel planks was installed in the coal mines of Lorraine in the mid-1500s to haul the laden trucks. Tracks with flanges were developed shortly after. Rails made of iron came much later.

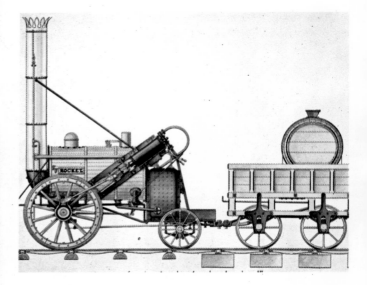

Above: George Stephenson's famous "Rocket" could achieve the then terrifying speed of 29 mph. A modern locomotive can comfortably manage speeds in excess of 120 mph. The steam engine that brought the railroads into being was an adaptation of the earlier stationary engines by Newcomen and Watt.

plate iron rails soon cracked and broke up. Once the rail assumed the form of a substantial deep girder, the way was open to build the system that has lasted until today.

George Stephenson, one of the pioneer steam railroad engineers, selected a gauge (distance between the rails) of 4ft 8½in. He and other Britons built the first railroads in most of the world's countries, so a uniform gauge was established. To show what might have happened, the states of Australia each selected a different gauge (4ft 8½in, 5ft 3in or 3ft 6in) and jealously resisted any change. Everything had to be unloaded at the state border and put into a different train. In recent years there has been a painful rationalization costing over a billion Australian dollars. If this had happened on a world scale the cost and problems would have been dreadful.

By 1830 the flange was on the wheels and the rails were secured by various means to crossties or sleepers. Methods had been worked out to build rail junctions, at first operated by hand on the spot, then laboriously by levers and long push/pull rods from a signal cabin. Today, a powered electric or electrohydraulic actuator signalled from up to 100 miles away is employed to "switch the points." Rails were laid in increasing length as the steel industry's rolling capacity increased. Continuous welded rail with the welds ground smooth now give a near-perfect ride with no bangs or clicks caused by the wheels passing over joints. Rails still occasionally suffer from "burns" caused by slipping wheels on one spot, from fatigue cracks, and even from fragments falling off. But these dangers are extremely rare and most rails last many years despite immensely increased traffic weights and speeds.

The sleepers, which are massive beams usually

Above: Differing gauges between one area and another can make running a railroad very difficult. This locomotive is having its bogie changed to fit standard Australian gauge. **Below:** Australia's railway gauges were rationalized in 1968. *Left*, broad gauge, *center*, broad/narrow, *right*, standard gauge.

Left: the continuous welded rail is becoming common on many national railroad systems. The rails are welded together on site and supported by concrete sleepers. Continuous welding makes the ride smoother and safer at high speed and reduces noise.

about twice the length of the gauge, have in the past usually been creosoted wood in western Europe, heavy steel pressings in the tropics, and materials like concrete. They have usually been deeply embedded in ballast, a layer of granite chips or other durable material. All traditional track needed frequent inspection and repeated maintenance. The rail fixtures had to be inspected, the ballast cleaned, and the track repeatedly resurveyed and adjusted by armies of men shoveling ballast into particular places to raise either one short section of line, or just one rail in relation to the other. Since 1960 extensive research has led to new forms of track laid either on continuously cast concrete, a thick strip of asphalt or any of several other materials which subsequently need infrequent inspection and, it is hoped, no maintenance. This will reduce delay to passengers and freight.

Essential to the running of a modern railroad is control and communication. At first, rail police signalled trains with flags, and a red lamp was hung on the back of the last vehicle so that if the couplings broke, leaving rolling stock blocking the line, the fact would be evident to signalmen or station staff even at night. Today railroads are almost as electronically equipped as the air routes. The entire network is linked by electric power and communications which in its simplest forms merely serves color-light signals and telephones. In the most advanced systems electronics exert automatic control over the trains, so that drivers spend most of their time passively obeying the orders of the computer.

For over 50 years there have been simple "automatic train control" systems. Some have boxes beside the rail which, when a signal ahead is at danger, extend a small lever to apply the brakes of a train passing overhead whose driver has failed to do so. Others have devices between the rails which ring bells in the cab, sound a buzzer or make a partial or full application of the brakes. The latest systems rely mainly on electrically conductive co-axial cable (like that connecting a home TV set to its aerial) or other conductor. These are laid in complex patterns between the rails and fastened accurately in place in a way that will need no attention for many years. Each train cab is served by conductive cells which pass very close above the patterns of track cables. These coils in turn feed a small digital computer, which in turn sends signals

Left: As the railroads spread in Britain, it was realized that a signalling system was needed as early as the 1850s. Old-style signal boxes like this changed the switches (points) mechanically by means of levers.

Below: the computer controlled signal box at London Bridge station. This has to monitor the progress and position of trains on the whole of the Southern Region system of British Rail. Obviously, a simple mechanical signal box could not cope with such a task. Every train's behavior is controlled.

to a bright display like a small TV in the cab.

The most basic function is that of signaling. By inserting loops or other patterns into the track coils and then transmitting particular signals from a central control, the interaction between the track and the coils on the speeding train can light a green display in the cab (or a bright red one), and sound an audible warning. If no action is taken, it will apply the brakes. By measuring the speed at which the train coils pass over the patterns in the track the speed of the train can be either measured and displayed to the driver or the central controller, or it can be controlled directly. Temporary speed restrictions can easily be built into the track system and later removed. Small sensing boxes can be added at strategic points to measure the temperature of the bearings supporting every wheel passing along the track, sounding an alarm if there should be a single "hot box" (perhaps without lubricant) needing attention. At any point where the train speed is always low, such as at the end of a station platform, it is possible to make the track itself weigh each axle passing over a particular point, and feed the system with the digital data. Identification panels beside or under each vehicle can be automatically scanned and the information – such as vehicle serial number, type, and date when next inspection due – transmitted to a central computer for possible action.

With little extra complication, messages can be sent to trains and either emitted aurally from a loudspeaker or, more probably, displayed line by line on the cab display. One of the difficulties with attempts at radio communication on railroads has been the poor or nonexistent reception in cuttings or tunnels. With the track conductor systems now available this difficulty disappears. Of course it is even simpler for the driver and guard to converse by each having his own set of coils running over the track, with a special facility for preventing their conversation from being transmitted elsewhere in the system. Not the least of the advantages of automatic control systems is that they ensure that all trains are controlled as efficiently as possible, in a way that sticks to schedule exactly without wasting propulsion energy. But eliminating the human crew is not anticipated on most routes. Even in the railroads that are part of the way along this path to automation there is usually an operator in the cab. On London's Victoria and Jubilee subway trains, he actually serves as the guard and opens and shuts the train doors. But his presence is reassuring to passengers. The only routes where unmanned trains are socially acceptable are those carrying freight only. The first was London's Post Office underground railway which handles bulk mail between sorting centers. One of the newest is a

totally automated line serving open-cast coal mines in the American midwest.

This new all-freight route operates on electricity supplied by overhead cable at 50,000 volts. It is generally agreed that electrification is the best form of motive power for railroads, the main drawback being the initial capital cost. The earliest electric railroads used direct current at about 400–800 volts, but by the 1930s many nations were turning to more efficient alternating current at much higher voltage. A common choice was 15,000 volts at 16.7 cycles. Today the world standard, if there is one, is 25,000 volts at 50 cycles. Though it might be thought this high voltage is dangerous, it is so widely used with so few problems that it offers an almost perfect solution. The higher the voltage, the fewer the substations needed to feed current into the line and the lower the heat losses in the cables. A few systems, like the midwest coal mine route, have adopted doubled voltage of 50,000 volts, though it is unlikely this will be exceeded this century. All modern electrification supplies the current via an overhead conductor cable called a catenary. This is held level and horizontal above the center of the track. The speeding pantograph on the train slides past underneath causing the minimum of vertical movement. Much research was necessary to find cheap suspension methods that avoided dangerous undulation and vibration of the catenary as trains pass at speeds up to 200 mph.

Since 1960 research, mainly in Britain, has discovered the exact mechanics of how two flanged tapered-tire wheels on a live axle (one fixed to both wheels) actually rolls along today's rails. The problem was surprisingly complex, and its solution assisted the design of wheel-sets and swivels which ride perfectly smoothly, without shaking from side to side or oscillating directionally. The startling result is that, whereas in 1960 it was thought doubtful that existing track could safely handle any trains faster than about 120 mph, with freight trains limited to about half this speed, today both these limits have been roughly doubled. This means that the world's rail network will remain competitive for as far ahead as can be foreseen.

Rail Propulsion

Above: a steam locomotive on the Darjeeling Railway. Steam locomotives still dominate India's railroads. Many of them were inherited from the days of the British Raj and still show no signs of wearing out.

There are many, including motive-power engineers, who regret the near elimination of the steam locomotive. Its demise was mainly on the grounds of its extremely poor overall efficiency, which hardly ever exceeded nine percent. (Ninety-one percent of the energy released in burning its fuel was wasted, most of it passing out of the funnel). The diesel engine has an efficiency from three to five times higher. Electric locomotives, considered in isolation, have an efficiency well over 90 percent, but this does not take into account the much poorer efficiency of most of the turbogenerators that supply the current. The answer of the steam enthusiasts is that these figures ignore what could have been achieved with intensive development of steam motive power over the past 25 years. Certainly railroads will soon have to find alternatives to oil-based fuel.

One of the basic handicaps of steam engines, never fully solved and partly responsible for failure of modern steam automobiles, is that they tend to need time to become ready for use. In the case of railroad engines this time was seldom less than an hour. Once "in steam" the fire kept burning, often wastefully, until it was extinguished at the end of the day's work by even more wastefully raking the unburned coal or other fuel into an ash pit. A diesel starts in seconds, and can be switched off even for the shortest halts. This is also true of the gas turbine, which has increasingly come to the fore in rail propulsion in recent years. The newer forms of power are widely socially acceptable. None makes severe physical demands on a human operator, and though some diesels are noisy and smoky this can be significantly reduced by proper design and maintenance to give a prime mover causing little offense in cities or tunnels. Emissions from gas turbines are even less noticeable, and the electric locomotive is virtually pollution-free, although fossil fuel may have been burned to generate the current. A unique advantage of any train with electric drive is that "regenerative braking" can be used. To slow the train the traction motors are converted at the flick of a switch into generators. These not only provide a powerful and controllable resistance to the train's motion, but also feed current back into batteries or into the supply cable.

Trains can be hauled in various ways. Most have a locomotive at the front; the heaviest freight trains have sometimes as many as eight at the front with still more distributed along the length of the train. A

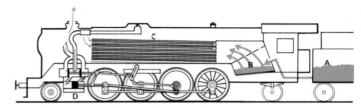

A steam locomotive gets its power from coal or oil (**A**) fed into the firebox (**B**). Hot gas and smoke pass through tubes in the boiler (**C**) and out of the funnel. High-pressure steam from the boiler drives double-acting pistons (**D**) coupled to big driving wheels. Exhaust steam also escapes through the funnel. Though powerful, steam locomotives are no longer built.

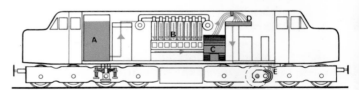

A diesel-electric locomotive converts oil fuel (**A**) into power in a diesel engine (**B**); the engine drives a generator (**C**) from which electric current flows to resistors (**D**). These control the speed of electric motors (**E**) geared to driving wheels. Because a gearbox on the scale needed would be unwieldy, the electrical system is needed to act as a variable-ratio drive giving a smooth ride to the train.

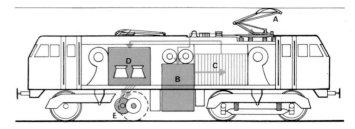

The all-electric locomotive draws power from a high-voltage overhead line (**A**). The current, stepped down by a transformer (**B**), passes through control resistors (**C**) to the rectifier (**D**), which converts it from alternating to direct current. This current drives motors (**E**) coupled to the main wheels.

Above: British Rail's 125 mph HST. These are diesel-electric trains with a power car at each end. The introduction of the HST in 1976 provided the British with one of the quickest and most comfortable rail systems in Europe.

Left: electric trains in St Anton, Austria. Since 1945 the electrification of Europe's railroads has proceeded at a rapid pace. Electric trains have been proved to be able to carry large loads with minimal damage to the environment. Diesel power coexists with electric.

few trains, including some fast passenger trains in Britain, are permanently coupled to a locomotive which pulls the train to its destination and then pushes it back. At one time it was thought that propelling from behind would be dangerously unstable. This proved not to be the case. A few trains, including some on the very earliest railroads, are pulled by cables from power sources at each end of the route. An increasing number of passengers have, since the early 20th century, been transported either in self-propelled single railcars or in multiple-unit (MU) trains in which the propulsion is built into a proportion of the coaches. Though French electric locomotives hauling ordinary passenger coaches set a remarkable speed record in 1955 at over 205 mph, it is significant that almost all today's fastest passenger trains are of the MU type. Britain's High-Speed Train (HST), which cruises at 100–125 mph, is not quite an MU. It has a locomotive at each end to provide streamlined ends with both-ways driving capability without putting propulsion under the passenger coaches. Both-ways driving does away with the need for laborious shunting operations once necessary to prepare a train for its return journey.

Many railcars and MU trains are driven directly by diesel engines in the same way as a bus. The engine is mounted on its side under the floor and coupled through a gearbox or automatic transmission to one or more axles. This arrangement is not practicable when greater power is needed. Designers have instead turned to electric or hydraulic drives. The hydraulic drive, which was popular between 1955 and 1975 but has been on the wane since, connects the diesel engine to a variable-output hydraulic pump. This pumps oil round a closed circuit in which is a hydraulic motor. The hydraulic motor is somewhat like the pump, but operates in reverse. Its output shaft is connected to the wheels. The diesel-electric, which is by far the most common type of locomotive in the world, has each diesel engine driving an electric generator directly. Current from this is supplied to traction motors geared to the driving axles. Surprisingly, a diesel-electric sometimes uses a different electric generator or even an oil-fired steam boiler to heat the train.

Electric locomotives merely convert the externally supplied current into wheel rotation. Their large batteries, switchgear, transformer and rectifier units, and control systems combine to make them hefty contraptions. Nearly all collect current via a roof-mounted folding pantograph, often with a carbon strip, sliding along the high-voltage supply catenary. The earliest railroad passenger coaches were simply modified stagecoaches, while freight trucks were

43

similar to the traditional chaldrons used in mines. It is a far cry from these primitive vehicles to today's rolling stock in which the structural principles and materials are nearer to the jet aircraft than the coach and four. But all are linked by their compatibility with the same duo-rail infrastructure.

When the railroads were first built, the gauge was usually the same around the world. But the loading gauge varied. This is the envelope of extreme dimensions for width and height of all rolling stock. It had to be specified to ensure that all stock would pass safely through tunnels or clear signals, and other lineside fixtures. In Britain the loading gauge was set at a height (at the center of a rounded top) of 13 feet 6 inches and width of 9 feet 6 inches. But other countries accepted larger vehicles and in North America the standard is 15 feet 6 inches by 10 feet 9 inches. Such limits obviously have their effect on the design of rolling stock. In North America, for example, it has proved simple to build double-deck passenger cars for suburban use, and to use equally lofty long-distance coaches either fitted with vistadomes on the roof, or else with passenger accommodation at an upper level, with lower-deck toilets, baggage, air-conditioning, and similar ancillaries. Double-deck usage is not economically viable in Europe, as maximum tunnel heights are too low.

While the loading gauge fixes height and width, other factors determine the length of rail vehicles. For over 100 years Britain was inclined to build small

Above: Canadian Pacific's "Canadian" diesel expresses provides its passengers with vistadomes. These would never clear Europe's tunnels.

vehicles. Virtually all British freight cars and many passenger coaches had only four wheels. For passenger transport the longer swivel truck car rides better, and in recent years weight has sometimes been saved by articulating adjacent cars to single swivel trucks. This process reaches its ultimate in the Spanish Talgo lightweight train. Each car rides on a single axle at the rear, the front being pivoted to the car in front. It is now known that for the best ride at high speeds cars should be as long as possible. The limits are set by the curvature on station platforms and the clearance past signals on the sharpest curves. Again, Britain differs from the majority in having high platforms from which passengers board the trains. In most other countries the "platform" is at, or very near, ground level. The train often has steps at the ends of each car which passengers climb up to floor level.

Interurban coaches invariably have traditional hinged doors, but urban stock usually has powered sliding doors controlled, at least partly, by the driver or conductor. For commuter use the interior provides mainly standing space. There are doors at frequent intervals to allow large numbers of people to enter or leave the train with minimum difficulty. A few subway trains run on noise-reducing rubber tires. This is not economical at present, despite a long list of claimed advantages.

Above: passengers climbing aboard the "Turbotrain," Canada's HST. Most countries in the world have low set platforms on their railroads. Britain has high platforms, which make boarding easier, especially with heavy luggage.

Above: the switching diesel is familiar on most railroads (this one belongs to CP Rail). To make a train ready for a return journey requires elaborate shunting. Many modern trains have a power cab at each end to avoid this complication.

Nearly all suburban and many long-distance passenger trains remain permanently coupled into one unit. Freight cars, on the other hand, are continually marshalled into different trains, which always have to be hauled by locomotive. The couplings have to be very strong to take the pull of as many as eight locomotives on one end of the train. Modern couplings incorporate springing to absorb axial shocks (for example, when a long train starts from rest), connectors for electric or pneumatic supplies, and also grip each car rigidly to prevent it from overturning in a derailment. Compatibility between vehicles of differing national rail systems has become a very important factor in Europe, where inside the European Economic Community (EEC) there is great "foreign" use of freight vehicles. It will take many years to eliminate all the old nationally designed incompatible vehicles.

There is always a rule in freight transport that costs can be reduced by moving loads in bigger units. Freight stock has in recent years been built up to the limits of the loading gauge and reasonable length. The standard British tank car has jumped from 12 tons to 90 tons riding on four axles instead of two. Clearly, if the big wagon costs twice as much to make (and it does not cost twice as much to pull) it pays if it can carry more than seven times the load of the small wagon. Virtually all freight vehicles today have vacuum brakes throughout the train. Until recently it was common to have no brakes except on the locomotive and the rear brake van, and whatever hand-cranked brakes on the wagons could be reached.

One vital factor is axle loading. The weight that can be supported by each axle has tended to rise in recent years, partly as a result of many railroads shutting down minor routes, and concentrating on the main trunk network where the traffic pays for a higher standard throughout. Modern track on major routes is suitable for very high speeds despite axle loadings which often approach the 30 or more tons common in

North America. Many railroads today use two-axle freight cars with a greater capacity than the swivel trucks of 30 years ago. Increasingly there is a move away from open wagons toward either special-purpose carriers, such as for cars, bulk cement or other defined commodities, or to long flat cars which have anchorages for standard containers. In North America there are "piggy back" cars on which are secured the giant semitrailers of articulated road trucks, which at their rail destination are coupled to tractor units and driven off on the road.

Below: an iron ore train at the Llanwern steelworks in South Wales. Railroads are ideal for carrying heavy loads like iron ore and coal. Road haulage interests leave this type of work to the railraods. Nearly all the world's railroad systems are now state owned.

High-speed Trains

Above: a high speed electric locomotive of the French railroad SNCF. Engines like this pull ordinary railroad coaches, and in March 1955 two types achieved a speed of 205 mph in trials. Passenger trains do not yet approach this speed.

Like coal, the railroad industry in many parts of the world went into decline in the 1950s and 1960s. It was popularly thought out of date, and in almost every developed country suffered drastic closures of up to three-quarters of entire national systems. Today, again like coal, the railroad is widely accepted to be of the greatest long-term importance. Thanks largely to British research it is known that, apart from the exciting but possibly costly promise of totally new systems, existing track can serve for future running at 200 mph.

The first time speeds remotely approaching this level were achieved was in an important series of trials on the SNCF (French Railways) between 1953 and 1955. Using ordinary passenger coaches hauled by electric locomotives, modified only in having higher ratio gearwheels in the drive from the motors to the axles, speeds were progressively pushed up until, in March 1955, two types of locomotive each reached a speed of 205 mph. Even though the section of line chosen was straight and of high quality, this was a daring exploit. At that time little was known of the behavior of locomotives or coaches at anything over 100 mph. Few regular trains travel above this speed. Powerful electric and diesel traction, however, have made progressive increases in the average speeds of ordinary expresses possible, until today 100 mph is not uncommon.

In the United States competition from the low-fare bus and the swift jet has almost eliminated long-distance passenger travel by rail. Most passengers now make such a journey for its scenic attractions without regard for economic considerations. In other countries trains are increasing their traffic, and increasing speed is vital if they are to compete with the airlines.

In very few instances the bold decision to lay down new rail routes has been taken.

The first of these was the New Tokaido line in Japan. This links some of the world's largest cities, including Tokyo, Nagoya, Yokohama and Osaka, with a total population of well over 20,000,000 people, on a 320-mile stretch of line. The existing line was winding, slow, and clogged with traffic. Despite severe

congestion on the air routes nearly all passengers were switching to the airlines. The New Tokaido line was the first new route in a developed country for many years, and it was built for speed. The civil engineers avoided all sharp bends and speed restrictions. Signalling and automatic train control is designed for average speeds above 100 mph. All trains are electric, drawing overhead current at 25,000 volts, and the standard passenger trains are MU streamliners with traction motors rated at up to an aggregate of 12,000

Right: Japan's New Tokaido Line links that country's largest cities. The first new rail route to be opened in a developed country for many years, the New Tokaido Line was built for speed – the run normally averaging 100 mph.

hp. Much was learned with this line about how to make it possible for trains to pass each other at a relative speed of 300 mph, how to plunge into a tunnel without causing "popping" of the eardrums, and even how to design proper toilet and washing facilities. To some degree the New Tokaido is a brute-force system, and it is costly to run. Each night the "Super Trains" are in for maintenance, such as replacement of brake pads and regrinding of tire profiles. Most countries are more concerned with what can be done on existing track.

British Rail long ago discovered many new things as a result of extremely detailed research and computer studies. One was that, though it is sensible to try to eliminate the worst curves, bottlenecks and causes of "permanent-way slacks" (such as colliery subsidence or old bridges), it is far more cost-effective to put available capital into new trains rather than new track. While some research continued on new track, mainly to eliminate maintenance, extensive study began to solve the equations of motion of rail vehicles. Soon complete answers were being obtained. One of the first fruits of this research was a new standard design of four-wheel freight wagon. This was able to run with precise directional stability at 145 mph, whereas previous wagons had lurched uncontrollably at less than half this speed, and were prohibited from exceeding 45 mph.

In planning an Advanced Passenger Train (APT) the new knowledge confirmed the need for a very high power/weight ratio in order to make the steady running speed almost the same as the maximum, no matter what the uphill gradient might be. Fast acceleration is also a requirement, though there is a clear limit to power/weight ratio set by the slipping of the wheels. There was additional research on how the slippery film on the rails, caused by oxides, ice, oil, and other extraneous material, could most effectively

be removed to give near-perfect adhesion. An especially far-reaching conclusion was that the speeds that would be possible with an APT would subject passengers to considerable discomfort on bends. There was no question of danger, but rounding the sharpest bends on the mainline routes at speeds close to 200 mph would fling passengers sideways and cause alarm. The answer was to tilt each coach as it rounded the curve.

In Canada the rather slower Turbo-Train was built with lightweight coaches supported at each end on high pivots. On curves, the cars swung outward automatically. Though this went part of the way to improving comfort it did not give the optimum rotation of the body of the coach, which ideally had to rotate as the train entered the curve, not be flung out as a result of it. The APT was therefore designed with extremely long coach bodies mounted on pivoted low-level supports. These were coupled to actuators able to tilt the body nine degrees to either side. Lateral-acceleration sensors in the leading swivel truck of each coach detect any bend in the line, and command the tilt needed to keep passengers sensibly "upright." Testing of the first APT began in 1973. By 1978 both gas-turbine and electric versions had been built, the latter matched to a 25,000 volt track. The tilting system worked well from the first, and was soon tailored to give an ideal smooth response. General directional stability of the new swivel trucks is so perfect that there is no false lateral input which could cause the coach bodies to keep rolling slightly even on straight track. Other features of these trains include stressed-skin aircraft-type construction and hydro-kinetic brakes in which energy is put into fluid circu-

Below: Britain's Advanced Passenger Train (APT) is designed to keep passengers comfortable while the train takes the sharpest bends at high speed. The APT will probably replace the 125 mph HST in the course of time. Such trains conserve energy.

Above: a Metroliner leaves Philadelphia on the high speed run from Washington DC to New York. Operated by the publicly owned Amtrak, the Metroliner has actually succeeded in tempting people away from the private automobile and the airlines.

lating in pipes. These revolutionary brakes are so powerful that the APT can run at full speed with the signals still at their existing spacing.

A less-advanced relative of the APT family is the SNCF's Turbotrain. Like the New Tokaido, it "cheats" by having a special high-speed track laid for it from south of Paris to Lyon – bypassing Dijon. Powered by Turbomeca aircraft-type gas turbines, the four-car Turbotrain has two power cars and can cruise at 186 mph, a speed handsomely exceeded many times on trials since 1972. The scheduled time over the new, more direct, line from Paris to Lyon will be two hours at an average of 158 mph.

The astronomic capital sums of money required for

new routes and trains need plenty of traffic to justify. For many years the most important single route in the world has been that from Boston to Washington, DC. This passes through densely populated areas including Providence, Hartford, New Haven, New York, Philadelphia, Wilmington, and Baltimore. There have been prolonged studies of this "North-East Corridor." The Metroliner cars show what can be done with conventional technology. They are traditional products of the Budd Company, extremely large – 85 feet long, and weighing 75 tons each. They can be coupled together to make a train of any required length up to 20 cars, and each is provided with no less than 2560 hp from its electric traction motors. Acceleration is tremendous. Two minutes from rest the train is at 125 mph with 160 mph attainable. They seldom reach

Below: the gas turbine powered Turbotrain introduced in France by SNCF in 1970. Turbotrain covers the 230 miles between Paris and Cherbourg in three hours.

Above: the French Aérotrain system was designed for travel in cities, between suburbs, or at airports where the distance covered would not be more than a few miles and may be as little as a couple of hundred yards. Self-powered air cushion vehicles run along a lightweight overhead track. The vehicle capacity can be as many as 100 or as few as four passengers according to requirements. Up to 15,000 passengers an hour can be carried each way on the system. Such a system can offer a flexible and economic transportation network.

their top speed but their quick braking and acceleration makes it possible to sustain high average speeds, despite frequent stops.

One of the oldest systems, other than the duorail, is the monorail. This carries cars or trains either riding astride or suspended from an overhead "rail." But the monorail does not offer any prospect for increased speed, and is chiefly seen as overhead routes in urban areas.

Chronologically the next unconventional system, and the first of the high-speed guided transport (HSGT) systems in which methods other than traditional rails are used to steer the vehicle, was the French Aérotrain. This was also the first tracked air-cushion vehicle (TACV) though it is not considered the optimum design. It rode on a track or guideway in the form of an inverted "T." The vehicle was supported by air-cushion pads (each resembling a miniature hovercraft) running on the horizontal surfaces of the guideway, and guided by additional pads running along the vertical faces of the central wall. The first test car reached 190 mph with an air propeller driven by a small aircraft piston engine. Later it was given a small jet engine and achieved speeds of up to 264 mph. Various other Aérotrains were built with more advanced propulsion systems. In Britain a very much better Hovertrain was tested. It runs on a rectangular-section track assembled from cheap but rigid concrete box units giving high accuracy, and avoiding the problems of the inverted "T." The vehicle straddled the track and was powered by the best form of propulsion there is for the HSGT vehicle – the linear induction motor (LIM). In this the vehicle picks up current by any of the usual methods and passes it through coils running close beside an aluminum strip recessed along the guideway. The magnetic field induced by the inductive interaction between the coils and the strip serves as a propulsive force. The LIM is almost 100 percent efficient, and it gives perfect nonslip propulsion without mechanical contact, and therefore with no wear and no sound. There have been many other TACV research vehicles and experimental prototypes, the majority tested in the United States. Contrasting forms of guideway have been researched, and very high speeds reached. The results have demonstrated that the seemingly odd policy of leaving the wheels off a train can help it to be faster and smoother.

For the more distant future there are ideas which often seem like science fiction. The most thoroughly proved is a close relative of the TACV, the magnetic-levitation vehicle (maglev). In this the weight is supported by magnets interacting between the train and the track. Leaders in maglev research are a series of industrial groups in West Germany and Japan. All of these are studying superconductivity as a means of improving efficiency. Some maglev plans lift by mutual repulsion and others by attraction. But in all cases the train rides just far enough from the track to avoid touching it. Speeds can easily exceed 300 mph.

Below: The Transrapid Maglev in Munich is powered by the linear induction motor (LIM). The modern LIM consists of two elements, a "rotor" and a "stator", free to move past each other with a constant small air gap between them.

Bridges

Nothing more perfectly mirrors man's talent as an engineer than the bridge. The first was probably just a fallen tree. Over 4200 years ago man built a great bridge of stone and wood across the Euphrates river. In 1779 the first metal bridge, an arch of cast iron that still stands, was completed at Ironbridge, England. Today there are many structural categories of bridge,

Above: Ironbridge, Coalbrookdale, England. Designed by Thomas Pritchard and built over the Severn river by John Wilkinson and Abraham Darby between 1797 and 1799, the bridge was open for traffic until the 1950s.

made of many materials. Some bridges are simple beams, supported at each end and able to bear loads by being stiff in bending. To support greater weights or span, wider gap bridges of arch or truss design are adopted. Arch bridges are traditionally made of stones or bricks which wedge themselves together, and push outward on the support at each end. Modern arch bridges are often metal. The deck on which the road or railroad is laid is hung from the arch by ties or cables. A truss is a rigid frame made by riveting or welding. Cantilever bridges are made strong enough to hang well beyond the vertical support. Some bridges are beams supported by cables from a tower. Many beautiful bridges are made of reinforced concrete. The widest single spans are suspension bridges. A tall tower is built on each side of the gap. Cables are passed

over the tops and anchored at each end, to provide support for a bridge deck too long to be a plain beam or cantilever.

Some bridges are unusual in design and do not at first appear to fit these basic categories. One of the oddest was the Britannia tubular bridge across the Menai Strait between Wales and the island of Anglesey. Built by Robert Stephenson to carry the railroad to Holyhead, and opened in 1850, it was actually a beam bridge. But the beams were long iron boxes through which the trains pass. It was destroyed by fire in 1970. London's famous Tower Bridge carries the road on two true cantilevers, each hinged at the end. To allow large ships to pass underneath (today extremely rare) the cantilevers are pivoted upward. A bascule bridge, the girder beam across the top was merely a footpath, playing little part in supporting the structure. An alternative is the swing bridge, with the

Above: This drawbridge on the Welland canal, Ontario, Canada, is drawn up to allow ships to pass beneath. One of the most famous of the type is Tower Bridge across the Thames river in London.

main section pivoting on enormous bearings. Another way of taking traffic across a river used by large ships is the transporter bridge. Traffic is loaded onto a platform suspended from a high deck, such as a girder beam or suspension bridge, above the tops of the tallest ship masts. When the platform is loaded it is driven slowly across the river by powered wheels running on tracks along the supporting deck. Another is the vertical-lift bridge. Here, the entire bridge deck – usually a girder beam – can be raised or lowered by hoists inside a tower at each end.

In the late 19th century two of the most famous bridges were built to carry railroads across Scottish rivers. The Tay bridge resembled a viaduct in having a large number of short spans supported on vertical struts. Most viaducts, from Roman times, were stone arches, but the Tay bridge was iron. Because of unsuspected flaws, it was blown over sideways in a storm in 1879. The Forth bridge, completed in 1890, is really a cantilever with three identical giant trusses each supported at the center and linked by two short beam girders added to complete the bridge. Another superficially related group of bridges are called bridle-chord structures: their decks are long beams, supported by cables or diagonal tie-rods from a tower at the center. Yet another very modern structure is the portal bridge. Reinforced or prestressed concrete is used to make a series of cantilever beams which are then pivoted together to make arches.

Cantilever bridges can cross very wide gaps. The last record-span held by a cantilever was the Quebec rail bridge. In 1929 this was beaten by the Ambassador bridge in Detroit, a suspension bridge. To build such a bridge a very firm foundation must be laid on each side of the gap. A tower is then built on this. Though the height of the tower needed for a given span has gradually fallen as structural materials have become stronger, the towers for the largest bridges are among

man's tallest structures. Several suspension bridge towers exceed 650 feet in height. All are of high-strength steel, and some are clad in stone or concrete. The next, and most crucial, task is to "spin" the wires. Thousands of miles of extremely strong wire are needed. This is galvanized to prevent rust, and is about the thickness of a pencil. First, a supporting cable is laid from end to end, passing over the tops of the towers. This carried a large pulley which passes back and forth across the bridge pulling a double wire on each trip. Eventually thousands of wires are formed in two groups, making two giant cables supporting the two sides of the bridge. Verrazano Narrows bridge, New York City, had the world's record span in 1978 – 4240 feet. It has 18,750 miles of wire in two cables each over three feet in diameter and composed of 26,108 wires. These vast cables are securely anchored at each end, the wires being splayed out in groups and fastened

Above: two high lift bridges, the work of the McDowell Company, and a utilities bridge. **Left:** the Verrazano Narrows Bridge being built. 20,000 miles of wire was used in the four main cables.

to enormous anchorages with steel arms by the score set in highstrength concrete. To complete the bridge, tie cables are hung from the two main cables. These support the deck, laid one welded-steel section at a time.

Much has been learned in recent years how to build better suspension bridges. The Tacoma Narrows disaster in 1940 showed that a seemingly strong bridge can be vibrated by the wind like a violin string until it collapses. The old heavy bridge decks have become lighter and streamlined so that the wind has much less effect. Modern steel is so strong that the suspension cables can bridge larger gaps, while passing over shorter towers. The towers themselves have become much simpler and, most observers would agree, more aesthetically pleasing. Today nonmetallic fiber of carbon or graphite, very much stronger and stiffer than steel yet less than one-quarter as heavy, may make possible bridges more than three times the span of any yet built.

Tunnels

Again it was the Babylonians who built the earliest known tunnel – under the Euphrates river. It is known to have been in use well over 4100 years ago. Since then tunnels have been cut through previously impenetrable mountains, shortened other overland routes, burrowed under rivers and seas, and honeycombed below the cities of the world. Man is not an

Above: the longest road tunnel in the world connects France and Italy through Mont Blanc. It is 7½ miles long and saves a detour of more than 100 miles. Work began in 1960. Two teams – one French, one Italian – started digging towards each other. Picture shows the Italians at work in April 1960. Digging took 2½ years – the two teams meeting up on August 6, 1962.

Left: the French at work in the Mont Blanc tunnel. Road tunnels are expensive to build and there must be considerable advantages to be gained before the go-ahead is given for the work to proceed. In the case of the Mont Blanc Tunnel the time and distance saved has had a significant impact on freight costs as well as making life easier for the automobile driver and passenger.

underground animal by nature, yet he has burrowed beneath the Earth's crust to an astonishing degree. Apart from road and rail tunnels there are tunnels for hydroelectric plants, water supplies, electricity supplies, gas, petroleum, air raid shelters (especially in modern China), and tombs. Modern mining often involves tunneling on a large scale, and an exhausted mine often serves as a handy place to dump refuse or dangerous radioactive wastes. But concern is mounting over the extent of this dumping.

Tunneling is simplest through clay and soft rock, but such material easily collapses and is liable to flooding. Hard rock is safely excavated and is usually impervious to water. But it is much harder to cut. Another factor that determines how the work is done is the depth. The modern mountain tunnel is invariably begun at both ends. With an accuracy of much better than one inch the two halves meet in the center. But where the tunnel is shallow, the cut-and-

cover method is usually cheaper and quicker. A trench is dug, the tunnel built in it and the top covered over. This is how highway underpasses are built, and it has become common for long undersea tunnels. In the latter case the tunnel is built in the form of enormous prefabricated sections of reinforced concrete, each weighing thousands of tons. These are lowered onto a foundation in an undersea trench and made watertight.

The earliest roads wound across mountain passes, and in many places were safe only for a nimble goat. But the early railroads could not follow such tortuous routes. Tunnels were necessary to give gentler gradients and less acute curves. Other early tunnels were built for canals. Only just over 50 years ago a canal tunnel was opened in France which was big enough to take ships from Marseilles to the Rhone. Whenever a tunnel is dug through clay or sand a major problem is preventing the newly excavated sections from caving

Above: work in progress on the Victoria Line of London's Underground. This North–South line was finally completed in 1969.

Above: the Chesapeake Bay Bridge-Tunnel. This picture shows a portal of one of the two 1½-mile tunnels built to provide navigation channels between trestle sections of the bridge-tunnel. These man-made islands at the junction of trestle and tunnel are built of sand, stone, and concrete. The tunnels are made in sections of double-skinned steel tubes, and were sunk in position on site. The tubes can withstand enormous pressures.

in. Until 100 years ago the work was laborious and dangerous. Hundreds of men dug and burrowed into the spoil while others kept pace with heavy timbers to put up temporary formwork to prevent collapse while the permanent brick or stone tunnel was being built. In 1824 Sir Marc Brunel built a large tunnel under the Thames river in which he used a massive iron shield. This was a three-story frame on which the diggers worked, protected at the front by heavy timbers keeping out the sand, mud, and water except where digging was taking place. This shield was moved forward at intervals. Later shields have been giant drums of thick steel. These often carry cutters and can be rotated and pushed forward by hydraulic rams.

Rock tunneling is invariably done with explosives. Drillers work on a "jumbo," a large multistory frame the same diameter as the tunnel, and with tungsten-carbide drills penetrate up to 10 feet in a carefully arranged pattern of holes in the rock face ahead. These are then packed with an explosive, such as dynamite. The charge is fired electrically from a safe distance, and the loosened spoil is removed by a large digger/conveyor. There are still hazards. It cannot be assumed that rock will not collapse. While building the Simplon road tunnel in the 1960s one jumbo was crushed by a rock fall and 17 men were killed. And there are always dangers from water, which may rush in under great pressure (sometimes at near-boiling temperature) and poisonous gases which may be released from underground cavities. It is essential continuously to pump in ample cool fresh air, and of course to dispose of the gases from each blasting explosion.

Sometimes special methods are needed to keep out water. One technique is to pressurize the inside of the tunnel so that water can no longer flow in. But this means that every worker becomes a diver and has to undergo a diver's gradual decompression after each

spell at work. Another is to pump grouting compound or bentonite, a kind of liquid cement that sets under water, to form an impervious barrier.

Still another is to use gelling salts or bitumen compounds, while in a few cases the area around the tunnel has had to be frozen by pipes carrying liquid nitrogen.

Though there are many mountain tunnels that are far deeper, the longest single trunk-route tunnel is the Seikan linking the main island of Japan with Hokkaido to the north. Though the two shores are 13.7 miles apart, the tunnel naturally has to be much longer, and is 22.6 miles in length. Extra sections were drilled sloping down from the deepest central section to avoid any catastrophic flooding. Chesapeake Bay in the United States is crossed by a slightly longer combination of tunnels, man-made islands and bridges. The proposed "Chunnel" linking Britain and France would have had a length of 32 miles.

Travel in Cities

The human population is still rising rapidly, and people are everywhere congregating in the cities. The growth of cities thus poses problems so great as often to seem insoluble. Aside from the question of making a city pay its way – many, like New York, are in desperate financial straits – the difficulty of maintaining good communications is taxing planners and engineers to the limit. The answer is usually that much could be done, but cities cannot afford it. Money is the stumbling block, not technology.

It would be simpler to build an efficient modern city if today's cities did not exist, and everything could start from scratch. Many grandiose plans for urban renewal, advanced "rapid transit" rail systems, and attractive plans for computerized "personal transport" networks have foundered on the rocks of existing cities which are the result of many years, sometimes centuries, of generally unplanned growth. The worst manifestation of unplanned growth is usually the road system. In the case of more than half the world's major cities the road system is totally unable to meet the demands made on it. In many cases the layout is much the same as that laid down one, two, or more centuries ago.

One way of attacking the problem is to build new urban highways. A second is to try to tailor the vehicles to suit the roads. A third is to build wholly new transport systems, if possible either above or below ground. In almost every major urban area – which often forms one continuous "agglomeration" with several cities growing into each other – new highways have proved enormously costly and wreaked large-scale demolition, causing vociferous discontent. City cars (cars designed for short urban journeys only) though on paper extremely attractive, have either never reached the market place, or proved marketing flops. New systems have hardly ever left the laboratory. In many cases they have not even progressed beyond the paper stage. Most cities have done little beyond pick around the edges of the problem with trivial local improvements to intersections and the worst bottlenecks. Liberal quantities of white and yellow paint have been used up marking the roads. Legislation of an almost wholly negative nature has been erected (for example, forbidding private cars from certain streets or even from entire city centers.)

Urban streets have borne various forms of personal transport for centuries. The Roman litter, the Sedan chair, the Hansom cab and the modern taxicab have all had their uses in their time. The public transport vehicle plying a fixed route arrived in the 19th

Above: traffic congestion is not a new phenomenon. This engraving depicts Ludgate Circus in London completely jammed in the 1850s. In the picture **below** is its modern equivalent – a traffic jam in Rome. In many cities the average speed of traffic through central areas is yearly getting lower. And the exhaust pollution of the modern traffic jam makes matters much worse.

century. The horse-drawn bus was followed by the motor bus, the tram, and the trolley bus. Urban railroads were also built. Today, the favored personal transport is the car, with various two-wheel vehicles also enjoying popularity. In 1945, the end of World War II was accompanied by optimistic paintings showing a personal helicopter on every front lawn. This did not happen. The only vertical urban transport is the elevator. But many designs exist for transportation systems that operate in both the vertical and horizontal plane.

Below: a West German *C-Bahn* (Cabin Taxi) system in operation. At a hospital near Kassel such a system is working well. Some cabs are for private hire, while others are available to people waiting at stops.

Such systems are stillborn by the existing infrastructure of most towns and cities. While they could be incorporated without difficulty in a new city they cannot in an existing one. All that can be done is conduct local experiments whenever a large new complex of buildings is constructed. Several large hospitals have recently been built with integrated horizontal transportation systems, and in a few cases these can function on different levels. But no example yet exists of the three-dimensional guided transportation system that so many inventors and companies have proposed.

Nearly all advanced engineering companies involved in transportation have sketched and patented city transportation systems. All carry people in small vehicles which sometimes run on wheels, sometimes on air cushions or magnetic levitation, and sometimes are linked to moving cables or belts. Some are suspended from overhead monorails, some straddle monorails, and a larger number use ordinary pneumatic tires on a reinforced concrete guideway. Typical of the more conventional systems, *Cabtrack*, a British proposal, uses four-seat electric vehicles which run on an overhead guideway in the form of a carefully planned network covering a whole city center. One or more computers control the system. Patrons walk to the nearest cab stop, pay a fee into a slot machine

and dial the destination. The central computer would control each of the many hundreds of cabs, slowing them on corners, avoiding bottlenecks or collisions and always selecting the most direct route. A later system in West Germany, the *C-Bahn* or Cabin Taxi, has actually moved beyond the paper stage. A local system at a hospital near Kassel has worked well in 1977–78. Much larger systems have been studied for several West German cities. Cabin Taxi has been planned with various sizes of standard vehicles hung from an overhead guideway. Three passenger cabins are designed for private hire, running direct on whichever route the customer selects. The 12 passenger size would run on fixed routes with "request stops," while 12 to 40 passenger cabins (seated or standing)

Above: electric cars may provide the answer to pollution-free private travel in cities. This one has been tested in Britain. But so far no one has developed an electric car to match the speed and performance of the internal combustion powered automobile. **Below:** a bubble car of the 1950s. The driver was seated too close to the ground for visibility and safety.

are planned for internal routes at airports, large industrial complexes, hospitals and other local networks.

Another alternative is the city car belonging to the user. A big step in this direction was the development in West Germany 30 years ago of a series of miniature cars, popularly called "bubble cars," which appeared ideal for urban use. Seating two adults, they were extremely small and could park in half the space needed for conventional cars. They had adequate performance and carrying capacity for commuting or shopping, and were extremely economical. Where they failed was in their incompatibility with buses, giant trucks and even ordinary cars in city streets.

Below: the Tokyo monorail system in operation. Providing an efficient public transport system has become essential in many cities as a means of discouraging the polluting use of private automobiles. The underground railway is the most popular.

They were often too low for other drivers to see them, and they were sometimes unwittingly crushed in heavy traffic. Another disadvantage was that they were incapable of serving a family, and inadequate in speed and payload for long holiday or business journeys. Owners usually had a large car as well.

Since 1960 more than 50 prototypes have appeared of commuter cars, shopper's cars, and other small cars tailored to urban use. The need to conserve energy and avoid pollution has recently given an added spur to the development of such vehicles. Added pressure has also been put on electric propulsion. This has never proved adequate for normal family cars, but is becoming acceptable for purely urban use. In the long term, the solution is probably to provide an electricity supply from a central source for urban vehicles. In the absence of such a supply the answer must be to fit a battery in each city car, and the extremely poor performance of electric batteries has been the biggest

single obstacle to the urban car. The standard battery is the lead/acid type. This is unable to store more than one percent of the energy stored in a tank of gasoline of the same weight. More advanced batteries exist but all are expensive, and in most cases use dangerous toxic or corrosive materials which may spill in a crash. There is abundant experience with electric commercial vehicles, mainly milk delivery vans, but the electric (or gas) city car has consistently proved unpopular.

This is in stark contrast to the various forms of electric train built for urban use. Many large cities have a subway, metro, or underground system covering at least the central business district, and in some cases

Above: a moving walkway in an airport near Copenhagen. Passengers and visitors carrying heavy baggage can stand and rest while the walkway does the work. Baggage-handling facilities at major airports have long been designed on the same principle, and the moving walkway is becoming increasingly common. People can travel distances easily that would be much too far to walk.

Right: a possible answer to the increasing congestion of the world's cities. Sea City was suggested by the British glass firm Pilkington. Their model would provide facilities for about 30,000 people. An outer floating boom would act as a breakwater. Sea City would have a high wall, profiled to deflect strong winds. Such suggestions, although interesting, are not yet becoming reality.

including personal hire vehicles with driver or computer control. A few involve personal baby cars which are driven from home to a station, put aboard a train or other public system, and driven off at the city center station. Other cities forbid private transport altogether and reserve central highways for buses and professionally driven taxi cabs.

One transportation system seldom seen outside the city is the endless-belt method such as the escalator, Travelator, moving sidewalk, Transcab, and many other low-speed systems. Numerous inventors have sought for years to solve two difficult problems. The first is how to make a single integrated system serve both horizontal "along" travel and vertical "up and down" travel in tall buildings. The second is how to achieve speeds much higher than the 2 mph typical of today's continuous-web systems without making it difficult for very young or old customers to step on or off in safety. Patent literature is full of cunning arrangements which either use belts of different speeds side-by-side, or else arrange for high-speed belts to slow down at points where travelers get on or off. Some plans use boxlike compartments running on the continuous web and are often hard to distinguish from other guided transport systems. Very few have successfully solved the problem of true flexibility in all three dimensions, which is anyway too difficult to incorporate into the streets and buildings of existing cities.

serving dozens of stations extending far out into the suburban areas. Some routes run at least partly at street level, often along the central reservation of highways and sometimes taking in former tram routes. Several cities, notably in Japan, have extensive elevated monorail systems. These are cheaper than going underground and use up much less ground-level space than surface railroads. The United States alone has proposed over 100 urban guided-transport systems, mainly operating on fixed routes, but often

A number of eminent architects, engineers and urban planners have studied the pros and cons of building entirely new cities on new sites, such as on man-made offshore islands. This solves many problems but creates fresh ones. For example, they have not yet suggested what should be done with today's sprawling inefficient cities. In any case, most people seem to prefer living in the inefficient old cities that they know, than in the few new towns that have been built since World War II.

CHAPTER 3

MARINE TECHNOLOGY

Ships, of a primitive type, were man's earliest vehicles after make-shift canoes and rafts. For thousands of years they have also been his largest vehicles, with the capacity to carry all his artifacts other than fixed structures. Modern water transport is changing with unprecedented speed. Nowhere else can the economies of scale – the reduction in unit costs brought about by using bigger vehicles carrying bigger loads – be realized so effectively. There is still no visible limit to the size of bulk carriers, but the larger the ship, the more restricted its choice of routes and ports. Two alternatives to the conventional displacement vessel are becoming increasingly available. At present, air-cushion vehicles and hydrofoils are used only for high-speed passenger and vehicle ferries, and for naval purposes. But there is no reason why these fast craft may not eventually become economic carriers of containers and other urgent cargo. Cargo submarines, on the other hand, pose severe problems.

Opposite: a production platform in British Petroleum's oilfield in the North Sea. Extracting natural resources from the sea is one contribution marine technology is making to our survival.

A Brief History

The first water craft were probably floating tree trunks, and their riders were unlikely to be able to swim. Simple boats were being built more than 40,000 years ago, and even at this stage men had probably learned to propel and steer them – first with their hands and then with paddles. But it was more difficult to build ships that could withstand the rigors of the open sea, and to learn how to navigate from one place to another.

But both these skills were mastered more than 6000 years ago in various parts of the Near East, including modern Iran, Iraq, Syria, and Egypt. By 4000 BC ships had been built over 325 feet long and propelled by sails, or large numbers of rowers using long-bladed oars. By the year 1000, Arabs were using the lateen

sail. For the first time this allowed ships to run across the direction of the wind. By zigzag motion, or "tacking," a ship could make progress in the face of the wind.

From Roman times until the 18th century, large ships tended to be divided into two types. The first were fast rowed galleys, with a narrow beam width at waterline in relation to length, and usually propelled by galley slaves. The second type were fat-bellied merchant ships with sails on one or more masts, often with a grotesquely high poop and stern from which the vessel could be defended if attacked. Generally, the larger the ship, the greater the number of oars or the area of sail required. Slaves and convicts gradually disappeared as potential rowers. This brought about the disappearance of the multi-oar

Above: lateen sails on Arab boats on the Lower Nile river. In use for nearly 1000 years, the lateen sail is still an efficient means of propulsion.

Above: the *Statsraad Lemmkuhl* competing in a Tall Ships race. Totally non-polluting, many tall ships can outpace some modern diesel-powered vessels.

Left: a Roman Bireme. An early galley, it was propelled by slaves – usually convicted criminals.

galley, but the labor-intensive nature of saled ships meant that large vessels had a crew numbered in the hundreds. The largest barks had two or three masts with square-rigged sails mounted widthways. As many as five or six sails were rigged to each mast. As many as 300 men were needed at once just to trim the sails of such ships. The Dutch-built schooners with fore-and-aft sails, needed fewer crew, and the most efficient of all was the single-sail Chinese junk which is still in use.

By the 1820s the steam engine was beginning to

Right: the famous *Great Eastern*, designed by Isambard Kingdom Brunel, pictured in 1858. This early steamship was 600 feet long and had five funnels. It also had 6500 square yards of sail. Designed to accommodate 4000 people in comfort, it proved too costly to run on passenger routes, and was converted to lay Atlantic cables. The *Great Eastern* was scrapped in 1889.

Below: one of the world's oldest craft to have remained basically unchanged – the Chinese "junk." Junks come in many shapes and sizes, though all have in common unstayed masts and fore-and-aft lug sails of matting spread on horizontal bamboo battens.

Above: the *Savannah*, the world's first nuclear powered merchant ship. It is likely that, as the oil crisis deepens nuclear power will be used more and more, although there are worries about the risks.

By 1840, the screw propeller was beginning to prove its superiority, except in shallow waters where it was prone to damage. Some vessels, such as the gigantic *Great Eastern* of 1858, had propeller, paddles, and sails. But by 1880 the future lay squarely with the steel-hulled screw-propeller ship with no sails. Coal firing was then universal, and from the beginning of the 20th century the steam turbine vastly increased the speed of the largest ships. The first motor ships made their appearance after 1911. Oil or diesel engines gradually gained dominance. This class overtook the

rival sail, and iron was recognized as an alternative building material to timber. The competition from the iron steamship forced changes in sailing ships. These culminated in the sleek Clipper ships of the 1850s, with three square-rigged masts, and an iron (later steel) hull. Many people refused to believe that iron could float, but in fact the reduced thickness in the structure of the metal ship allowed much more room for cargo. Also, the policing of the seas by the Royal Navy, and navies of other countries, eliminated most pirates and with them the heavy high poop and stern.

Throughout the 19th century shipowners argued about the merits of steam and sail. Many used ships equipped with both. All the early steamships were propelled by lateral paddlewheels or, in the case of American riverboats, a single full-width sternwheel.

steamship in total worldwide tonnage in 1960.

In 1959 the first nuclear-powered merchant ship was launched (the US-owned *Savannah*). Though such propulsion is technically proven, it has never been economically viable. Studies were carried out in the 1970s for extremely large nuclear-powered bulk carriers. None appear to be viable unless there are major rises in the price of petroleum. Meanwhile the aircraft gas-turbine engine has had a massive influence on marine propulsion. Three companies – Rolls-Royce of Britain, General Electric, and Pratt and Whitney of the United States – provide engines for nearly all the largest and fastest container ships. Not only are gas-turbines efficient and reliable engines but their greatly reduced weight and bulk, compared with any alternative means of propulsion, allow for considerably larger payloads to be carried.

Seagoing Ships

Ships' masters have been used to handling cargo for thousands of years. They learned how to stow it so that the ship was correctly "trimmed," riding exactly level in the water. Some commodities had to be kept well separated from each other, and whenever storms were likely everything had to be securely lashed down to stay in place while the ship heaved and tossed. So many ships foundered in the 19th century through overloading that in 1876 ships using British ports were required by law to have a "Plimsoll line" painted on their sides. Named after Samuel Plimsoll, the British member of Parliament who brought about its introduction, this line shows the maximum safe water level for different densities of salt or fresh water. Soon ships the world over had to be marked with a Plimsoll line. Four years later ships began to be equipped with refrigerated holds to transport meat and other perishable goods. Some vessels were designed to transport only one commodity, like coal, while the tramp was a privately owned ship that called at any port and carried any cargo available.

Today cargo is grouped into four classes. General cargo is the traditional kind, made up of sacks, boxes, bales, drums, and large items such as locomotives. Refrigerated cargo is invariably either foodstuffs or pharmaceuticals. Containers are internationally standardized boxes. Bulk cargo are raw materials such as oil or iron ore that can be carried without any form of packaging.

General cargo predominated a century ago but today it is rare, mainly because of the advantages of containerization. The container revolution has transformed all freight transport, especially by sea. Packing goods into a container at the start of its journey means it can be sealed against theft until it arrives at its destination. It is much better protected against damage. Throughout the journey only one item – the container, with its known serial number – has to be loaded and unloaded, saving much costly labor power. Similar savings are made in the paperwork. Containers, eight feet square and 20, 30, or 40 feet long, exactly match the lifting grabs of special cranes, or the locating rails of container ships and the automatic locking systems that secure them in place. Many shippers have discovered that containerization cuts total costs in half and reduces losses from theft and damage almost to zero. Nearly all urgent mass-produced goods travel in containers. Container ships are among the largest and fastest vessels afloat – only bulk carriers such as oil tankers being bigger. A growing number have compact and economical gas-turbine propulsion, giving speeds of about 26 knots.

Containerization has meant channeling cargo through fewer and fewer ports, each lavishly equipped with giant traveling cranes, computers, warehouses, mechanical handling systems, and extensive road and rail connections. Thousands of smaller ports have declined, and handle only infrequent sailings using old labor-intensive methods. The same has happened with bulk carrying. It is fundamental to all transport that if vehicles can be made larger, and still be filled, costs are reduced. The most obvious group of bulk carriers are the tankers built for transporting crude petroleum. In 1950, a headline read "Monster tankers exceed 25,000 tons." Today, they have reached 500,000 tons, the largest vehicles ever made. The largest are 1244 feet long, have a beam (width) of 204 feet, and a draft (depth in the water) of 92 feet.

Below: containerization has revolutionized merchant shipping. The container is transferred from a railcar to a truck by means of a gantry crane. The containers are lifted aboard ship by other gantries. When loaded, the ship's deck and hold will be packed with containers.

Below right: an aluminum sphere for a liquified natural gas tanker. Built in Quincy, Massachusetts, this was the fifth sphere for the first LNG tanker to be built in the United States. Completed in March 1977, this vast sphere is over 12 stories high and weighs about 850 tons empty. Such tankers will become commonplace.

Above: a newly completed bulk carrier being towed out of an enclosed shipyard at Sunderland Shipbuilders Ltd. Many of these carriers are so large that they never enter port and deliver their cargoes to offshore jetties.

Such enormous sizes severely restrict the routes open to these "supertankers." Other bulk carriers, almost as large, handling gravel, ores, salt and many other commodities, face the same problem. Usually they do not come into harbors, but load and unload at purpose-built jetties often far offshore, where the cargo is handled by pumps and pipelines, vacuum piping for powders or grain, or mechanized bulk-handling installations for solids. Despite their size these colossal ships are built inside enclosed factories, from which emerge finished sections weighing hundreds of tons. These are quickly welded together by a small labor force out on the slipway.

There are many special kinds of cargo ship. Liquefied natural gas (LNG) is carried in giant refrigerated spheres built into the LNG ship in a row. Ro-Ro ships carry road vehicles, which drive on board complete with their cargo, the name coming from "Roll-on, Roll-off." The LASH (Lighter Aboard Ship) concept attempts to speed up sea transport by dropping off cargo lighters at each port of call.

Right: oil tankers are by far the largest vessels afloat – some weighing as much as 500,000 tons. Oil spillages and tankers running aground are giving increasing cause for concern round the world.

Though it is difficult accurately to predict the distant future, is likely that the ocean-going passenger ship will prove to be one of man's most short-lived vehicles – with an active life of a mere 150 years. Before 1820 there were very few overseas passengers, except on short journeys by "packet" ferry between, for example, Britain and France. Almost the only people who traveled abroad in large numbers were soldiers. Likewise, since 1970 the number of passengers has dwindled almost to vanishing point, except on short ferry routes. In the years between some of man's greatest vehicles plied the oceans. They were floating hotels of the most lavish kind. These survive today only as cruise liners. These are not vehicles but floating holiday resorts.

The earliest commercial passenger ships offered little accommodation, ranging from a stretch of wooden deck in a slave carrier up to a small but ornate first-class cabin in the best ships voyaging between Europe and North America, or the Far East. Most ships carried mainly cargo, with passengers as a mere sideline. Only the larger or better "liners" (ships belonging to a recognized company plying a regular route) had cooks and servants to minister exclusively to the passengers' needs. Even on a well-appointed liner there were occasional bad crossings when bad weather or adverse winds would cause seasickness, injury, damage to the ship, and, most frequently, an extra week or two at sea – sometimes leading to shortage of provisions. Some ships, carrying large numbers of passengers, made no provision for them at all, beyond allowing them aboard. Between 1835 and 1895 poor immigrants from Ireland and other European countries had to bring all their own food. The ship supplied only fresh water in small allowances each day.

Throughout the 19th century the volume of passenger traffic increased remarkably. So did the expected standard of accommodation. In the 1880s the majority

Below: the *Mauretania*, one of the famous Cunard line which included the *Aquetania*, and the *Lusitania*. These magnificent ships offered those who could pay the price every luxury that a first class hotel on land could offer. Even cabaret was provided.

Below: at the end of the 19th and beginning of the 20th centuries, wave after wave of immigrants crossed the Atlantic to settle in the United States. They came by steamship and paid low fares. In return they were allowed a few square feet of deck and fended for themselves on the journey.

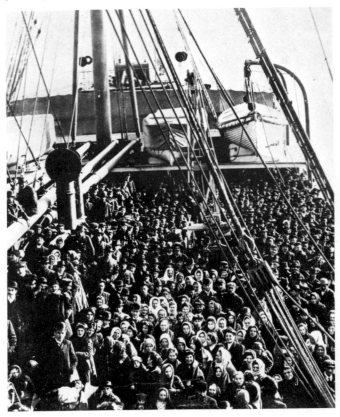

of non-immigrant transoceanic passengers at last had individual or family cabins. The largest ships offered sumptuous suites for complete families and their servants. Thousands of British families voyaged to such places as India, Singapore, Hong Kong, and southern and West Africa to administer the Empire, while the steadily growing system of worldwide trade produced a new class of businessman who spent months of each year aboard ships. The time was whiled away spending money, as several hundred card-sharp and confidence tricksters spent their entire lives (when not in prison) "riding the tubs," on legitimate tickets, swindling all the rich passengers they met.

On most routes ships seldom exceeded a displacement of 15,000 tons. But on the North Atlantic, competition and the influence of prestige produced exceptions. The Blue Riband, one of the most famous trophies in all transport, was a silver cup awarded to the ship making the fastest time between Sandy Hook light vessel, New York, and either Southampton or Cherbourg. The first turbine liner, *Mauretania*, won it on her maiden voyage in 1907, and held it for over 20 years. The last holder was the *United States*, fastest of all liners, at an average speed of 35.59 knots. Largest passenger ships were the British "Queens" (*Queen Elizabeth*, 83,673 tons, and *Queen Mary*, 81,237), and the French *Normandie* (83,423). All had

Left: the sumptuous splendor of the *Queen Mary's* dining saloon. The *Queen Mary* and the *Queen Elizabeth* were at their peak when ships were the only practical way of crossing the Atlantic. But not everyone traveled in such style.

steam boilers, but the Queens had conventional turbines, while the *Normandie* was equipped with turbo-electric propulsion. Longest liner of them all was the *France* at 1035 feet. Commissioned in 1962 it made heavy losses throughout its short career, and was finally laid up in 1974. Sole survivor of the once proud species of North Atlantic liners is the *Queen Elizabeth II (QE2)*. It was launched in 1967.

Why has *QE2* managed to survive? One factor is that it was designed after it was accepted that passengers in a hurry traveled by air. Speed no longer mattered, so an economical vessel was produced. Second, its light aluminum hull carries more passengers, in great luxury. Shallower draft (depth of hull under the water) allows the *QE2* to visit more ports without scraping its bottom, or, worse, running aground. As a result, it has become a most successful cruise liner. Just over 2,000 passengers can be carried on trips to all the favored holiday resorts, and often on round-the-world voyages. But these trips are enormously expensive, so the *QE2* does not exactly provide efficient transport.

Below: the *Queen Elizabeth II (QE2)*, the only remaining liner of its size. The few who can afford to cruise in it are charged enormous fares, and it still barely pays. The scarcity and rocketing price of oil has made oceangoing liners uneconomic.

Ships for Canals and Rivers

A large number of types of ship never go to sea at all. Privately owned pleasure boats are the most numerous. They include classes with sails, engines, oars or paddles. Hulls can be made of wood, glass-fiber, or inflatable rubber fabric. The traditional canal barge, or "narrow boat," is familiar to most people in Europe. It was once a vital link in the economies of many countries. Today, it is more often used for pleasure trips by holidaymakers. Other inland ships include harbor tugs, barges and lighters, police launches, fire floats, urban water buses, dredgers, and large vessels for use on great lakes or rivers. Lightships and weather ships leave port, but remain anchored at a fixed spot – never journeying anywhere.

Canal and river traffic is still of commercial importance in many countries. The world's busiest river is the Rhine. Every day it carries barges of several thousand tons each, as well as large groups of un-

Above: the harbor at the Rhine river port at Basle in Switzerland. This inland port is very important to distribution in Europe.

Left: a riverboat on the Mississippi. Paddle wheels were the favored method of ship propulsion in the early 19th century. They began to be replaced by screw propulsion in the 1850s.

powered barges, tied one to another, and propelled and maneuvered by a single tug. In the world's lumbering regions vast "islands" of floating tree trunks, with a combined mass of thousands of tons, are pushed and steered along rivers by tugs.

The ships that travel along "seaways" and sea canals are the same as those found on the oceans, but ships designed for true inland waterways continue to be built. Many of the large freighters used on the Great Lakes of the United States and Canada were delivered in sections, and assembled and launched into the lake. Most look like tankers but with a small deckhouse at the bows, and all have to have hulls strong enough to

plow through thick ice. A number of ships on the Great Lakes and other waterways are train ferries. They are fitted with rail tracks onto which complete railroad trains can be run, and delivered onto a different rail network at the destination port.

In the 19th century steamers on the Mississippi river formed the transport backbone of the United States, and a few still carry passengers for pleasure. Others ply such rivers as the Nile, Danube, Rhine, and Congo. Most have very low freeboard, the hull hardly showing above the water. But on top is a high superstructure of cabins and promenade decks. Some of these boats still use side paddlewheels or a sternwheel, which are

Above: water transportation is an economic way to shift heavy loads. Here, one of the largest tugboats in the United States pushes dozens of barges up the Mississippi river with ease. **Right:** A ferry crossing on the Trans-Gambia Highway. Ferry crossings are common in many parts of the world where a bridge is not justified.

better in shallow water than a screw propeller. When a screw propeller is used, it is often surrounded by a tubular duct called a Kort nozzle which increases efficiency and protects the revolving blades. A few inland ships use the Voith-Schneider propeller, with downward-projecting blades. This can push the ship in any desired direction for perfect maneuverability. But it is no good in very shallow water.

Most of the tonnage passing along the Rhine and other great inland waterways travels in barges. These are often little more than tough steel boxes shaped roughly like a ship. They are open at the top for easy loading – for example coal is poured in from giant overhead hoppers. The top often remains open on the journey. Some large inland ships have built-in ballast tanks so that they can adjust their draft for salt or fresh water, and the need to pass under city bridges often results in either a complete absence of super-structure in the so-called flat-iron shape, or masts and exhaust stacks that can hinge flat when necessary.

Some ferries carry passengers and vehicles across a river. A few have no engine but are fixed to a chain anchored on each bank, and crank a wheel against the chain to work their way across. In some places a num-ber of pontoon barges are moored side-by-side and a bridge laid across the top. Armies use this technique to build a large bridge quickly, and at Coblenz after World War II a pontoon bridge had a large central section which could be driven out of line to allow river traffic through, afterwards being moved back into place as part of the bridge.

Some tugs go to sea, towing drydocks, oil rigs, and disabled ships, but most are used for work in harbors and inland waterways. Some pull or push barges or barge trains. As many as six may work together in docking or undocking the largest oceangoing ships, especially when high winds make handling difficult.

Hydrofoils

Ships are at their most efficient at low speeds, because they can convey very large loads with small expenditure of power. The most cost-effective speed for the majority of ships is about 10 knots, and they remain reasonably efficient at up to twice this speed. But the resistance (drag) of the water increases approximately as the cube of the speed, and the cost of operating the 35-knot ship is enormous. For over a century inventors have tried to find ways of reducing the drag. The only one to work so far is the hydrofoil. A hydrofoil is a water blade or sea wing, but the term has come to be applied to the whole craft.

The earliest successful hydrofoils had ladder foils. Like all hydrofoils, the craft was a conventional displacement ship at rest. When it began to move forward, it was thrust upward by water passing over numerous curved steel foils fixed to vertical members in the form of ladders. As speed increased, the upthrust gradually overcame the weight of the projecting parts of the ship, and the whole craft began to rise out of the water. This reduced its buoyancy but also reduced water resistance. Speed, therefore, continued to increase. This in turn increased the lift, thrusting the ladder foils and boat higher out of the water. But as each foil emerged from the water it virtually ceased to lift (its trivial lift in the atmosphere could be ignored).

Eventually a self-stabilizing point was reached with the foil ladders roughly half in and half out of the water. The hull of the boat itself was completely clear of the water. With hull resistance virtually eliminated, hydrofoils could travel at several times the speed of displacement ships.

Ladder foils had the disadvantage of excessive depth, so they could only be used in very deep water. They were also clumsy, heavy, and prone to damage. But the potential was obvious. As early as 1918 a ladder-foil boat achieved 60 knots. Between the world wars a Swiss, Baron Von Schertel, perfected the simpler surface-piercing or vee foil. This looks like a broad letter "V" in front view, fixed to the boat by low-drag struts. As the speed of the boat increases, the foil rises out of the water, starting at the tips and working inward until at full speed the weight of the boat is taken by the deepest central part of the V. This foil is again self-stabilizing. The lift depends on the amount of foil in the water, which depends on the speed, the two balancing automatically. If the craft were to heel over, more foil would be submerged on one side than the other, automatically turning the boat upright once more. Banked turns at speed are as effortless as on a cycle, and the whole system is robust and practical. Most hydrofoil ferries use surface-piercing foils, with a horizontal foil at the stern taking a small part of the weight. A "sea autopilot" is fitted, as well as a simple pipe system through which air can be admitted to the foil and allowed to "bleed" out through holes in the upper surface. Both refinements make the hydrofoil usable in rough seas.

In the Soviet Union large numbers of foil ferries run on the Volga and other rivers and inland seas on "depth-effect" foils. These are broad horizontal foils that run just below the surface. They are excellent for smooth water, but unsuitable for open sea operation.

The best foil system for sea going is the fully submerged type. Here the boat rides on three vertical struts, either one at the front and two at the back or, more often, the other way around. These struts in turn ride on small separate horizontal foils whose angle (incidence) can be varied. An autopilot system constantly adjusts the incidence of the three foils according to the speed and motion of the water in waves to provide a near-perfect ride in rough seas, far smoother than in a displacement ship floating on the waves. All current production hydrofoil warships are believed to use submerged foils. They are compact

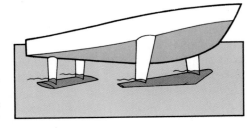
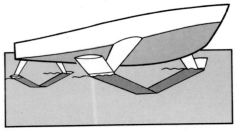
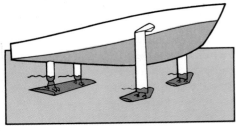

Left: the four most common types of hydrofoil in use today. *Moving clockwise:* the foil ladder, the depth effect, the submerged, and the surface piercing. The hydrofoil works on the same principle as the wing of an aircraft. It is surrounded by the medium in which it operates (water). It achieves its lift mainly by causing a reduction in the pressure on its upper surface. An inclined plane can be made to provide lift if it is moved horizontally through air or water.

Above: a commercial hydrofoil in service in Copenhagen. Hydrofoils are in use around the world, usually on fast passenger ferry services. A number of public hydrofoil services are in operation, notably "Seajet" between Britain and France.

Below: the Boeing research hydrofoil *Little Squirt*. Capable of speeds of up to 50 mph, Little Squirt is propelled by a waterjet. Water for this enters a scoop at the base of the aft strut and is forced out at the rear by a double-action centrifugal pump.

vessels able to travel at roughly twice the speed of most surface warships when foilborne, but also able to fold up their foil struts and proceed at modest speed as displacement ships when necessary.

One purely geometrical difficulty with hydrofoils is how to drive the propeller. The plain angle drive results in the propeller thrusting diagonally upward. The Z-drive needs two sets of gearboxes, while the V-drive is often used on surface-piercing foils but still needs a gearbox and two long shafts. The best means of propulsion may eventually be the water jet, in which the diesel or gas-turbine engine drives a powerful pump, discharging a high velocity jet above the water.

Below: the 320-ton USS *Plainview* was built for the US Navy by Lockheed Shipbuilding. It is powered by two 14,000 hp gas turbines.

Air-Cushion Vehicles

Air-cushion vehicles (ACVs) can look like ships, trucks, or trains. Not surprising, people find difficulty in giving them a name. Apart from air-cushion vehicles, they are variously called surface-effect vehicles, ground-effect machines (GEMs), and hovercraft. Some operate over water, some over land, and some on rigid land tracks. Most can negotiate almost any surface – water, concrete, soft sand, mud, snow, or ploughed fields. Most of the large ACVs are amphibious but could not reasonably travel over land in developed countries, and are used as seagoing ferries. They are much faster than ships, and give a smoother ride across waves.

Inventors tried to make boats ride on bubbles of air in the 19th century, but it was not until 1955 that British electronics engineer Christopher (later Sir Christopher) Cockerell patented the first successful

promptly driven about the Solent between the English mainland and the Isle of Wight. Nobody was previously sure that it would work or even float on its cushion. It was found to have exciting possibilities. But the clearance was only about four inches, so it could not clear a roadside curb, let alone a tree stump. In the early 1960s the vital flexible skirt was introduced.

The skirt is a flexible wall of tough rubberized fabric all around the craft. Air is pumped out through inward-sloping ducts, called fingers, to sustain the cushion. On the largest craft the cushion measures eight feet high allowing them to be driven across hedges, parked cars and other obstructions, causing no damage to either the ACV or the obstacle. For overwater use the skirt is at least as important as on land. It enables the ACV to run at high speed across large waves, preserving its cushion and giving a smooth ride. At about the same time the skirt was introduced, an alternative presented itself for pure overwater vehicles in the form of the rigid sidewall. This is a narrow wall projecting down along each side of the craft, just far enough to maintain a seal along the sides of the cushion. As sidewalls need no pumping power the total installed power is reduced, but their water drag makes extra power necessary to cruise at any given speed.

ACV propulsion and control offers endless scope for variety. Some simple craft have one engine driving

Below: the inventor Christopher Cockerell (*center*) with a working model hovercraft. His original model was fashioned from bits of an old vacuum cleaner, parts of a motor boat, and a couple of coffee tins. He tested it on the kitchen scales.

Below: a Bell SK-5 on river patrol in South Vietnam. United States and South Vietnamese forces used a number of these craft. Based on the British SR.N5, the Bell SK-5 proved particularly effective in operations in the Mekong Delta area.

scheme. He proposed a craft with a flat bottom containing a narrow slit all around, out of which air could be blown diagonally inward toward the center. This "peripheral jet" produces a cushion of air under the craft pressurized slightly above the pressure of the atmosphere, yet able to support the vehicle's weight. The SR.N1, the first ACV, was launched in 1959, and

a single fan whose output is used to feed the cushion, and also propel the craft via aft-facing nozzles equipped with some means of thrust-vectoring for control. Marine ACVs often use swivelling air propellers, ducted air propellers with exit control surfaces, marine screw propellers, or water jets. All the largest have gas-turbine power, the biggest being warships.

Above: British air-cushion-assisted transporters are designed to carry heavy equipment like electrical transformers over roads and bridges that might otherwise need costly strengthening. This transporter is moving a 155-ton load. The cushion cut the load by 70 tons.

Above right: the British SR.N6 is the workhorse of the hovercraft industry. It is employed in a wide range of civilian and military roles round the world. SR.N6 carry 38 passengers or 3.4 tons of freight.

The big question mark over the large overwater ACV is its commercial viability. Only two designs have extensive commercial experience. The French SEDAM N500 has three turboprops for propulsion. Its two lift fans, each driven by its own turbine, eliminate transmission shafting. But its first year demonstrated poor reliability. Britain's larger SR.N4 is the only completely reliable large ACV, and since 1978 these monsters have been lengthened by adding a new 55 feet center section which increases capacity to 416 passengers as well as 60 cars. Four swivelling propellers of larger, more efficient type, and a low-pressure "large-finger" skirt giving a wave clearance of nine feet have also been introduced recently. With 70 percent higher earning power for 15 percent greater costs, the Super-4 is the first ACV in the world to run at a genuine profit. An incidental and unexpected bonus of the "stretching" process is an improved ride and higher speed in adverse weather.

This strongly suggests that all over the world the displacement ship will gradually be replaced by the large ACV where there is a market need.

Below: the enormous SR.N4 ferries passengers and cars between Dover in England and Calais or Boulogne in France. It makes the crossing in a fraction of the time of the traditional ferry service. Lack of official interest nearly killed the original hovercraft research.

Navigation

Above: radar, originally developed in the mid-1930s to signal the approach of enemy bombers, is an invaluable aid to navigation at sea. The relief of the nearby seabed can be plotted as well as the position of surface obstacles. Radar is now a vital tool of navigation.

Sailors learned rudimentary open sea navigation several thousand years ago by heading their vessel correctly in relation to the changing position of the Sun. More than 1000 years ago the Arabs laid the foundation of astronavigation when they worked out detailed navigation charts based on the stars, as well as the Sun and Moon. This is still a required accomplishment of a ship's master. Using an optical instrument called a sextant, equipped with mirrors and a

Above: an early compass of the early 14th century housed in an ivory box. Pieces of iron have been used to guide ships to within sight of land since about 1000.

two-axis spirit-level, the exact direction (azimuth) and height above the horizon (elevation) of a heavenly body can be compared with the precise time to allow the ship's position to be read off a table of latitude and longitude. An even more basic method is pilotage. This is based on measuring the direction of coastal objects or other fixed marks such as buoys or lightships. From these calculations, dead reckoning can be used to give approximate positions plotted at intervals on a chart.

In the early 20th century radio stations began to be used for navigational purposes, at first by rotating large aerials in the form of pivoted rings of wire until that rotation brought them on to the correct wave-

length and the received signal was either zero or at a maximum. Each direction-finding (D/F) bearing gave a "position line," and two or more – hopefully crossing at some point – gave a "fix." After World War II radio navigation aids (navaids) were greatly refined. There now exists a widely used family of systems emitting either pulses or continuous waves from a "master" and interlinked "slave" stations at fixed land sites to give a stationary pattern of interference in the atmosphere. Decca Navigator is the most widely used precision system for inshore and general use to distances of about 1000 miles, while the pulsed Loran gives slightly less accurate guidance in midocean. The most useful radio navaid is Omega, coming into general marine use in the early 1980s. Omega combines extreme precision with coverage of the entire globe.

The accuracy of gyroscopes had improved to such a point by the early 1950s that the way was opened to a totally self-contained navigation system, the inertial navigation system (INS). Initially confined to aircraft and missiles, it was later installed in submarines and space vehicles. INS will be used increasingly in surface shipping during the 1980s because of the increasing value of individual hulls and the high incidence of navigation errors. The system is also becoming gradually cheaper and especially tailored for marine use. The basis of INS is a small platform which is held exactly level and parallel to the horizon by high-precision gyros. No matter where in the world the ship travels, the platform is always totally horizontal. On the platform are mounted accelerometers which record every acceleration of the platform with impressive precision. Once told its position at the start of the voyage the INS can, by multiplying the acceleration changes by time (to give ship velocity), and then multiplying by time again to give distance, continuously indicate the current position. Its accuracy falls off over a period, so on a long voyage the INS has to be corrected by other methods. One of these is satellite

Above: the flight-deck of an SR.N4 hovercraft showing the Decca Navigator equipped with a moving map display. The three phase-meters are above the map. Navigational aids like this make it safe to travel, even in very bad weather.

Right: the radar installation on HM Frigate *Charybdis*. Radar installations are now an essential tool in nearly all military operations. The radar equipment of this frigate will signal all potential hazards, and is vital to accurate navigation.

navigation. This is based on radio links with navigation satellites orbiting the Earth, and can give exact position information to all ships equipped with the appropriate receivers.

Today virtually all oceangoing ships, and in developed countries most inshore and inland commercial vessels, are equipped with radar which presents the navigating officer with a maplike picture of the ship's surroundings. Refinement of the radar has eliminated returns (echoes) from waves, while those from coastlines, and especially from other ships or objects projecting above the sea surface, have been enhanced. Though only indirectly a navigation aid, radar is of great value in avoiding collisions or running aground

or onto rocks, especially in fog or other bad visibility. Of course, soundings still have to be taken as in the old days to check the depth of water under the ship whenever necessary. But despite this, large numbers of ships continue to run onto rocks or collide with each other – even in broad daylight. Compared with aviation the record of shipping is extremely poor. In the busiest bottlenecks like the English Channel, Strait of Malacca, and Gibraltar the frequency of collisions and near-misses is a matter for grave concern. This is despite the technical ease with which potential collisions can be displayed, for example by providing the radar with circuits that indicate any ship posing a future hazard.

Ports
and Canals

Only the smallest boats can be pulled ashore on a beach or up a slipway. For thousands of years the main ports have provided a deepwater harbor (kept from silting up by dredging), strong quays alongside which ships can berth, and wherever the sea can be rough, the protection of a breakwater or sea wall to give calm water inside the harbor mouth. In the past 100 years further requirements have been added. Road and rail links are brought right alongside the ships. Warehouses must be provided to store cargoes, often bonded (sealed) by customs officials so that nothing can be removed without payment of a national tax (duty). Cranes are needed to supplement those of the ships,

and in some cases special handling gear is necessary for particular cargoes. Coal and iron ore, and similar bulk solids, require built-in bucket/conveyor systems, or vertical grabs on swinging arms. Grain and other granular solids are handled by gravity chutes for loading and vacuum pipes for unloading. Exceptions are salt and sugar, which must be packaged and sealed against damp. Flour and cement can be pumped by air pressure through pipelines – often for great distances.

The modern container port is a vast network of fixed installations, which are linked by road and rail to a country's transportation network. Each berth is served by at least two cranes. These have a long horizontal girder which extends over the ship, and out across the landside. Both roads and rail tracks pass beneath this girder. Each crane has a special four-cornered grab which can be set to handle any of the three standard lengths of container. The serial number and weight of each container are recorded and the

Below: with an efficient import trade, Bremen docks control the supply of most of the imported products to the German and bordering countries' raw material market. Considerable storage capacity is available so that goods can be stockpiled.

information fed into a central computer. Sometimes one crane loads each ship while another unloads. The most efficient method is for each container to go straight onto its correct road or rail truck. But this is usually difficult to arrange and most are taken by a large straddle carrier or container transporter to a park for subsequent collection. Many of the busiest container ports are fairly new sites, not built on existing harbors. Where this is the case models are used to investigate the hydrodynamics of the new harbor to make sure the proposed design of quays, breakwaters, and adjoining beaches will be the cheapest that will handle the traffic without silting up or suffering damage from the worst storms. Many long-established harbors are badly sited, and require constant protection against the sea – for example, by piling heavy concrete tetrapods against the outside of the breakwaters, or building extensions to them.

Modern ports need a traffic control center, with comprehensive radar coverage as far as the horizon, and radio communications with all ships. Where possible, lights are fitted to navigational obstacles, and sea lanes are marked out by distinctively shaped and colored buoys which often carry radar reflectors, lights, and/or gongs or other aural warnings. Many harbors also have dry docks. Once a ship is inside, the gates are closed and the water in the dock drained to expose the normally submerged parts of the vessel for maintenance. Floating docks, which can usually be moved from one place to another when necessary, admit a ship and then pump out water from their own tanks to increase their buoyancy. They can rise until the ship is completely out of the water.

There are a few inland ports accepting large deep-water ships. These are connected with the sea by man-made canals. To change the level of the water large locks are required, each with two sets of gates spaced far enough apart to admit the largest ship. Each pair of gates is equipped with controllable valves, called sluices, through which water passes to adjust the level inside the lock when both sets of gates are shut. In hilly areas ships may have to pass through five or more locks in a little over a mile. A few canals include tunnels, and in southern France a new tunnel big enough for large cargo ships links Marseilles with the Rhone river. Ship lifts or inclines are increasingly replacing locks, because they are quicker. Standard Europa cargo ships are lifted or lowered by such installations as the French Ronquières inclined plane or the German Luneburg elevator. The Garonne Lateral canal has a sharply inclined section along which ships pass in a pool of water trapped by a giant sliding gate pulled by powerful locomotives at each side.

Underwater Vessels

It is a sad reflection on mankind that the entire early history of the submarine was concerned with warfare – discounting isolated individual experiments. Even today the overwhelming majority of all submarine funding is for military applications. But economic pressure forces man to make fuller use of the oceans and the raw materials that lie within their still little-explored depths. The greatest single spurs to man's modern and growing capability to operate deep within the oceans were two United States military disasters, one in 1963 and the other in 1966. The first was the

Below: William Beebe and Otis Barton being released from the Bathysphere after their record breaking dive of 2510 feet on August 11, 1934. Crew members on the stand-by barge *Ready* have just opened the hatch. Spherical shapes can best withstand deep sea pressure.

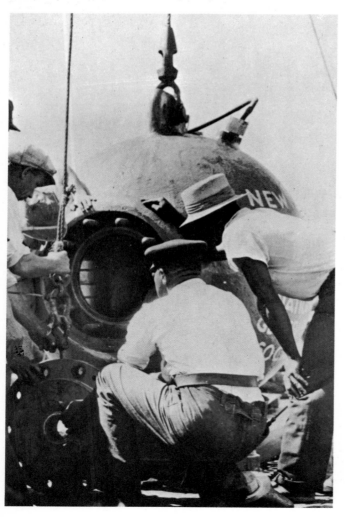

uncontrolled descent and fatal crushing of the submarine USS *Thresher* in the Atlantic, in depths where no rescue vessel could operate. The second was the loss of a hydrogen bomb in the sea off Palomares, Spain. These events brought a revolution in research and advance in deep-ocean technology.

There have been divers equipped with breathing apparatus for at least 2500 years, but it is only in the 20th century that they have ventured to any significant depth. At the greatest depth the water (hydrostatic) pressure reaches well over 1000 tons per square foot. A sphere is the only possible shape for a container of humans. The first "bathysphere" was developed by Otis Barton in 1930. In 1933 he descended a short distance in the first to be built, accompanied by William Beebe. More recent deep explorations began in 1948 with Professor Auguste Piccard's "bathyscaphe" FNRS II. A bathyscaphe usually comprises an extremely strong sphere. Inside are the crew and instruments. The sphere is attached to a much larger buoyant hull filled with a fairly incompressible liquid of a lower density than seawater, such as petroleum. It is impossible to make the sphere buoyant by itself. Suspension from the surface was tried, but with depths of up to 36,000 feet it was totally impractical. Variable buoyancy tanks and vertical screw propellors were adopted instead.

Man has learned a great deal about how to support human life in the oceans, how to navigate in its depths, how to work on the seabed, and how to extract valuable resources from the sea. Already there are a remarkable variety of submersible vehicles. They include rescue craft, many kinds of research and exploration platforms, routine commercial undersea work boats for numerous jobs, undersea habitats (most for fixed-base operation), seabed based remote-control working devices, and even such things as bulldozers and similar seabed travelers.

Most undersea vehicles designed for depths greater than 1500 feet do not attempt to copy the military submarine. Instead, they are built on the bathyscaphe principle with a sphere housed in a buoyant and navigable hull. Almost all use electrical power exclusively. Other than nuclear energy – which is practical only in large vehicles – there is no other source of energy that does not demand a supply of oxygen, fuel, and possibly other materials which have to be carried on board. Electric batteries are compact and reliable. They also help ballast the hull outside the sphere, and ways have been found to make them withstand the severe pressures.

The submersible is usually equipped with several systems for propulsion and control. As mentioned earlier, vertical position is controlled by variable buoyancy tanks and vertical screw propellers. The sitting of the vessel in the water (fore and aft trim) is adjusted by liquid transfer between front and rear ballast tanks. Propulsion is provided by screw propellers, usually equipped with a fixed or rotating peripheral shroud ring. This is either fitted at the stern

Right: Westinghouse built *Deepstar* in 1965 for oceanographic research. The US Navy's bathyscaphe *Trieste* had already descended nearly 36,000 feet to touch bottom in the deepest of the Pacific Ocean trenches. Such equipment has served greatly to extend humanity's knowledge of the seabed.

Below: this science fiction-like contraption is *Beaver IV*, one of a series of submersibles for the offshore oil and gas industry. With its manipulators, it can undertake maintenance work on wellheads 2000 feet underwater. Submersibles like *Beaver IV* also pick up samples.

Left: the submersible *Aluminaut* in shallow waters. Visibility just below the surface is good, but no light at all penetrates below 1000 feet. At *Aluminaut's* design depth of 15,000 feet it is pitch black. Also, particles in the water can severely restrict vision even at modest depths. This presents a serious problem to marine geologists studying the ocean floor. About 85 per cent of our learning is gained through seeing. Powerful lights can be used, though their beam is restricted under water. Infrared film offers a possible solution.

with a rudder, or one at each side with variable speed on either for immediate directional control. The forward speeds of submersibles are low – only about three knots. A few also have bow and/or stern lateral thrusters. In some operations these can push the whole vehicle bodily sideways, or can counteract a small lateral current in the surrounding water.

Virtually every manned submersible has both direct vision ports and at least one television camera. Powerful lights penetrate the darkness of the deep sea where sunlight has never reached. Some submersibles are designed to operate by themselves, with the capability of traveling distances in excess of 60 miles. Others are incapable of independent operation but must remain linked to a parent ship on the surface above. Depending on the mission, there may be some form of "landing gear" for resting on the seabed, as well as a wide range of external manipulators.

The most common manipulator is a pincerlike "hand." There are several types, and it is essential in some jobs to be able to vary the force of the grip. In a few submersibles the gripper is very large, having a diameter matched to that of sections of pipeline, but usually piping is lowered from above and all the submersible has to do is position it. A few more sophisticated working submersibles have provision for underwater drilling, cutting, and welding. But there are severe limits on the depths at which most operations of this kind are possible. Oceanographic and prospecting vehicles usually have claws and a box fitted nearby for gathering samples of rocks or other material.

Submarine rescue craft vary considerably. No internationally agreed design has yet emerged for rescues from civilian submersibles. Standardized designs have been agreed by major users of naval submarines, which have escape hatches onto which a rescue vehicle can be sealed even at depths exceeding 1200 feet. A few rescue craft, in the United States Navy for instance, can operate at several times this depth, though this is well beyond the design limits for naval

Above: the research submersible *Alvin* being maneuvered by cranes aboard USS *Fort Snelling*. A submersible is basically a pressure hull accommodating the crew, surrounded by an unpressurized streamlined fairing containing power supplies and other gear.

Above: the robot submersible CETUS (Computerised Exploration and Technical Underwater Surveyor). It can operate on the seabed to a depth of 1500 feet. Developed in Britain, it steers itself automatically along buried pipelines to find blockages and breakages. Picture shows it about to enter the North Sea.

submarines. The original Deep-Submersible Rescue Vehicle (DSRV) could clip onto a stricken submarine and bring out 24 of the crew at a time, repeating the operation as often as necessary.

One of the earliest and deepest-diving of all working submersibles was the American *Aluminaut*. This 72-ton commercial vehicle had a thick aluminum hull, lighter than a steel hull, but strong enough for safe operation at depths of 15,000 feet and more. *Aluminaut* played the central role in the final successful recovery of the Palomares H-bomb. In 1969 it made a unique and successful salvage of another submersible, the United States Navy's *Alvin* DSV-2. This had been dropped into the water with its hatch open, and sank in a depth of over 5000 feet. After being recovered the *Alvin* was fitted with the world's first submersible titanium hull. It went into service for work at depths of up to 12,000 feet.

Since 1960 sporadic attention has been paid to the idea of a cargo-carrying civilian submarine. Many existing submersibles carry small cargoes, but for bulk transport purposes very much larger vessels are needed. There is no evident problem in building such a vehicle. Missile submarines already have displacement close to 20,000 tons. The great advantage of the submarine as a transport vehicle is that, contrary to what might be expected, it has less hydrodynamic drag than a surface ship. It has greater "wetted area," and therefore higher skin-friction drag caused by shearing the adjacent layers of water molecules in the surrounding "boundary layer," but there is no wave-making drag. This is the largest single component of the resistance of most ships at full speed. Instead of dissipating energy in making large waves at the sea/air boundary the submarine moves totally in one medium, seawater, and makes no waves at all, other than trivial eddies behind projections from the hull. The submarine looks an interesting possibility as a bulk carrier of crude petroleum or other materials, even though it would need exceptionally deep water and probably nuclear propulsion. On some routes the depth of water needed would lengthen the journey, for example by prohibiting passage through a canal. But on others the ability to pass under the Arctic ice cap would dramatically shorten the journey.

In the 1980s undersea habitats are certain to have an active role to play in harvesting the underwater resources, but until the end of the 1970s all had been mostly research tools intended to find out how man can best live and operate in the subaquatic environment. Captain Jacques Cousteau has been one of the best-known pioneers of undersea living. His *Conshelf* (Continental Shelf) series of habitats have operated for several weeks at a time in a depth of well over 1000 feet in the Mediterranean and Red Sea. West Germany, Japan, the United States and to a lesser degree many other countries have built and operated habitats at modest depths. Over a long period, they have laid a solid foundation of experience on which man's future work in the oceans will be based.

The Technology of Diving

For thousands of years humans have dived to the seabed in search of such items as fish living on the bottom, pearls, and sponges. They reached depths of nearly 120 feet. It takes time to descend to such a depth and regain the surface. Any prolonged exposure to high pressure causes trouble when the pressure is released. Nitrogen bubbles are released in the blood, causing excruciating pain and sickness. Commonly called "the bends," such an attack can be fatal. To dive deeper or stay down longer, a diver has to have special breathing equipment and a life-support system controlling the rate at which the pressure is released. To return from prolonged work at the 330 foot level a diver needs a full day to return to atmospheric pressure. From 500 feet the recovery time should ideally be three days. There are other difficulties with breathing high-pressure air, and since World War II intensive research has tried to produce the best gaseous mixture for humans working underwater.

Nitrogen in the air has long been identified as a source of trouble, quite apart from "the bends," and it has gradually been eliminated from breathing mixtures. A helium/oxygen mixture has been widely used, though helium is difficult to obtain. It also makes it almost impossible for humans to communicate by voice. Once the gas reaches the throat it alters the behavior of the vocal cords so that the spoken word comes out as a meaningless gabbling. Some breathing mixtures still use nitrogen, to which the human body is attuned, and reduce its emergence as bubbles by administering partially depolymerized hyaluronic acid (PDHA) to reduce the body's fat content. As far as possible, large-scale underwater operations of the future will attempt to preserve a "shirtsleeve environment" for human beings, with sea-level atmosphere and temperature. Only a few specially trained and equipped aquanauts will be required to work outside these life-support chambers.

Apart from the self-supporting naked diver or aqualung-equipped diver, the simplest diving method of the past was the diving bell. This heavy bell-shaped air container was usually supplied with air from a pump at the surface. For many years suited divers operated either from diving bells or surface ships. For the deepest dives extremely strong pressure-tight suits were used. Today, all deep diving is done with

Below: in the past the breathing apparatus of divers gave them a very restricted time underwater. Developments like this rebreather Aqualung "scrubs" exhaled carbon dioxide so that it can be breathed again as fresh oxygen. Diving time is greatly extended.

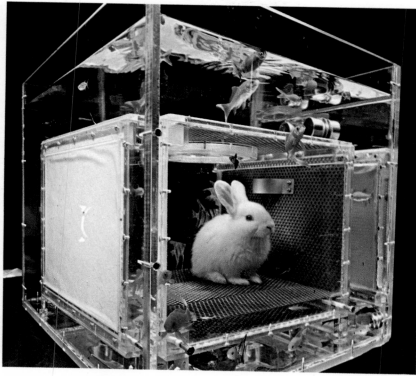

Above: a diver using a powered underwater wrench. Deep underwater it is impossible to exert the muscular power necessary to operate conventional manual tools. The pressure around the diver is intense. This makes a task, which on dry land would be quite straightforward, extremely arduous – even for the strongest of people. Divers must be supremely fit.

Above: completely surrounded by water, this rabbit is breathing oxygen passing through the membrane-covered sides of its perspex cage. The silicone-rubber membranes act similarly to the membranes of the lung.

Below: a diver employed in the Forties oilfield in the North Sea in a decompression chamber after a dive.

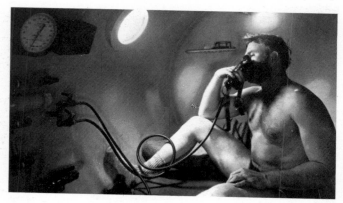

the people inside some kind of pressure-tight environment, such as a submersible or habitat. Free-swimming aquanauts stay below 500 feet, wearing only a foam rubber insulated wet suit and carrying one or more self-contained underwater breathing apparatus (Scuba) packs on their backs. When it is necessary to cover considerable underwater distances, the aquanaut's swimming ability is enhanced by providing him with a powered tug which he holds with both hands. These are equipped with electric batteries and a screw propeller.

In nearly all undersea operations there is some means of communication between the divers and a support team on the surface. This may be a telephone using a cable. Since 1960 both sonar and lasers have been used to open up additional possibilities. Neither needs a physical link, and the depth to which communication is possible is governed only by the available electric power to energize the transmitters. Sonar, using ultrasonic sound energy, has been brought to a very high standard as a result of its crucial role in antisubmarine warfare. A modern system can transmit not only coded speech but a valuable line-by-line TV-type picture of the underwater scene. Lasers can be selected to operate at a blue-green wavelength giving maximum transmissibility through seawater, and though the range may be less than with high power sonar, the rate at which data can be handled is greater.

There are several diving machines which form a direct link between the surface and the seabed. They can attain depths of about 350 feet, which is typical of the important continental shelf on which most of man's underwater commercial exploitation is based. This shelf surrounds all the Earth's land masses. In many places the shelf is well over 1000 miles wide and includes, for example, the entire area surrounding the British Isles and Indonesia. Even in these areas of relatively shallow water commercial operations are still in their infancy.

One unusual machine is a displacement ship, or floating instrument platform (Flip). It can be up-ended so that one end rests on the seabed. Another is the POE vertical platform.

81

Extracting the Sea's Resources

The largest commercial harvest from the sea continues to be fish and other edible marine life. But it has now been recognized that the oceans are the Earth's greatest storehouse of raw materials, surpassing even the crust of the land masses. The sheer technical difficulty of gathering a significant amount of this huge reserve of useful elements and compounds has until now made the land far more important. But the future commercial harvesting of the seas is being carefully managed so that no harm is done either to the environment or to life in the oceans.

Today the most economically important ocean commodities are oil and gas. These are probably widely distributed over the Earth, but existing prospecting techniques make exploitation possible only in the shallow waters of the continental shelf. These shelf areas are believed to contain at least half the Earth's oil and gas resources. Prospecting or production are in full swing around the coasts of every continent.

Oil and gas are closely related. In fact, natural hydrocarbons range from gases through a wide range of heavy or light liquids to solids. Crude oil deposits are usually overlain by natural gas, which is nearly always 80 to 90 percent methane. Natural gas is also often found where there are no oil deposits. As on land, prospectors search with seismic methods, which plot the reflections of intense shock waves from strata in the crust. But because of the overlying water, trial drillings are very important. Sometimes a hollow drill is used to extract a "core" – a rod the length of the drill's penetration showing the types of submarine rock present. Drilling ships have taken cores through the rock under the sea to depths of over 26,000 feet – adequate for all today's commercial wells. Another aid to prospecting, and to subsequent seabed operations like the laying of pipelines, is the side-looking sonar which by reflecting ultrasound pulses from the bottom can paint a high-definition picture like that achieved by radar on land.

Drilling for either prospecting or production can be tackled in a number of ways. The simplest drilling platform is a ship, but the ship needs to have special equipment to pinpoint its exact position, as well as side thrusters to hold position directly above the hole. Most rigs are large platforms, which either float, or rest on the seabed. Some are of the jack-up type with retractable legs which are extended on site to rest firmly on the bottom. Many drill rigs for open sea operation are among the largest of man's movable structures – well over 300 feet wide and tall. The seabed can also be drilled by a manned compartment on the bed itself, often with a controlled sea-level atmosphere, or by an unmanned remote-controlled drill rig resting on the bottom and electrically powered via cables from the surface.

The hole may be vertical or inclined. A single platform or rig may drill several holes from one location to form a fanlike or conelike series of penetrations covering a large area at the gas or oil bearing stratum. Muddy liquid is pumped down to lubricate the diamond-tipped drill bit, and provide hydrostatic pressure to prevent a sudden gush of oil or gas. If the hole uncovers a workable deposit of gas or oil it is lined with pipe and concrete and capped with a "Christmas tree" of valves. These are joined to transport pipes leading to an ocean storage tank or a shore terminal.

Left: this view from the derrick of the deep sea drilling ship *Glomar Challenger* shows the automatic pipe racker. There, in precise order are the thousands of feet of drill pipe needed to reach the floor of the deep ocean. Drilling beneath the seabed is even more demanding of skill than dry land operations.

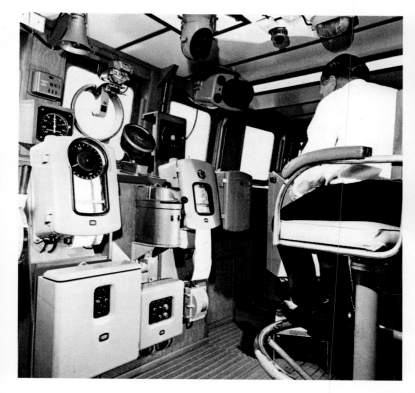

There are numerous other products from the sea. One of the most obvious is salt, sodium chloride. This is obtained by evaporating seawater in large shallow pans so that only the salt remains. Bromine and iodine are related elements obtained from the sea. In the latter case seaweed is allowed to concentrate it first. The bulk of magnesium, one of the most important and lightest of metals, comes from the sea where there is an almost limitless supply. Many other metals are found in some regions, including silver, gold, copper, zinc, lead, beryllium, and nickel in alluvial or sedimentary deposits on the sea floor. Other metals, including gold, are present in vast quantities in the form of fine suspensions throughout the depth of the sea, but no economic way has yet been found to exploit this rich resource.

In 1872 the British research ship HMS *Challenger* dredged up a small lump of material. This turned out to be almost pure manganese, one of the most useful of metals. It was gradually realized that vast areas of the ocean bed are littered with these small "nodules." Further exploration of the nodule beds revealed that some are mainly of nickel while others contain variable proportions of iron, cobalt, vanadium, copper, and other metals, usually in the form of oxide. The more scientists have learned about these nodules, the more puzzling they have become.

Careful study has shown that some are many millions of years old, with gradual growth like a tree or on a much shorter time scale a hailstone. Layer upon layer is formed – like an onion. Nodules vary from small granules to objects nearly a foot across. Most are a few centimeters (say, one inch) across. This enormous resource contains most of the Earth's known reserves of many metals. Scientists do not yet

Above: a desalination plant in water-starved Kuwait. Removing salt from seawater is a very expensive process. Even so it is the only way that a country like Kuwait can provide sufficient fresh water. The country's oil wealth means that it can afford the massive investment necessary.

Above: a manganese nodule from the Pacific Ocean. Manganese nodules contain manganese, nickel, copper, and cobalt. There is thought to be enough in the Pacific alone to last for 400,000 years. **Below:** harvesting manganese nodules. A wire dredge basket is trawled along the seabed. Its content is emptied into a barge alongside the trawler.

understand why these elements are always found on top of the seabed deposits, instead of becoming buried.

Although of enormous commercial value, it is only since 1960 that the technology to "sweep the ocean floor" for these deposits have become available. The first commercial operations on a large scale began in 1979. Such an endeavor has meant the solution of many problems, and the development of impressive new engineering. One of the major stumbling blocks remains the legal one. Never before has it been necessary to decide who, if anyone, owns the ocean floor. No precedent exists for a nation or company to lay claim to a piece of it. In 1978 one of the several giant international consortia formed to gather the seabed minerals laid claim to a carefully designated rectangular area in the Pacific ocean to test the response of the lawyers.

This area was chosen after prolonged exploration and studies with computers. Nodules and other mineral deposits vary from place to place. Mineral consortia have studied the content of each nodule, their individual size, distribution density, seabed state, year-round weather, and ocean depth before deciding where to mine, and how. And the hundreds of research laboratories and companies involved have come up with many novel methods for reaping the harvest.

The most obvious methods are mechanical, using dredger-bucket loops or similar digging systems. Crawler tractors have been designed to travel along the bottom gathering nodules into scoops. Large scoops can be towed by ships, gathering nodules into various kinds of transport and lift system. Equipment resembling powerful vacuum cleaners have been studied. Several kinds of reusable containers have been tested. The two-sphere technique, for example, uses one sphere to suck in the minerals while the other provides the lifting buoyancy. As the stern test of actual oceanic mining approaches, attention is being focused on large towed sledgelike devices. Each will cost more than £1 million ($2 million). This form of mining is not without risks, as the equipment can be lost on the seabed. Considerable research effort has gone into finding ways of guiding collectors at depths of over 13,000 feet. They are towed with cables that will only break under enormous stress, and are accurately navigated to avoid such obstructions as sunken wrecks.

The leaders in this modern "gold rush" are Japan, West Germany, the United States, Canada, and Belgium. Smaller parts are played by the United Kingdom and France. Nobody expects the vexed question of who actually owns these minerals to hold up operations, though the legal wrangles are likely to make the business less commercially attractive. For example, there have been various proposals by the less-developed nations that they should, because of their greater need, have a major share in the profits even if they play no part in the actual harvesting of the resource. Such questions are harder to answer than the technical ones.

Right: side-scan sonar is one of the most advanced types of echo sounding equipment. It can produce acoustic "pictures" of the seabed. This is 32 feet long side-scanner GLORIA. It was developed by Britain's National Institute of Oceanography. From a depth of 600 feet it produces clear seabed traces to a range of 12 miles.

Below: fish farming is already being carried out on a large scale around the world. Using the most sophisticated fish farming techniques it is possible to provide a reliable supply of fish – in quality and quantity. Fish farming on the ocean bed itself is likely to be necessary if the ever-increasing number of mouths are to be fed. This impression of a fish farm shows biotrons, or fish hives (*bottom left*) containing spawn. This is carried by divers to hatcheries (*left background*) where the eggs are fertilized.

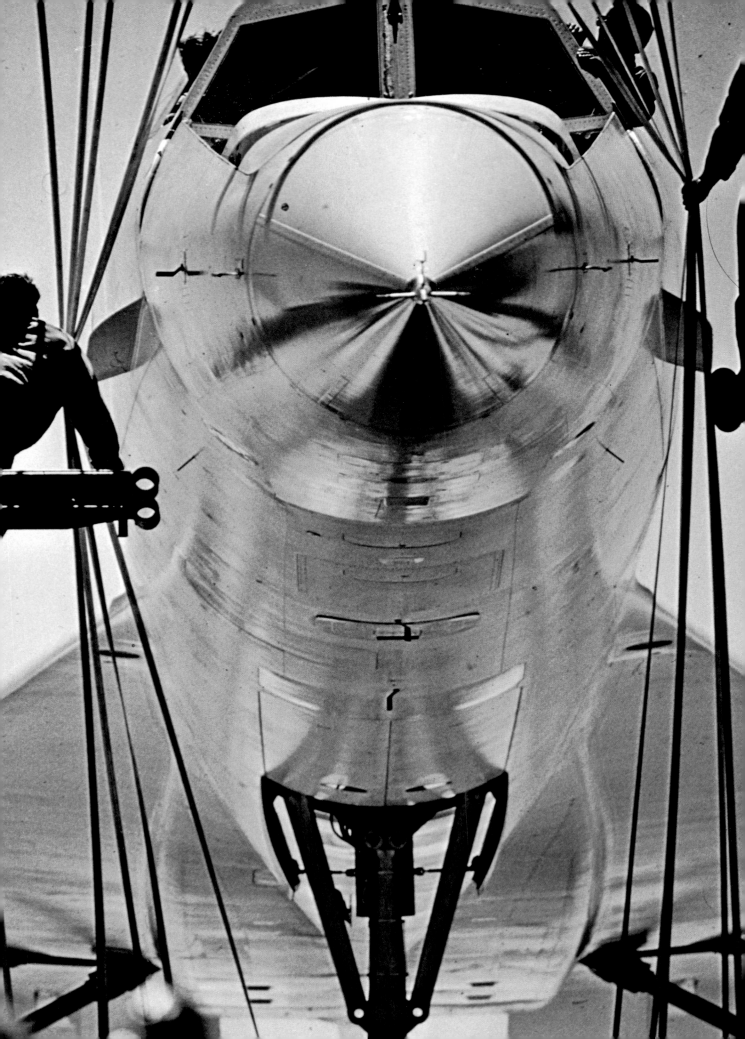

CHAPTER 4

TECHNOLOGY
OF FLIGHT

Few branches of technology offer such diversity as aviation. From 1783 man-made balloons were at the mercy of the winds. Dirigible (steerable) airships steadily improved capability. The controllable airplane dates from 1903. All three of these aircraft families have attracted the attention of military customers. Avaition technology has to an exceptional degree been lent to the waging of war. This tends to mask its enormous accomplishments in unifying and helping mankind. In more than half the world's countries aviation is the chief means of transport for all but local journeys. By far the fastest growing sector of aircraft operation is general aviation, which excludes all airline and military aviation. Aviation has used the advance of technology chiefly to improve efficiency, reliability, and safety. As a direct result, costs have been reduced. Air transport is cheaper today than it has ever been. Taking inflation into account air transport costs roughly one tenth what it did in 1930.

Opposite: the world's only supersonic airliner in use, the Anglo-French Concorde, under test at Toulouse. The runway was fitted with a safety net to prevent damage to this prototype by overshooting. Air France and British Airways Concordes went into service in January 1976.

A Brief History

On November 21, 1783 Jacques-Etienne and Joseph-Michel Montgolfier became the first men to fly in a hot air balloon when they launched one over France. Subsequently balloonists, or aeronauts, tried rowing with oars and rudder. But proper travel through the air did not begin until another Frenchman went aloft in his airship on September 24, 1852. By this time experimenters were toying with heavier-than-air flying machines, and the German Otto Lilienthal was the first to make glides down hillsides in order to learn how to control such a machine. But it was the American brothers Wilbur and Orville Wright who by methodical research solved the vital problem of control. From 1900 they flew gliders, and having come to certain conclusions, built an engine and flew the first airplane on December 17, 1903.

By 1910 the German Count von Zeppelin was building airships able to carry freight and passengers, and by 1914 his airline, Delag, had carried 33,722 passengers and crew without injury on 1588 flights throughout central Europe. In the same year a flying boat operated a short-lived scheduled service across Tampa Bay in Florida. Aviators had already flown seaplanes and flying boats, gliders, primitive helicopters, four-engine airplanes, and made successful landings and takeoffs on platforms mounted on ships.

Scheduled passenger flights can be traced to August

Above: a hot-air balloon designed by the Montgolfier brothers. It was in a balloon of this type that Pilatre de Rozier and the Marquis d'Arlandes stayed aloft for 4 minutes 24 seconds on October 15 1783.

25, 1919 when erratic and unreliable services began between Hounslow, London, and Le Bourget, Paris. Converted bombers were used – carrying up to four passengers swathed in greatcoats, scarves, and goggles. Operations were extremely difficult and grossly uneconomic. But the number of airlines was gradually rationalized and equipment improved. The advent of reliable aircooled radial engines – the Bristol Jupiter in 1922 and the American Wasp and Whirlwind four years later – transformed reliability and economics. By 1930 the network of air routes had spread from Europe to the Far East, southern Africa, and South America. Networks were extending within Europe, Asia, and North America.

By far the greatest boost to air transport was the development in the United States of modern cantilever monoplanes with all-metal stressed-skin construction, retractable landing gear, flaps, and variable-pitch propellers driven by 1000 hp engines in low-drag cowlings. As early as 1920 Fokker in Holland and Junkers in Germany had produced superior monoplane transports. But the new United States models not only reduced cost per seat-mile, but almost doubled cruising speed from 100 mph to 190. The most famous of the new breed was the Douglas DC-3, first flown on December 17, 1935, which by sheer chance had a structure able to resist fatigue and remains operational without major maintenance for 50,000 or more

Above: a bomber triplane converted into the Bristol Pullman for civil use. The enclosed cabin was luxuriously appointed to carry 14 passengers. Civil aviation was not allowed in Britain until April 1919.

flight hours. For this reason the old DC-3 is still in active service. Over 3000 were still flying in 1979.

During World War II, large and powerful troop transports such as the Lockheed Constellation and Boeing Stratocruiser pushed cruising speed up to 300 mph and the number of passengers to 100. Soon after the war, new methods of rocket and turbojet propulsion began to replace piston engines, initially in combat aircraft, but in Britain in civil transports. The United States flew the first supersonic aircraft beyond the speed of sound (about 750 mph) on October 14, 1947, and Britain sent the jet-propelled Comet airliner on its maiden flight on July 27, 1949, and opened scheduled jet services on May 2, 1952. The Comet roughly doubled both the cruising speed and altitude of passenger air travel, and provided a wholly unexpected bonus in a much smoother and quieter ride. Most airlines doubted the viability of the jet, and the turboprop was first to win wide acceptance in the short-haul Viscount which first flew in July 1948 and entered service in April 1953. Piston engine aircraft had to plow through bad weather at low speeds. But these fast new transports had air-conditioned pressurized cabins for comfortable flight high above the elements.

In 1955, while air forces around the world introduced supersonic warplanes, the French flew the Caravelle, the first short-haul jet. To the surprise of many it proved to be a best-seller, and its successors – the Tu-104, BAC One-Eleven, Trident, DC-9, 737, Tu-124, 134, and 154 and, above all, the 727 – made jet engines almost universal in passenger air travel by the 1970s. The classic Boeing 707 and Douglas DC-8 did the same

Above: a Lockheed Constellation. This was the most advanced airliner developed before World War II, with 2000 hp engines, and a pressurized cabin.

over long hauls, and in 1969 the Boeing 747 "Jumbo jet" transformed the scale and economy of air transport and opened up a new era of mass travel. Carrying upward of 400 passengers, the 747 was quieter than its predecessors, and in terms of seat-miles much more economical. Air travel is now within the grasp of the ordinary family.

Above: the French *Sud-Aviation* (now *Aerospatiale*) go into production with their Caravelle. Its engines were mounted at the rear of the fuselage. This reduced noise for the passengers, but brought those on the ground no relief.

Propellers versus Jets

All aircraft propulsion is jet propulsion, in that forward acceleration is obtained by a jet of air providing backward thrust. The largest jet is that of a helicopter's lifting rotor. This gives a mainly downward thrust, but can be tilted slightly to provide forward thrust. The next largest jet comes from a propeller. Early propellers were carved from laminated strong wood. Modern blades are made separately from the hub. Densified wood, plastics, light alloy, hollow thin-wall steel or glass and carbon-fiber reinforced plastics are all commonly used to make the blades. The blades can be set to a "fine" pitch for high power on takeoff, and adjusted to "coarse" pitch for economical engine operation at high speed.

Below: wooden propellers being manufactured for the German airline Lufthansa. These early propellers were individually carved from laminated hardwood. Later propellers were metal, with separate hub and blades.

In the event of engine failure they can be "feathered," and set to reverse pitch to brake the aircraft after landing. Propellers have always been in some degree dangerous, heavy, clumsy, and a source of intense noise and vibration.

They also impose severe limits on aircraft speed. The rotational speed of the blades has to be added to the forward speed of the aircraft. The tips of the blades can reach the speed of sound while the aircraft speed remains modest. This results in severe noise and loss in efficiency. Altering the propeller by increasing the number of blades, their width, diameter, or their rotational speed, incurs penalties. To this day the speed record of a propeller airplane is little over 500 mph.

The turbojet removed the restriction on aircraft speed, but early jet engines were extremely noisy, consumed huge quantities of fuel, and were inefficient except at very high forward speeds. In the early 1950s the jetpipe of many turbojets was enlarged, fitted with additional fuel burners and converted into an afterburner, with a variable-area nozzle. With the nozzle opened wider, tremendous quantities of additional fuel could be burned in the afterburner. This boosted thrust for takeoff or for supersonic flight, but at the penalty of even poorer fuel efficiency and even greater noise.

Such engines were ideal for warplanes, but commercial transports needed greater economy. The

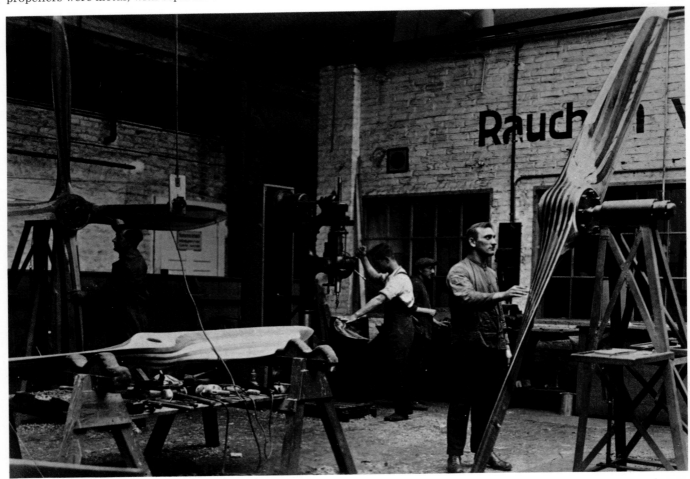

thrust reverser to brake aircraft after landing. Engines can be hung under or above the wings or on the sides of, or inside, the rear fuselage. The choice depends on the number of engines, aircraft range, and other factors. The whole engine pod is anti-iced, fire protected and fitted with comprehensive instrumentation

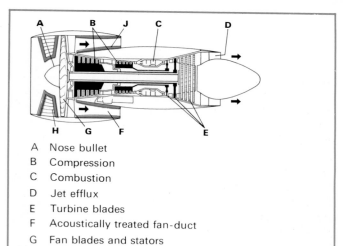

A Nose bullet
B Compression
C Combustion
D Jet efflux
E Turbine blades
F Acoustically treated fan-duct
G Fan blades and stators
H Sound-splitter plates
J Sound-absorption material

Above: the modern turbofan engine is significantly quieter than earlier jet engines. This sectional diagram shows the construction of an acoustically treated turbofan of the type powering wide-bodied airliners.

Above: an employee of the United States airline TWA stocktakes the company's supply of propellers in the 1950s. These replacement propellers were fitted to existing hubs, making spares cheaper to provide.

turboprop seemed to be a reasonable compromise, combining the power and reliability of the gas-turbine engine with the propulsion efficiency of the propeller. But it was subject to the unavoidable speed restrictions of propellers and has been used mainly in short-haul passenger airliners. The key to the future lay in more advanced gas-turbine engines which allowed the use of much higher pressures and temperatures, dramatically cutting fuel consumption. Today these surpass the most economical diesel. This has not only allowed the gas turbine to replace piston engines in many major markets apart from aviation but it has transformed the economy of air transport.

The development of the high bypass ratio turbofan in the 1960s increased efficiency still further. Earlier turbofans merely bypassed a little air from the compressor to rejoin the flow in the jetpipe, but the high bypass ratio engine was almost like a turbojet driving a multibladed ducted propeller. These huge engines, used in all the giant wide-body airliners and now scaled down to power the newest business jets and STOL (short take-off and landing) airliners, combine the extreme economy of the latest high-pressure high-temperature gas turbines, the propulsion efficiency and low jet velocity of the turboprop (but without the turboprop's limitations on flight speed), and the jet's freedom from vibration and noise. The new crop of "fan engines" are the nearest to perfection yet attained by any engine designer.

Like the propeller engine the jet can be fitted with a

Above: a Rolls-Royce RB.211 turbofan secured to a gantry while engineers position microphones to measure its noise level. The RB.211 is much quieter than its predecessors.

to control and record the health of the engine inside. Though the modern turbofan may appear large in diameter, its "drag" (air resistance) is close to zero. The engine sucks in air at the front and discharges it at the back, most emerging as cool air from the fan.

Subsonic Flight

Nearly all civil air transport is subsonic. Over the past 25 years the average size of the aircraft used has quadrupled. Speed and range have both doubled. The proportion powered by piston engines has fallen from about 98 percent to below two percent, while in recent years the quiet and efficient turbofan has steadily replaced both the turbojet and turboprop. Airframes have been developed that give no structural trouble for 60,000 or more flying hours. Airplanes have improved in reliability to a remarkable degree.

significantly larger than previous long-haul passenger carriers, and carried a load of kerosene fuel roughly equal to the entire laden weight of contemporary piston-engine airplanes. They were also original in engineering design, with four-wheel bogie landing gears to distribute the weight. These retracted into bays in the capacious lower section of the fuselage, under the main floor. The four engines were hung well below and ahead of the leading edge of the wing in "pods" which also incorporated thrust reversers and noise-suppressing nozzles. These pods were deliberately widely spaced along the wing to serve as antiflutter masses to damp down vibration – a common cause of metal fatigue.

In the late 1950s the choice by the French of rear-fuselage mounting of the two turbojets of the short-haul Caravelle was copied on two, three, and four-engine airplanes. Putting the engines at the back generally reduced noise in the passenger cabin and had other advantages. For example, failure of an engine did not cause such pronounced asymmetric

Above: an *Aerospatiale* Caravelle airliner at Helsinki airport. Introduced in 1958, the Caravelle had a cruising speed of 510 mph, carrying between 80 and 140 passengers. Other aircraft copied the rear-engined layout.

Airliners were divided into jets and nonjets in the 1950s. Jets had sharply swept-back wings, first used on military aircraft, to increase cruising speed closer to the speed of sound at the expense of higher takeoff and landing speeds and various other penalties. Such pioneer jets as the Boeing 707 and Douglas DC-8 were

thrust problems. But the helpful masses of the engine pods were taken away from the outer wings, making the wing heavier. The tail was also made larger and heavier. "Deep stall" control problems arose. These were caused by the ineffectiveness of the tail in checking very steep descents in a flat nose-up attitude. A number of fatal accidents have occurred. Man now knows how to build safe rear-engine jets. But the newest airliners such as the Boeing 757 and 767 and all Airbus versions have underwing engines only.

In the 1960s aerodynamic research produced a new family of wings with "supercritical" section profile. These have flatter tops than earlier wings, with more bulged undersides and rather more bluff leading edges. The trailing edge is often curved gently downward. It might be thought these would suffer increased air resistance (drag), but in fact the reverse is the case, especially at cruising speeds only a little short of the speed of sound. Older wings are what aerodynamicists call "peaky." Nearly all the lift is generated in a band of intense suction just a little way back from the leading edge on the upper surface. The supercritical wing spreads the suction evenly across the top. The result is a more efficient wing which, compared with traditional jet airliner wings, has less sweep (20° to 25° instead of 35° or more), greater thickness, and greater slenderness in plan form (higher aspect ratio). Higher aspect ratio means a given amount of fuel will carry the aircraft further. With increased thickness the wing weighs less and can accommodate more fuel.

Nowadays almost all new airliners, except STOLs

Above: engineers examine a model of a Lockheed TriStar before deciding to extend the wingtips. They went ahead with the extension and achieved a 3½ per cent cut in fuel consumption.

Left: Supercritical wing fitted to an F-8 aircraft at the NASA Flight Research Center. Supercritical wings are thicker, lighter, and accommodate more fuel.

Below: the interior of a Boeing 767. The model shown here can accommodate 18 first class and 197 economy class passengers.

designed for use where there are no good airfields, have supercritical wings with only slight sweep, and large-diameter fan engines. They are as fast as the old jets, much more fuel-efficient, and quieter even than the old propeller aircraft. And the development of engines with 55,000 pounds of thrust means that transports can be much larger than ever before. Economies of scale and bulk transport reduce the unit cost.

Once settled, the most difficult thing to change in a transport aircraft is the cross section of the fuselage. Older airliners had rectangular sections just wide enough for one seat on each side of a central aisle. The DC-3 of 1935 was wide enough for a double seat on one side, and in the denser seating of the postwar era this and similar sized airliners seat passengers two-plus-two. In the 1950s the triple seat made its appearance (some would say, a retrograde step), so that all the widest aircraft of the 1960 era seated passengers three-plus-three. By this time virtually all

Above: airliners queue for take-off on the tarmac at La Guardia airport, New York. The congestion of the world's airports and air routes has made strict traffic control vital to protect the safety of passengers and public.
Left: freight being unloaded from a Boeing 747 at London's Heathrow Airport.

airliners were pressurized, and though the lightest section for a pressurized fuselage is a circle, the best shape was usually judged to be a modified egg shape with a large circle above the floor and a deep smaller circle below to accommodate cargo, systems, and landing gears. In the late 1960s the so-called wide-bodies essentially doubled fuselage width at a stroke, so that two aisles became standard with first-class seating two-plus-two/two-plus-two and high-density seating three-plus-two/two-plus-three – in other words 10-abreast, but with nobody far from an aisle. The changing body sizes have also been reflected in a range of standard baggage and freight pallets and containers, tailored exactly to the available space.

Over the past two decades the development of the high bypass ratio engine has, almost fortuitously, prevented two crises which would otherwise have severely restricted the growth of air transport. One would have been caused by the cost of fuel, and the other by noise, which with growing traffic and noisy early jet engines threatened to become a major social problem. Today's large turbofans are the quietest aviation engines in history – quieter even than the engines of small lightplanes – and, installed in today's extremely efficient aircraft, they burn roughly one quarter as much fuel per passenger journey as the first jet airliners of the mid-1950s.

In common with other forms of transport, aviation today uses its complicated and extremely costly vehicles much more intensively and over a much longer period than in the past. In 1930 a typical utilization for a major airliner was 350 flight hours per year, but today it is considerably more than 10 times this amount. As an example, "Big Orange," one of Braniff's

Above: Braniff International's Boeing 747 "Big Orange" flies a daily round trip between Dallas and Gatwick, London – in the air 18 hours out of 24.

of components in the aircraft were either inspected at frequent intervals or were "lifed" and removed and thrown away when the allotted life had expired. Today the only part regularly inspected – and that only after intervals which might include 1000 ocean crossings – is the airframe, which is especially designed with smooth curves and polished surfaces, bonded joints, multiple load-paths, and protective surface coatings so that it will neither crack nor corrode. All the other amazingly complicated parts and systems are provided with automatic BITE (built-in test equipment) and "health-monitoring" systems, many of which spell out incipient trouble either on the flight deck or in the form of magnetic tapes which are played back after each flight. Nothing is torn apart or inspected unless the diagnostic equipment warns that trouble is coming. This has transformed reliability, engineering costs, and wasted effort in opening up items that are working normally.

This fantastic improvement in reliability has swept away previous statistical beliefs on aircraft design. There is no longer any need for multiple engines; a new British quiet short-hauler, the BAe 146, has four engines for particular reasons of cost and convenience, not reliability. Indeed, if passengers and thus airlines would accept it, the single-engine 200-seat short-hauler could show costs even lower than today's aircraft. The Airbus A300 was the first efficient supercritical-winged wide-body, and the first with only two engines. Probably this basic shape, with two underwing engines, will become by far the most common on the air routes in the final years of the present century.

By 1990 hydrogen, carried as the cryogenic (refrigerated) liquid, will have begun to replace petroleum-based kerosene as the standard aviation fuel. This will demand aircraft tankage of vastly greater volume, though gross weights will not change much. As liquid hydrogen is dangerous because it is so cold the tanks will be giant pods on the tips of the wings.

colorful 747s, flies the round trip from Dallas to London/Gatwick and back every day of the year, about 18 hours out of each 24 being spent in the sky earning revenue. This intense utilization is combined with a working life likely to exceed 30 years, compared with three years for the 1930 era. Thus, total flight time on each airframe has jumped from about 1000 hours to about 100,000 hours. To meet this severe demand, structures have had to be completely rethought, as have systems, components, and the whole business of maintenance.

Traditionally the structure was designed without thought of fatigue or corrosion, and all the thousands

Left: as the world's oil supplies dwindle the search is on for alternative sources of fuel. This is a model of a Lockheed L-1011 airliner designed to be fueled by liquid hydrogen. Financed by NASA, Lockheed engineers designed a 400-seat airliner, equipped with three decks – two for passengers and one for cargo. When built, it would fly 6300 miles non-stop at the same speeds as today's wide-bodied airliners. Liquid hydrogen has been used to fuel space rockets for years, and is a possible fuel for aircraft. In any case, alternative energy sources will have to be found before the oil runs out.

Beyond the Speed of Sound

Very little was known about supersonic flight until after World War II. This was partly because free-flying aircraft were incapable of reaching such speeds, 761 mph or more at sea level, and about 660 mph in the intense cold of high altitudes. Research in wind tunnels was made difficult by the fact that tunnels "choked" with shock waves at speeds well below that of sound, and could not operate over the important transonic region (speeds near that of sound). Eventually ways were found to make transonic tunnels. But long before that a United States Air Force (USAF) pilot, Charles "Chuck" Yeager, had flown a rocket-propelled Bell XS-1 research aircraft faster than sound on October 14, 1947. Soon Mach 2 (twice the speed of sound) was exceeded by various American

Above: airflow around a model of Concorde in a wind tunnel taken during the aircraft's development.

Right: Concorde's cabin is narrow but journeys are brief.

research aircraft. By 1956, fighters and even a bomber, the B-58, were capable of Mach 2.

Supersonic transports were studied, and a passenger version of the B-58 bomber was on the drawing boards by 1958; but difficulties are severe. The most intractable is that the L/D ratio, the ratio of aircraft lift to drag (a measure of its flight efficiency) is roughly halved when the speed of sound is exceeded. Thus, with a given load of passengers and fuel, a supersonic airliner will fly roughly half as far as a subsonic one. Moreover, the afterburning engines have to be installed in extremely complex and costly nacelles with both the inlets and nozzles fully variable in both area

and profile. The noise of such an engine on takeoff is high and difficult to reduce. In cruising flight all supersonic aircraft cause shockwaves. At Mach 1 (the speed of sound), these are perpendicular to the line of flight, and at increasing Mach numbers, lean back more and more acutely. With large aircraft the powerful shock waves form a cone extending down to the ground, leaving behind a "boom carpet" several miles wide within which the passage of the shock waves is heard as a dull multiple boom.

There were many other fundamental difficulties. No traditional shape would do for supersonic transport (SST). This meant bringing radical new aerodynamics, flight characteristics, and technology to the ultra-conservative airline industry.

The body cross-section, called the fineness ration, has to be extremely large in an SST. For any practical overall length this meant a slim body. Some passengers – especially after familiarity with the roomy wide-body aircraft – think SST accommodation cramped. Cabin pressure also had to reach record levels to preserve a comfortable interior in the near vacuum of flight at 60,000 feet. SSTs have therefore been fitted with very small windows. Many designers believe that windows should be eliminated altogether, though this would not be popular.

Though the United States spent over $1 billion on studies and engineering development, only two types of SST had been built by the end of the 1970s. The Soviet Union, conscious of the size of the country and the benefits of saving time, planned a large fleet of up to 75 SSTs, produced by the Tupolev bureau as the Tu-144. First flown in December 1968, this aircraft was totally redesigned and suffered many problems. It has never gone into intensive passenger service. Proving flights began in 1974, followed by Aeroflot services with special freight. This terminated in 1978. The reason for final withdrawal had not been made public in 1979.

Slightly smaller than the Tu-144 is Concorde, which was developed by Britain and France after the signing of a 50/50 intergovernment agreement in November 1962. The first prototype flew in March 1969, and scheduled services by Air France and British Airways began in January 1976. Apart from slight enlargement and various details the design changed little during prolonged development and route-proving in 5335 flying hours when the unrestricted Certificate of Airworthiness was awarded in late 1975. Powered by four 38,100 pound-thrust Rolls Royce Olympus engines, Concorde seats up to 128 passengers in four-plus-four rows, with provision for lavish meal service. Structure is mainly a heat-resistant aluminum alloy, and though cruising at Mach 2.2 (1460 mph) is possible, the Mach number is normally restricted to about 2.09 to delay fatigue. An unusual feature, shared by nearly all other SST designs, is a drooping nose with retractable visor to give perfect streamlining in cruising flight but good forward vision for landing and takeoff. The configuration is a tailless ogival delta, with powered elevons on the wing trailing edges. Leading edges are fixed, and a large spiral vortex formed on takeoff and landing increases lift by making the wing behave like a much thicker wing with a generously rounded leading edge. By 1979 all 16 production models were in service.

Above: the first SST to fly, the Soviet Union's TU-144. Aeroflot terminated its services in 1978.

Below: a Concorde takes off from Fairford in Gloucester. By 1979, 16 Concordes were flying with four airlines.

Cargo Planes

There was very little air cargo before World War II. Freight was restricted to small urgent items, which were expensive to send, and had to be accommodated in passenger airliners. Military transports carried only troops, with small items of supplies loaded by hand. Cargo carriers could have been built decades earlier, but the first really capable machine was the German Me 321 glider. This monster was designed for the planned invasion of England, Operation Sealion. It had an unobstructed level cargo hold. Two immense

most useful design features of cargo aircraft. The fuselage is an unobstructed cargo hold, with a flat level floor at truck-bed height. The interior is pressurized and air conditioned. Four low-pressure main tires, suitable for unsurfaced airstrips, are fitted, and the sprightly field performance is matched to very short or high altitude airfields. Four powerful turboprop engines provide this performance, and the C-130 cruises at speeds of over 375 mph. At the rear of the fuselage are doors which not only form a ramp for vehicles, but can also be opened in flight for dropping heavy loads by parachute, or, at very low level, without. By 1979 orders for the military C-130 had reached 1600, while many civilian customers have bought the L-100 Commercial Hercules. Most versions have a lengthened fuselage because with low-density cargoes the basic Hercules' space is limited.

Lockheed also built 284 C-141 StarLifters. Powered by four turbofan engines, they have considerably greater range than the Hercules. Another Lockheed product is the C-5A Galaxy. They built 81 of them. The

Left: the first really effective cargo aircraft, the World War II German glider Me 321. Otherwise known as the Gigant, this enormous craft was loaded with tanks and heavy artillery through double doors at the front. A glider's behavior is unpredictable.

Below: a C-130E Hercules at RAF Mildenhall. The C-130 series has been flying since 1954, and is still widely used in military and civilian roles. Its speed of 375 mph, and its ability to use unmade airfields make the C-130 suitable for ferrying essential supplies to remote areas.

doors formed the entire nose through which large loads like tanks or artillery could be loaded. By 1943 the powered version, the six-engine Me 323, was in service. This introduced multiwheel landing gear for operation from rough terrain. Other cargo aircraft followed, including the American C-82 Packet with twin tail booms, and a cargo hold with left and right rear doors.

Soon after World War II Britain produced the Bristol 170, which sold in both civilian and military markets. A simple 4000 hp high-wing machine with fixed tailwheel landing gear (which precluded a low level floor) and twin nose doors, the 170 is still used for such cargoes as cars or beef, but more modern and sophisticated models did not appear. Many freighters followed, for military and civilian use. The one that won the biggest market was the Lockheed C-130 Hercules. First flown in 1954, this for the first time combined all the

largest airplane in the world on some counts, the Galaxy carries a 100-ton load including two M60 battle tanks or 16 3-ton trucks yet can use rough airstrips with its 28-wheel landing gear. Another military need is for the aerial tanker, and Boeing delivered no fewer than 888 piston-engine KC-97 Stratofreighters and Stratotankers, and 732 jet- and fan-engine KC-135 versions of the civilian 707 passenger airliner. In the Soviet Union Antonov produced the four-turboprop An-10 and 12, resembling the C-130, and the much larger An-22 Antei used for military and civilian haulage of the heaviest loads, especially in opening up developing regions such as the oilfields at Tyumen.

Right: the world's biggest passenger airliner, the Boeing 747 "Jumbo", is also in service as a freight carrier. This 747-200F is being loaded at Tel Aviv. It carries more than 250,000 pounds of cargo.

Below: the hold of the largest cargo aircraft in the world – the Lockheed C-5A Galaxy. Originally used in the war in Vietnam, the Galaxy's enormous hold can easily carry 16 three-ton trucks.

tainers or palletized cargo. This is loaded and stowed under computer control to keep the center of gravity in the right place and enable items to be unloaded in the correct sequence at destination airports.

The largest and most capable commercial freighter is the Boeing 747-200F. This has an airframe similar to that of the 747 "Jumbo" passenger jet except for a

In 1958 the British company Armstrong Whitworth marketed a carefully planned family of Argosy civilian and military cargo aircraft, but failed completely through no fault of the aircraft. Civilian operators still fly cargo either under the floor in passenger airliners, or above the floor in combination passenger/freight (Combi) or quick-change (QC) jets, externally often identical to passenger aircraft. Almost all have one or more large freight doors, typically 8 feet 3 inches high and about 12 feet wide, in the left side. A strong steel floor is essential, with powered rollers and free-running "ball mats" for handling the con-

modified fuselage with upward-hinged nose for full-section cargo loading, computerized loading system, and powered conveyors controlled from various local push-button stations. A side cargo door and passenger windows are omitted. Up to 254,640 pounds of cargo can be loaded and exactly positioned by two men in 30 minutes. These aircraft are often assigned to carrying 100-ton loads across the Atlantic to such cargo terminals as Paris or Frankfurt.

An unusual large-scale cargo network in the United States is that of Federal Air Express, of Little Rock, which uses Falcon business jets.

General Aviation

In many countries general aviation (GA) has for many years been growing more rapidly than any other sector of the industry. GA includes all aviation other than civil airlines and military equipment. Light club and private aircraft number about 190,000 around the world. There are also single-seat gyroplanes, sailplanes, hang gliders, and an army of homebuilt machines which include replicas of early flying machines or famous types from the two World Wars. A vital sector of GA is aircraft operated for business houses, charter operators, photography, and other special operators. GA also fulfills police and paramilitary functions such as customs and border patrol, and plays a major part in the development of modern agriculture.

Privately owned aircraft are predominantly single-engine. Almost the entire world market is dominated by piston engines from two American companies, Lycoming and Continental. These engines have four or six horizontally opposed cylinders (called "flat four" or "flat six" types). They are increasingly being fitted with direct fuel injection instead of a carburetor. A small but growing proportion have exhaust-driven turbochargers to maintain full power at high altitude. As a result, increasing use has been made of pressurization even in the cost-conscious GA market. Engine powers usually range from 100 to 400 hp.

Apart from the often bizarre home-built machines and the replicas, virtually all nonagricultural GA airplanes are clean monoplanes with comfortable well furnished cabins. Increasing attention is paid to the provision of comprehensive instrumentation, avionics (aviation electronics, such as radio, navaids, and landing aids), and even autopilot for safe flight in instrument flight rules (IFR), such as cloud or bad weather. More and more people are using private aircraft not only for casual sport but also as an essential working tool, and need to fly whatever the weather.

Business aviation is a case in point. It has grown phenomenally in the past 25 years. Business aircraft enable executives to go straight to their destination without the delays and inconvenience often experienced on surface transport or commercial airlines. The majority of company-owned airplanes are piston-engine. A high proportion are twin-engine aircraft which can carry larger groups of passengers over greater distances than would be practical with most single-engine machines. The most produced single-engine lightplane is the four-seater 150 hp Cessna 172.

Above: A Cessna 172 parked by a private house in Canada. This four-seater is one of the most popular light-planes to be produced. The Cessna is easy to learn to fly, which is the main reason for its popularity.

Over 31,000 have been made in 23 years. Cessna built one twin, the Skymaster family, with push/pull engines at front and rear of the fuselage. Any pilot could fly it, while other twins need special instruction and pilot licenses.

Piston and turboprop twins normally have a wide instrument panel packed with controls sophisticated enough for a commercial airliner. Dual controls are usually fitted, though the right-hand front seat is generally occupied by a passenger. Behind are seats for a further four to seven people, and large quantities of baggage can be stowed in the long nose, behind the cabin, and/or behind the engines in the rear of the nacelles. Retractable landing gear, constant-speed and feathering propellers, flaps, electric and hydraulic systems, and, usually, pneumatic inflatable-strip de-icing are standard. Even the piston-engine twins often cruise at speeds around 280 mph. The more powerful pressurized turboprops are generally quieter despite cruising at about 325 mph. Range is adequate for almost any European journey, or a one-stop coast-to-coast flight across the United States.

By 1960 it was evident there would be a market for

small business jets. Though this still accounts for a mere one percent of the GA market, these 2000-plus machines cost from $1 to $5 million each, and therefore make up almost 20 percent of GA by price. Nobody buys or leases a business jet without a definite need. This need has been expressed by over 1000 companies, some of which operate several jets day and night. A few aircraft in this class are full-scale jetliners, the BAC One-Eleven being especially popular because of its low costs. Its beautifully tailored interiors are arranged not for the airline 75 to 100 passengers but for only 15 to 25, with lavish office facilities and special equipment. Some interiors are fitted with powered armchairs. The occupant can drive along seat rails in the floor and then swivel to face the other passengers. Early business jets were powered by turbojet engines, but today the quieter and more fuel-efficient turbofan predominates. The smallest models available are compact to the point of being cramped, the outstanding example being the baby Foxjet which is smaller than most lightplanes. The largest are the long-range Gulfstream III followed by the French Falcon 50 trijet and Canadair Challenger.

A highly specialized branch of aerial work is firefighting, particularly forest fires. Many types of air-

Top left: a number of companies use the BAC One-Eleven as a luxuriously appointed executive jet.

Left: a business jet is equipped much like an office suite.

Below: the popular private four-seater Cessna Skyhawk has a cruising speed of 140 mph.

plane and helicopter have been used for this work. This usually involves either "bombing" the fire with chemicals or "water bombing" in which a large mass of water is dropped on a carefully selected spot to blow out that part of the fire and leave the ground cool and wet. Canadair in Montreal designed a unique aircraft primarily for firefighting, but capable of many other duties. Called the CL-215 and first flown in 1967, it is an amphibian, with a deep flying-boat hull and retractable tricycle landing gear. It is the last aircraft designed to use large piston engines, with two 2500 hp Pratt & Whitney Double Wasps. Its method of operation is to swoop onto the nearest stretch of water near a fire at 69 mph and scoop up 1200 gallons in 16 seconds. Special fire-retarding chemicals can also be carried and mixed with the water. The CL-215 then dumps the load on the fire and returns for a fresh supply. Often water has been scooped up in waves as high as six feet, and single aircraft have dumped as many as 100 loads on a fire in one day.

With a nose radome the CL-215 is a versatile search and rescue aircraft. Large numbers of radar-equipped aircraft with piston, turboprop, and jet engines are used all over the world for coastal surveillance, fisheries patrol (to enforce controversial limits), marine environmental protection, and many other duties. But in terms of numbers the largest aerial work fleet in the world is agricultural aviation (ag-aviation). There are about 40,000 such aircraft in use around the world. Largest is again the CL-215, some of which have been equipped as wide-swath budworm killers to meet a major crisis. Most agricultural aircraft are single-engine, and unlike any others flying. A few are helicopters.

Below: a Canadair CL-215 being operated by the Royal Thai Navy. This amphibious design was built primarily to fight forest fires in remote areas.

Agricultural duties include spraying insecticides, fungicides and top dressing, dusting with a wide range of powders, seeding (for example virtually the whole of the North American rice crop is air seeded), and distributing weedkillers or defoliants. One aircraft can do as much work in a given time as at least 15 tractors, and use only about one tenth the fuel. The essential qualities of an agricultural aircraft are robustness, reliability, capacity, and weight-carrying ability for heavy loads of chemicals or seed. Good forward visiblity, and excellent handling at low speeds and very low altitudes are also vital. Accidents are more frequent than in any other branch of aviation, so special provisions are built in to insure that accidents are as far as possible minor, and certainly not fatal. Another requirement is total proofing against the chemicals which are often corrosive. The airfield is any convenient space on a farm. Some ag-planes are braced low-wing monoplanes. But a few have a high wing and many are biplanes. One unique jet biplane, produced in Poland, carries its load in two hoppers between the wings. Usually the tank or hopper is in the fuselage, which is generally tall and hump-backed with the cockpit at the highest point for optimum visibility. Piston or turboprop engines of 300 to 700 hp are usual, and there has to be either a power takeoff shaft or a ram-air windmill to power spraying pumps or other equipment. Chemicals are expelled through jets along the wings to give a swath width of about 115 feet at full extent.

As a complete contrast, gliders range from the highly popular hang gliders to the world championship sailplanes, surely the most beautiful aircraft in the sky. Sailplanes have slender (high aspect ratio) wings, the structure usually being of glass fiber reinforced plastics (GRP), often stiffened by carbon fiber. The pilot sits in a reclining position in the pointed nose under a giant molded canopy, with controls for tow-

Above: gliding is not only a popular sport. It was extensively employed during World War II to land troops behind enemy lines. The German invasion of Crete was accomplished with the help of gliders. This sailplane is flying along the flank of the Remarkables – a mountain chain in New Zealand.

release (usually either from a winch or a tug airplane), conventional flight controls, airbrakes, retractable landing wheel on the centerline, water ballast, and, in some cases, braking parachute and/or flaps. Standard class machines have a span of 49 feet 2½ inches and open class a span of at least 56 feet; many open class today have spans of well over 65 feet and have a glide ratio (distance traveled divided by height lost) as high as 55.

Above: aircraft are playing an increasingly important role in agriculture. Their major job is spraying crops with insecticides, fungicides, and top dressings. There are an estimated 40,000 agricultural aircraft in the world. One of these can do more in an hour than a tractor in a day.

Left: in recent years hang gliding has become a hugely popular sport, especially in the United States. The modern hang glider has its origins in the foldable wing developed by NASA for spacecraft re-entering the Earth's atmosphere. A hang glider has a tough Terylene sail attached to a strong tubular A-frame. The pilot hangs in a harness.

Helicopters

Though the first helicopters to lift their inventors – Paul Cornu and Louis Breguet – just cleared the ground as early as 1907, it was to be almost another 40 years before this type of vertical takeoff and landing (VTOL) flying machine was to become useful. Long before the helicopter matured the autogiro was to be developed, through the perseverence of the wealthy Spaniard Juan de la Cierva. Disturbed at the toll taken by airplanes, he sought a machine that could not stall (lose lift because of breakaway of airflow from the wing). He put a lifting rotor above an airplane and got it off the ground in January 1923. His autogiro is essentially an airplane propelled by a conventional propeller which instead of having a fixed wing has a set of slender radial wings on a free-spinning hub. This is rotated by the passing airflow. There are thousands of autogiros in use, but nearly all are sporting single-seaters needing only the area of an average lawn to take off. A few can clutch the engine to spin up the rotor for a vertical "jump" takeoff. All can make

almost vertical landings, the rotor "autorotating" as the air passes up through it.

In contrast the helicopter's engine drives the rotor system all the time. This system usually comprises one main lifting rotor above the center of gravity plus a small tail rotor to cancel the drive torque (and thus stop the helicopter from rotating in opposition to the main rotor), and privide directional control. A few helicopters have two equal lifting rotors with axes tilted to give stability and rotating in opposition to cancel torque. The tandem rotor machine is not uncommon, but the world's largest helicopter – the 231,500 pound Soviet V12 – has side-by-side rotors, and the intermeshing "egg-beater" has also proved popular. The Soviet Kamov bureau has adhered to the superimposed coaxial arrangement, but this may have been to make a powerful helicopter compact enough for use from small ships.

Unlike most autogiros the helicopter derives both lift and propulsion from its main rotor, and whereas that of the autogiro is tilted slightly backward, the

Right: the first helicopter to leave the ground – the 1907 Cornu. The contraption remained aloft for less than a minute with Paul Cornu at the controls. An unwieldy machine, it crashed and broke up on landing. **Below:** the world's largest helicopter, the Soviet V-12.

Above: some helicopters are ideal for maneuvering heavy loads in areas where a crane would be impractical. Here, a US Army Sikorsky CH-54A Skycrane repairs a bridge in Vietnam. Skycranes can position loads up to 40,000 pounds.

Right: maneuverability makes helicopters ideal for rescue work over mountains and seas. Here, a helicopter from the Swiss mountain rescue service is employed in aid of stranded climbers.

air flowing diagonally up through the rotor disk, the helicopter's main rotor is tilted forward in normal cruising flight, the air passing down through the disk. The pilot has two main controls. 1: the cyclic-pitch stick in front varies the pitch of each blade cyclically as it passes round the 360° disk to tilt the disk in any desired direction, so that the helicopter will travel in the same direction. 2: the collective-pitch lever low on his left side alters the pitch of all blades together to make the machine move up or down, as required. These two controls give considerable maneuverability.

The first helicopters had very limited capability, and the only ones to be mass produced were the small Bell and Hiller machines with piston engines of about 200 hp and a bench seat for three under a plastic "bubble" in the nose. In 1955 France produced the similar Alouette with a turbine engine, and gradually the lighter, more powerful, more reliable, and usually quieter turboshaft has eaten into the piston-engine market and made the helicopter far more attractive. Though still costly, compared with airplanes of similar capacity, the helicopter's ability to use small areas for takeoff and landing, its ability to hover, and its general versatility have made it a popular tool for more than 10,000 companies and a number of airlines.

In undeveloped regions, helicopters of all sizes are often the most numerous vehicle. In the Soviet Tyumen oil region, for example, they fly heavy equipment to otherwise inaccessible sites. Hundreds support the world petroleum industry, and a helipad is every oil rig's link with the outside world. A few cities have helicopter airlines carrying passengers and mail between downtown office roofs, airports, and suburban heliports, though costs are high and legislation on safety and noise is rigid. For such air carrier work, forthcoming helicopters such as the 44 seat Sikorsky S65C will greatly improve economic performances. At present nearly all commercial helicopters are 10-seaters or smaller, used for business or utility tasks such as patrolling pipelines or electricity grids, traffic control, or photography.

It is only recently that, partly by using ordinary auto parts, helicopter costs have been cut dramatically. In the 1980s, large passenger helicopters with greatly improved performance and economy will probably become commonplace.

VTOL and STOL

The acronym VTOL, from vertical takeoff and landing, came into use in the 1950s when it was realized that the light and powerful gas turbine engine opened up new possibilities of vertical flight. Over 80 technically novel experimental VTOL and STOL (short takeoff and landing) machines have been built since, and today there is a better appreciation of not only what is possible but also what makes economic sense. In particular, though the VTOL airliner, and even the large VTOL jet, have been feasible for many years, none has proved financially attractive and recently such ideas have sharply receded from the fever-pitch of excitement current before 1970. In contrast, the less radical STOL has amply proved its value.

Right: the Rotodyne, built by the British firm Fairey Aviation. This 60-seat convertiplane prototype was designed to achieve high cruising speeds. But financial constraints killed the Rotodyne project.

Below: the tilt-wing Ling Temco Vought XC-142A was a military transport. It was never built in quantity.

There are literally hundreds of ways in which engine power can be used to enable airplanes to use smaller airstrips. At some point the use of tilting wings, tilting propellers large enough to be called rotors, and many other devices becomes so unusual that the machine can hardly be called an airplane at all. Many, especially those fitted with some kind of lifting rotor (which was often hinged or even retractable), were called convertiplanes. Though impressive prototypes have been built none has entered service. One of the first large VTOL convertiplanes was the British Rotodyne of the late 1950s. In its final form it had a 60 seat airline cabin body, a small wing carrying two 5000 hp turboprops, and a large lifting rotor driven by jets at the tips. Bell's X-22A had four giant ducted propellers which pivoted to give lift or thrust. The Curtiss X-19 used four tilting propellers on the tips of

case, the sheer convenience of flying from the center of (for example) London to the center of Paris was a tantalizing prospect, especially as it could be done at jet speed. Rolls-Royce developed from proven lightweight lift-jets a series of lift-fans of much greater diameter. These could handle many times the airflow, and were therefore much quieter. Typical of the proposals that resulted was the Hawker Siddeley 141 of 1970. This resembled a normal 120 seat twin-jet with a very small wing and a strange fairing along the lower sides of the fuselage. Within this fairing were to be eight or more RB.202 lift fans, for VTOL. It was calculated that overall noise would be less than for conventional takeoff and landing (CTOL) airliners, but despite the claimed higher productivity the airlines showed little interest.

In contrast the STOL airplane exists in large numbers, and its size has reached the 50 seat level with the de Havilland Canada Dash-7, first flown in 1975 and put into airline service in 1978. The Dash-7 is a four-engined successor to the same company's Twin Otter, of which more than 600 are in use. Both have PT6 turboprop engines on simple wings of large span with generous double-slotted flaps to give high lift at low speeds. Tails are large, for good control at speeds down to about 70 mph. The tail of the Dash-7 is mounted atop a lofty fin. The Dash-7 is pressurized for high altitude cruise. The British 30 seat rival, the Short's 330, has a roomy unpressurized cabin of rectangular section and is consequently much cheaper.

In the mid-1970s the United States Air Force tested two extremely advanced STOL transports to replace

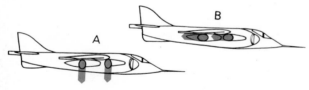

Above: the British Hawker P.1127, an ancestor of the Harrier jump jet, demonstrates its vertical take-off ability. Diagram **A** shows ducts and louvers directing the gas jets down for vertical lift off. In **B** the jets are deflected backward for conventional level flying.

Right: a USAF Boeing YC-14 on a test flight in February 1977. The YC-14 is a STOL aircraft designed for delivering troops and supplies to restricted sites. Similar machines could be employed in many civilian roles.

small wings at front and rear. The LTV/Hiller/Ryan XC-142A had four turboprops with large propellers on a wing that pivoted through 90°.

By the 1960s designers of civilian jetliners were studying the jet-lift schemes promoted by Rolls-Royce. Most suffered from very high noise level, to an extent that threatened to rule out VTOL from city centers and possibly even from airports, where severe congestion was generally believed to be a prime reason for hastening VTOL operations that would not use the runways and would occupy different airspace. In any

the C-130 Hercules as a standard medium airlifter. Though budget restrictions prohibited buying either, the technology of these aircraft will strongly influence advanced civilian STOL transports of the 1980s. The Boeing YC-14 has USB (upper-surface blowing); the two large turbofans are mounted high on the small wing and blow back across the upper surface. And when the high-lift flaps are lowered the augmented jet effluxes are turned down almost 90° for lift. The McDonnell Douglas YC-15 achieves the same performance by engines blowing straight into the flaps.

Safety in the Air

From 1920 until 1940 most air services were operated from grass fields, with rudimentary equipment and facilities. Partly because of the restricted size of airfields, many of the largest transport aircraft were flying boats, operated from marine bases in the same way as ships. But the worldwide development of major airfields in World War II provided the foundation for the modern airline industry based almost entirely upon airplanes operated from land.

Such aircraft need strong paved runways from 5000 to 12,000 feet long. Until the jet era it was common to provide three runways aligned in different directions so that airplanes could take off into wind. Modern aircraft are less sensitive to wind direction, and most airports use multiple parallel runways to increase the rate at which aircraft can land and take off. Many important airports still have a diagonal crossing runway. This interferes with traffic on the main runways and poses a collision hazard. Research has shown the safe capacity limit for each runway layout, and the generally accepted configuration for future major airports is two pairs of parallel runways, one of each pair used for landings and the other for takeoffs.

To avoid collision hazards either on the approach or in the departure pattern flown by outward bound aircraft, the distance between takeoff and landing runways has to be considerable. Today all incoming traffic has to fly a long straight-in approach along a glide path that usually slopes down to the runway threshold at about three degrees. This is the usual inclination of the instrument landing system (ILS). With two radio beams (one, called the glide-slope, gives vertical guidance and the other, called the localizer, gives directional guidance) it influences a cross-pointer instrument in front of the approaching pilot to steer him safely down to the runway. Few airports have an instrument runway (one equipped with ILS) of sufficient accuracy for truly blind landings. But when a flight does terminate at such a runway a suitably equipped jet can make an automatic landing in zero visibility. Autoland facilities include precise control of flight path, automatic control of engine power, autoflare to check descent as the ground approaches, and automatic "kickoff drift" to cancel the effect of sidewind while landing level.

Below: the German Dornier DO-X Flying Boat in New York harbor. When it was built in 1929, the DO-X was the largest aircraft in the world. It once carried 129 passengers at a speed of 134 mph.

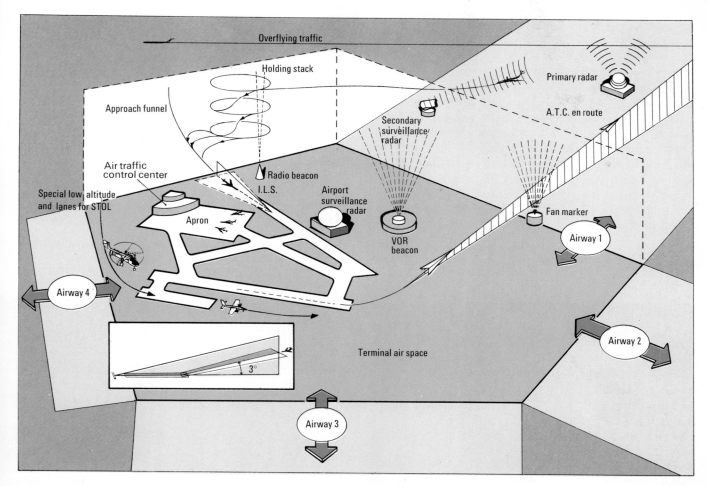

Overflying traffic

Holding stack

Approach funnel

Air traffic
control center

Special low altitude
and lanes for STOL

Apron

Airway 4

Radio beacon
I.L.S.

Airport
surveillance
radar

VOR
beacon

3°

Terminal air space

Airway 3

Primary radar

A.T.C. en route

Secondary
surveillance
radar

Fan marker

Airway 1

Airway 2

In the 1980s existing ILS will gradually be replaced by MLS (microwave landing system), in which the radio beams are different in character and sweep the sky from side to side. One of the advantages of MLS is that airplanes do not need a long straight approach. By chaneling in traffic from all directions, the safe capacity of a single runway will be increased. It will

Right: the weather radar system of a DC-9 installed in its nose. "See and be seen" is the doctrine of safety in the air.

Below: the microwave landing systems (MLS) enables aircraft to touch down in dense fog to an accuracy of within two feet on a two mile runway.

Above: an airport's terminal airspace (*blue*) and four airways (*pale blue*) adjoining it. Taxiing and take-off instructions are issued from the traffic control center.

be possible to use two close parallel runways for simultaneous landings.

Air traffic control is handled by different groups of controllers. Some of them are in airport towers to

handle arrivals or departures, and others are in control centers for upper airspace. The upper part of the usable airspace is used by military aircraft, supersonic transports, and long-range business jets. The bulk of subsonic airliners use the middle airspace (about 25,000–40,000 feet), and the lower levels are used by propeller aircraft and helicopters. In good weather, aircraft at low levels can fly without ground control and often without a radio – following the doctrine "see and be seen." All airspace above 18,000 feet in North America and 24,500 feet in Europe is strictly controlled. Aircraft using it must carry many electronic aids, file a detailed flight plan, and remain in contact with various ground controllers. When

"interrogated" by ground radar, each aircraft automatically sends back its identity, and often its exact altitude.

Each airport has different tiers of controllers, often on different floors in the tower. Ground control is responsible for everything on the airport, including vehicles, and handles the often urgent and congested ground traffic that directs incoming aircraft to the right stand (the departure gate, or parking place), and the hords of vehicles that swoop on it for quick "turnaround" for the next departure. Radar control of incoming traffic works alongside departure controllers, who all have details of every flight on computer printed strips. At the boundary of the airport

Below: the interior of an air traffic control tower. This controller is watching a radar screen. Two radars are used in air traffic control, one for surveillance and the other for precision approach.

Bottom left: a surveillance radar screen in an air traffic control tower. It shows the precise positions of aircraft approaching and leaving the control area of the airport, and facilitates accurate guidance to pilots.

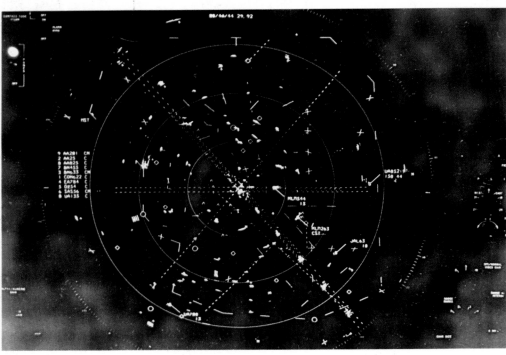

control zone or terminal area, control is handed over to the controller in the next area sector. In this way, an airliner is constantly monitored from the ground. When aviators first faced the problem of navigation they had a choice of three methods – direct contact flying (watching the ground and following their track on a topographical map), dead reckoning (using a plotting chart and knowledge of cruising speed and time on a given compass course), and astronavigation (using a sextant to measure the inclinations of stars as mariners had done for centuries). All had grave deficiencies, and the first really helpful navigation aids (navaids) emerged in the United States to assist pilots flying the mail through bad weather. Bonfires were the first, and most primitive navaids, and pilots tried to keep one in sight to the rear until the next showed up in front. By 1929 radio beacons were taking their place. In the early 1930s the Radio Range

began to cover the United States with a network of air routes linking the Range stations. Each station sent out a radio beam heard by the pilot as a letter N in Morse code on one side of the airway and a letter A on the other. Left or right depended on whether the pilot was approaching the station or leaving it. Along the centerline the two signals merged into a continuous steady note.

This helped greatly, but had the unfortunate psychological effect of carving up the vast expanse of sky into narrow highways. In course of time, these became greatly overcrowded. Also, such aids were useless for military purposes. In 1940 the German *Luftwaffe* found its bomber targets using radio beams

Below: an aerial view of Heathrow Airport, London. The runway pattern is clearly visible. One of the problems of mass air travel is that airports have to be close to major cities where land costs are astronomic.

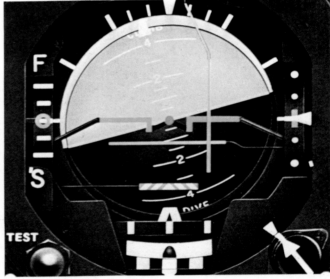

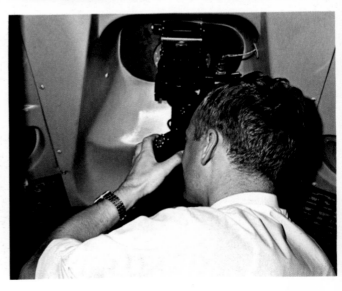

sent out over the enemy territory, with crossing beams to indicate when bombs should be dropped. This again was useless except for individual targets, and the following year the RAF began using Gee, the first of a diverse family of area coverage or R-Nav systems. It used a "master" radio transmitter precisely keyed to three "slave" stations. All emitted pulses of radio waves together, so that the interference between the wave fronts had the effect of setting up a stationary pattern of waves in space. These patterns had the form of hyperbolae, so Gee was called a hyperbolic aid. By studying the different times separating the signals from the four stations a pilot or navigator could pinpoint his position on the hyperbolic pattern, and therefore in relation to the ground. Subsequently the continuous wave Decca, American Loran, and many other hyperbolic navaids came into worldwide service. Today the newest R-Nav system is Omega, which with a mere eight transmitters provides coverage of the entire globe. A mini computer on the aircraft analyzes triple signals emitted by the stations and instantly indicates position within less than two miles.

While the final fling of the narrow "airway" principle was very high frequency omnirange/distance-measuring equipment VOR/DME, which has been used everywhere since 1957 under international rules, new methods have helped aircraft to fly direct to their destination. One is nose radar, which can "paint" coastlines and give other details of the terrain beneath, besides warning of mountains or storm clouds. Doppler radar measures the speed and direction of the aircraft over the ground by sending radar signals to the ground ahead of and behind the plane, and measuring the signals' change in frequency.

By far the most important navaid carried by large

jetliners and all advanced military aircraft is the inertial navigation system (INS). This became possible as the result of totally new standards in precision manufacturing. The basis of an INS is a stable platform held by gyros so that it is always exactly parallel to the Earth's surface. In moving through 90° in latitude or longitude the platform rotates precisely 90°. On this platform are mounted sensitive accelerometers which sense and record all the accelerations experienced by the aircraft. When all these accelerations are integrated with respect to time the result is a continuous record of velocity. A second integration gives position measured from the known starting place. The navigator merely tells the INS its starting point, and it thereafter tells the navigator its exact position.

In the 1980s even newer systems will come into wider use. One is based on one or more lasers which send light continuously around a small triangular path with mirrors at the corners. This gives more accurate results than a flywheel-type gyro. Navigation satellites (Navsats) should by 1990 enable pilots to know where they are to within a few feet at all times.

Above: a Flight Management System built by the Lockheed Aircraft Corporation. This system equips the aircraft with a digital computer with a 32,000-word memory. It can automatically control the aircraft's trajectory and power to give maximum fuel economy.

Below: a navigator on the flight deck of a Super DC-8. Navigating over huge expanses of ocean presents special problems. Chief among these is the absence of short range navigational aids. Long range fixing systems and vertical navigation are necessary.

CHAPTER 5

LIGHT AND ITS USES

Light is the most vital of natural phenomena to man's survival, and is fundamental to the maintenance of life. Man has harnessed light to his own technical progress with astounding results. From a basic understanding of the workings of the human eye developed devices to aid industrial and scientific advance. Telescopes were devised to explore the stars. From their primitive beginnings massive computer-linked machines were spawned that can examine stars light years from Earth. Man can also observe the smallest microbes with modern microscopes, which magnify thousands of times. Industry, science and the dabbling layman have reaped massive benefits from photography – becoming daily more sophisticated. Man-made light, in its various forms, provides an essential service in the home, the office, the factory, and the hospital. Light is also put to work detecting the presence of chemicals and poisons. The development of the laser has been a boon to both industry and medicine, and, like many other good ideas, it has been perverted in the cause of war. Light can even be used to reproduce objects in three dimensions. What it actually is remains a mystery.

Opposite: the effect of difffraction on a finely grooved video disk. Diffraction patterns enable researchers to measure light wavelengths to a remarkable degree of accuracy.

What is Light?

For thousands of years man has pondered the nature of light. The earliest recorded references are found in the chronicles of the ancient Egyptians. But it was with the rise of Greek science around 500 BC with its fascination for abstract speculations that the first theories emerged. Today they seem wildly unlikely.

years before systematic investigations of light began to provide the first modern theories.

In 1666, the English physicist Isaac Newton performed the earliest really important experiments with light. He discovered that by passing light through a prism of glass, it could be split into a band of colors – red, orange, yellow green, blue, indigo, and violet in that order. From this, he deduced that ordinary light was a mixture of different kinds of light which individually affect our eyes so as to produce the sensation of color. But important though Newton's discovery was, his idea of the actual composition of light echoed the ancient Greeks. He proposed that light consisted of minute luminous particles, or "corpuscles," traveling at enormous speeds. Newton's reputation ensured that his theory would be taken seriously, even though dissenting voices demanded to

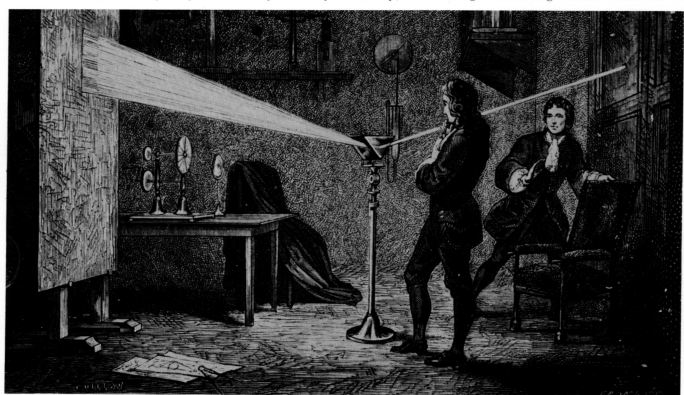

Above: in one of his experiments with light, Sir Isaac Newton positioned a prism so that a narrow beam of sunlight would strike one face of it. A rainbow of colors was reflected on a nearby wall. Ordinary white light is a combination of all the colors of light.

Pythagoras, for example, believed that the eye possessed invisible "tentacles" which reached out to illuminate distant objects. His contemporary, the philosopher Democritus, argued that we see by light emitted by the objects we look at. With typical ingenuity, Plato attempted to combine both ideas, suggesting that light was caused by emanations from both object and eye. Aristotle managed to undermine all three by asking with shrewd and devastating logic why, if either the eye or objects themselves generated light, are things invisible in the dark? It took nearly 2000

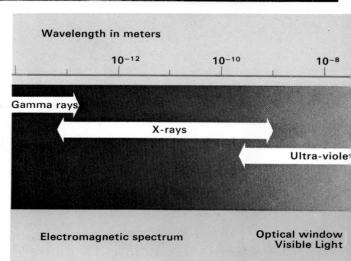

Wavelength in meters

10⁻¹²   10⁻¹⁰   10⁻⁸

Gamma rays

X-rays

Ultra-viole

Electromagnetic spectrum

**Optical window
Visible Light**

know why light beams could cross each other without the corpuscles colliding. A few years later in 1678, Dutch physicist Christiaan Huygens established a new and opposing school of thought, showing how light could be regarded as a wave. Huygens used his wave theory to explain many of the properties of light but, like Newton's corpuscular theory, it also had weaknesses. In particular, Huygens could not explain why light was always propogated in perfectly straight lines.

The two theories vied with each other for over a century. Then in 1801 English physicist Thomas Young discovered that light could be made to form patterns of bright and dark bands when passed through a narrow slit. The only explanation for this was a phenomenon called interference, something exclusive to waves. In 1818, Frenchman Augustin Fresnel dis-

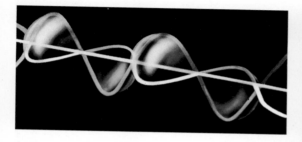

Above: the motion of a light ray presented in three-dimensional form. It is thought that light consists of a stream of electrical and magnetic pulsations. **Right:** a prism splitting light into a spectrum.

covered a similar effect, which also could only be explained by the wave theory.

In the face of this experimental evidence, the corpuscular theory foundered. But problems still remained. For example, if light was a wave, of what was the wave formed? The answer was provided in the 1860s by Scottish mathematician James Clerk Maxwell. He showed that light was just a small part of a wide spectrum of electromagnetic radiation. Such radiation consisted of waves of oscillating electrical and magnetic fields moving at great speed in a direc-

tion perpendicular to the direction of the oscillations. Therefore, light was seen as a kind of transverse electromagnetic wave to which the eye is sensitive.

By the end of the 19th century, the wave theory was almost universally accepted. But within a few years German-born physicist Albert Einstein had shown that the nature of light was far more complex than had been imagined. Einstein had used a detailed consideration of the behavior of light as one of the cornerstones of his theory of relativity. But it was some earlier research which gave him an important new insight into the nature of light. He investigated the phenomenon of photoelectricity by which certain metals exposed to light emit electrons. Einstein showed that light behaved like a hail of tiny corpuscles which dislodged electrons from the atoms of the metal by their repeated impacts on its surface. He

called the corpuscles photons.

So the long search for an answer to the question "What is light?" ends with a remarkable dilemma. Light seems to possess a strange duality. In certain circumstances, it behaves like a wave, in others like a flux of particles. But what it actually *is* remains a mystery. Scientists today might even argue whether the enquiry itself has a meaning. Faced with the question "Is light a wave or a corpuscle?" modern physicists give the simple, baffling but correct answer, "Yes"!

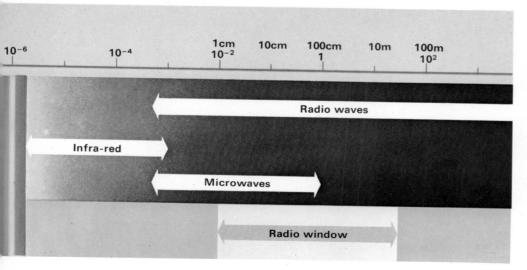

Left: an electromagnetic spectrum, ranging from short radiation (gamma rays) to long radio waves. Light is a form of electromagnetic radiation. Optical and radio wave radiation can penetrate the Earth's atmosphere fairly freely. Optical waves are those wavelengths to which the eye is sensitive. Ultra-violet and infra-red rays are invisible to the naked eye, but have a variety of industrial uses.

117

Vision

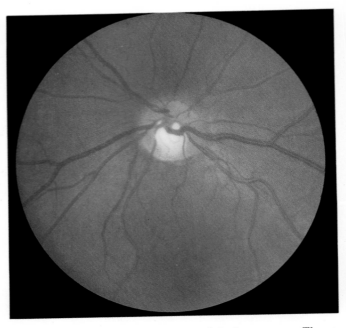

Above: an opthalmoscope's view of the human eye. The yellow spot visible in the iris is shown, along with the blood vessels of the retina and the blind spot where the vessels and nerves leave the eyeball. The retina converts images into signals for the brain.

Man's knowledge of vision is based on the simple idea that the eye is an image-forming device. No matter how much the apparatus for seeing varies from one kind of animal to another, the process of vision always starts with light entering the eye carrying with it information about the object from which it has been reflected. All the biological systems of the eye are then coordinated to bring the light to a sharp focus at its sensitive backwall or retina, much as a camera focuses an image on film. But producing a clear image is no easy matter. In the human eye the tasks of controlling and focusing light are performed by a series of delicately engineered biological components whose precision and versatility make the most sophisticated camera seem a child's toy by comparison.

The amount of light entering the eye is controlled by the iris, a word taken from the Greek for rainbow. It is the most distinctive part of the eye because of its variation in color from person to person. The iris is in fact a diaphragm of circular and radiating muscle tissue. By dilating or contracting it allows more or less light to enter the aperture in its center – known as the pupil – so that the image formed on the retina has the optimum intensity.

The focusing system in the eye has four main components – the cornea, a fluid called aqueous humor,

the lens, and a jellylike substance called vitreous humor. The cornea is a clear membrane curved over the outer surface of the eye, which bends light sharply inward. Next, the light passes through the aqueous humor between the cornea and iris. By far the most important job is done by the lens, suspended directly behind the iris. The lens, the size and shape of a small bean, is made up of over two thousand layers of tissue. It is remarkably pliable, and small external muscles vary its shape to ensure accurate focus for objects a few inches, or several miles away. These changes in focus take place instantaneously so that man can switch his gaze, near and far, without a moment's pause.

After passing through the vitreous humor, the focused light strikes the retina. Here light-sensitive cells transform the energy of the light into tiny electrical impulses which are relayed by the optic nerve to the brain. The information is then decoded into the details of size, color, distance, and texture which together make up the sense of vision.

The workings of the eye are completed by other safeguarding and manipulative systems. It can be pointed in different directions by six external muscles attached to the walls of its protective bony socket. The frontal section of the cornea is cleaned by a constant flow of tears from glands, and the eyelid has the dual role of protecting the eye from dust particles and shutting out stimulae to make sleep easier.

Although the eye is an organ of impressive capabilities, it does not always function perfectly. The most common faults affect focusing and are due either to defects in the cornea or the lens. For example, when the cornea is imperfectly curved, rays of light

Below: the human eye in the bony socket of the skull. Muscles connect the outer coat of the eyeball to the socket, enabling the eye to turn in a variety of directions. Weakened muscles can lead to faulty eyesight.

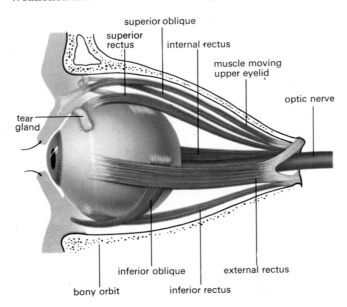

superior oblique
superior rectus
internal rectus
muscle moving upper eyelid
optic nerve
tear gland
inferior oblique
bony orbit
inferior rectus
external rectus

Above: this 14th century Italian painting depicts a scribe wearing eyeglasses. It is the earliest known picture of spectacles in use. Spectacles were originally cumbersome and uncomfortable to wear, and often provided little improvement in vision.

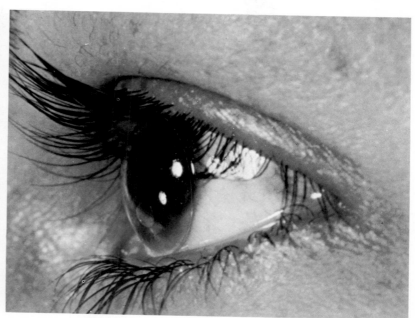

Above: a contact lens in place on the eyeball. Made of plastic, contact lenses come in two types – flexible and rigid. As there is no distance between the iris and the lens, near perfect vision can be attained. Lenses must be regularly cleaned to prevent inflammation.

may be refracted to a greater degree in the vertical plane than in the horizontal, or vice versa, causing a blurring of sight. This defect is called astigmatism. The commonest fault in the lens is an inability to stretch itself to bend light from distant objects sufficiently, or to compress and bulge enough to bring light from close objects into proper focus. These defects are commonly known as near and long-sightedness.

The use of glasses to correct such focusing defects dates back to the 13th century, long before the workings of the eye were properly understood. For centuries following their invention glasses remained crude instruments, matched to a person's eyes by trial and error. Moreover, most were so grotesque in appearance that they caused considerable social embarrassment. Today the situation has changed dramatically. Modern opthalmic optics – the science of diagnosing and treating eye defects – is able to measure accurately the nature and extent of a focusing fault. Lenses can now be ground and polished to the exact prescription needed to put things right. Today, glasses are a matter of high fashion. A wide range of frames are available, expensively styled in plastic and metal. Often large lenses are favored, which if made of glass would prove uncomfortably heavy. Technological advances in plastics have made it possible to mould optically perfect clear plastic to the precise curvatures required to compensate for an eye's defects. For those who are still embarrassed by wearing spectacles, tiny contact lenses are available to be worn in direct contact with the cornea, dispensing with the need for external frames altogether.

Left: people have always been concerned that spectacles might alter their appearance for the worse.

Telescopes

There are two kinds of optical telescopes: refractors and reflectors. Both collect light from distant objects and bring it to focus. The main distinction between them is that refractors use a system of glass lenses while reflectors employ large curved mirrors.

The earliest telescopes were refractors and the simplest form – called the Galilean – is still widely used. A large convex (outward curving) lens at the front end collects and bends the light to a focus. This lens is called the objective because it is closest to the object being viewed. A smaller convex lens is used as an eyepiece to magnify the image. One major drawback of refractors is their dependence on the quality of the lenses they use. The glass from which they are ground must be pure and bubble-free, and their curvature must conform to precise geometric requirements. They are also affected by an inherent problem shared by all lenses, known as chromatic aberration.

It stems from the fact that a lens bends light of different colors to differing degrees so that images formed are surrounded by multicolored halos. The problem was solved in 1733 by Englishman John Dolland who showed how chromatic aberration could be corrected by using a compound lens formed by combining a convex and a concave lens. Since then all refractors have used compound lenses based on Dolland's design.

Most optical astronomers want the maximum light-gathering and magnifying power consistent with a sharp focus. This means using a big objective lens in a refractor. But as the size of the lens increases so does the difficulty of manufacturing it to the rigorous standards necessary for top optical quality. In addition, the objective has to be supported in the telescope by its edges and large compound lenses are so heavy that they bend slightly under their own weight, distorting their carefully ground surfaces. Therefore, refractors are restricted to manageable sizes. The largest in the world is the 40-inch telescope of the Yerkes Observatory in Williams Bay, Wisconsin. Gathering 40,000 times as much light as the naked

Above: the reflector built by Sir Isaac Newton in about 1671. It had a one inch diameter mirror which collected light and directed it onto a smaller flat mirror, and from there into the side of the tube. Here, the image is magnified by an eyepiece. Many modern reflectors work on the same principle as this 300-year old device.

Left: by the end of the 17th century the telescope was in widespread use all over Europe. This painting by Donato Creti depicts two astronomers observing the Moon with a telescope of the type used by Galileo.

eye, it magnifies 3000 times, so that the moon's surface can be viewed as if it was only 80 miles away. But the Yerkes telescope was built in 1897. None of the advances of 20th century technology have extended the capabilities of refractors any further.

Because their light-gathering and focusing is done by mirrors, reflecting telescopes can be made much larger than refractors. Unlike lenses, mirrors can be supported from behind and need only be made with an accurate curvature – it does not matter if some imperfections exist in their surfaces. So, while making large mirrors is not easy, it is at least practicable. The world's largest operational reflector at Mount Palomar, uses a giant 200-inch mirror, providing 25 times the light-gathering power of the Yerkes refractor. In the Soviet Union a still larger instrument is being built using a 236-inch mirror housed in a massive 800-ton mounting.

The first reflecting telescope was built by Isaac Newton, and the system he designed still bears his name. Light from the main mirror of a Newtonian reflector is deflected by a small mirror through the side of the telescope tube. In the other main type of reflector – called the Cassegrain after its inventor – light is reflected back toward the main mirror and is viewed through a small hole drilled in its surface. Many modern reflectors dispense with secondary mirrors altogether by substituting a film-holder to make direct photography possible.

Right: the introduction of radio telescopes, like this one at Jodrell Bank, England, has provided a new way of observing the universe. Radio telescopes have picked up the emissions of quasars (quasi-stellar radio sources), which are from four to 10 billion light years away from Earth.

Below: the world's largest operational telescope at Mount Palomar, California. Picture shows the observer seated in the center of the 200-inch mirror. The Mount Palomar telescope is immensely powerful, and has detected stars and galaxies which are many millions of light years distant.

Optical astronomy today places greater emphasis on recording images by long photographic exposures than on astronomers making direct observations themselves. Because the time needed to register the faintest objects on film can be many hours, research has concentrated on trying to improve the sensitivity of photographic emulsions. The most important advance in recent years was made by an Edinburgh team who perfected a technique of "soaking" the photographic plates in a sealed tank of nitrogen. Impurities in the emulsion are chemically flushed out by the process. Plates of far greater sensitivity than ever before are provided.

Whether astronomers use reflectors or refractors, it is essential their instrument is mounted on a movable platform in order to compensate for the motion of the earth. Otherwise, the object they observe would seem to be moving slowly, making photography impossible. The simplest system is called equatorial mounting. The telescope is steered to compensate for the apparent "up and down" movement of the star field while the observer himself follows the east to west motion. The biggest instruments, however, use complex horizontal and vertical compensators. Linked to computors, huge mountings can be delicately maneuvered both to eliminate the apparent movement of the stars and to target the telescope on the precise point in the sky the astronomer wants to observe. Observations can take place with minimal disturbance.

Microscopes

Just as telescopes make it possible for man to study the remote mysteries of space, microscopes reveal the otherwise invisible world of the very small. The first probably appeared in the early 1400s. Simple instruments with a single lens of modest magnifying power, their use was confined to the study of insects and plants. Today's optical-microscopes are complex instruments using multiple lens systems with a magnifying power that reveals even the most minute bacteria in detail.

The optical requirements for microscope lenses are even more stringent than those for telescopes. Although the problem of chromatic aberration in telescopes was solved in the early 1700s, it was not until nearly a century later that scientists discovered how to apply the technique to microscopes. Even then their construction was a hit and miss affair, and it was not until the work of German physicist Ernst Abbe that microscope design was placed on a firm scientific

Below: a Zeiss microscope dating from about 1895, fitted with a large illumination apparatus designed by the German physicist Ernst Abbe. Carl Zeiss and Ernst Abbe began making microscopes together in 1866.

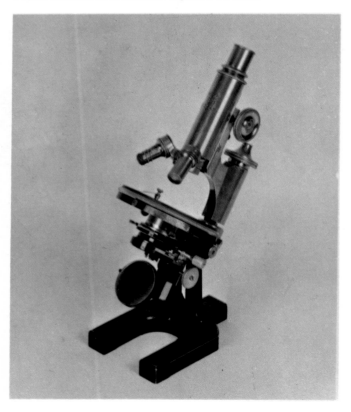

footing. Abbe formed a famous partnership with the German lens manufacturer Carl Zeiss. Together they produced the finest instruments of the 19th century.

Modern microscopes use a combination of lenses of various kinds to produce a highly enlarged image. The objective lens system – closest to whatever is being viewed – and the eyepiece can usually be moved in relation to the specimen being examined to achieve the best focus. The body tube, housing the lenses, may be single for viewing with one eye or double for binocular vision. A third tube is sometimes added, linked to a camera so that photomicrographs can be made to record specimens. The main tube is carried on a metal arm fixed to a horseshoe or tripod base. Attached to the arm and extending in front of the objective lens is a platform called the stage, on which specimens to be examined are placed.

In high-powered microscopes it is usually essential to illuminate the specimen with a brilliant, uniform light. Lens systems called condensers are fitted below the stage, focusing the light from a tungsten filament lamp onto the field of view. Modern condensers can also provide special kinds of illumination, used to maximize contrast in semitransparent specimens that would otherwise be difficult to see. For example, if a condenser is adjusted to provide illumination at a very oblique angle to the specimen, the rays of light do not enter the objective lens directly and only light reflected and bent by the specimen enters and forms the image. The result is a brightly illuminated specimen against a dark background.

Like astronomers, microscopists usually want the maximum possible magnification consistent with sharp focus. The fundamental limitation on the optical microscope's power to resolve details of tiny objects is the wavelength of the light being used to view them. The shorter the wavelength, the smaller the object resolved. But when the object's size is similar to the wavelength, light can pass around the object without being properly reflected. This leads to a blurred image with no visible details. Scientists have tried to extend the usefulness of optical microscopes by employing light of shorter and shorter wavelengths, eventually turning to ultraviolet light. The earliest ultraviolet microscopes were difficult to use. All results had to be recorded by camera. But recent advances in the design of television cameras have made them sensitive to ultraviolet light. So it is possible for scientists to see images from such microscopes displayed directly on television screens.

Television link-ups both for ultraviolet and normal light microscopes are being used increasingly. The first benefits were in teaching. A whole class can examine a biological specimen with their teacher, viewing its magnified image on a television screen. More sophisticated applications have followed. With the aid of image-analyzing computers, the electronic scanning process by which a television picture is built up is utilized to provide detailed measurements of the specimen being examined. Using this technique, scientists can view a television picture of the specimen, and simultaneously make a wide range of valuable measurements.

The search for ever-greater magnifications has led some scientists to abandon optical microscopes altogether. Now, instead of light, it is possible to use beams of electrons to produce magnifications of over 2,000,000 enabling the outlines of individual atoms to be discerned for the first time. The workings of such microscopes depend on a strange duality in nature. Just as light can sometimes behave as a flux of tiny particles, so particles such as electrons can show wavelike properties. It is actually possible to attribute a wavelength to electrons and because it is far shorter than the wavelength of ordinary light, electron beams can be used to resolve some of the tiniest objects in nature.

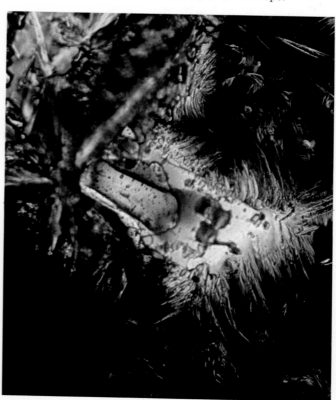

Below: a crystal of ascorbic acid under the microscope. This specimen is illuminated with a brilliant uniform light – dark areas can distort the image at great magnification. Pictures like this are produced by fitting a camera in the eyepiece position of a microscope.

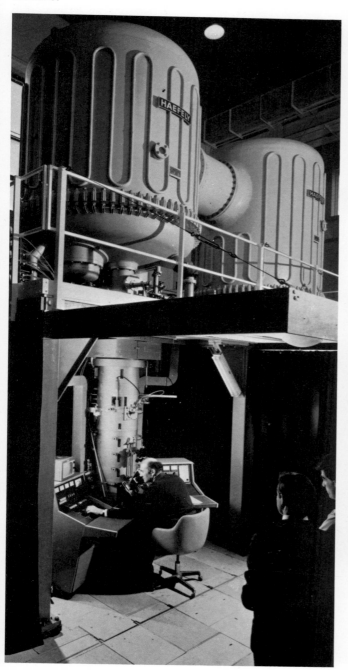

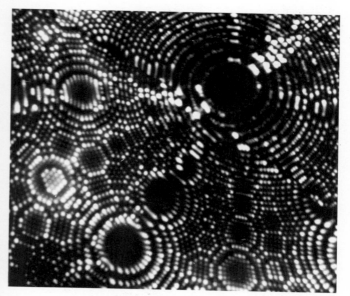

Below: a photograph of a tungsten needle taken through a field-ion microscope at a magnification of five million times. The bright spots are tungsten atoms. Pictures like this are of vital use in industrial research.

Left: a one million volt electron microscope installed at the Atomic Energy Research Establishment, Harwell, England. The first electron microscope of its size in Europe, it is intended to help research into neutron radiation damage in reactor materials. The extremely high voltage enables sharp images to be obtained from thick specimens.

123

Photography

Since the 1830s when William Fox Talbot produced the first dim photographic images, photography has become both a fast-moving technology and a massive industry. Most early cameras were bulky contraptions using delicate glass photographic plates. The light-sensitive emulsions coating these plates were slow to respond to an image and exposure times of several minutes were needed for even routine photography. Modern cameras are compact enough to fit into the palm of the hand and produce perfect photographs in a range of lighting conditions with exposure times as small as one thousandth of a second.

In 1888 Eastman Kodak introduced a revolution in photography when they marketed a small box camera. Instead of plates, it used a roll of photosensitive paper. Together with the company's cheap and widely available processing service, the first roll-film camera transformed photography into an amateur hobby of enormous popularity. Although plate cameras continued to be used by professionals intent on top quality results, greater technological breakthroughs doomed their bulky equipment to obsolescence in the mass market.

In 1924 the Leitz Company of Germany introduced the Leica, the first compact camera to use film in a 35mm format. With the Leica, a large number of frames could be taken on a single roll of film and the film itself offered high quality results at an economic price.

Today amateur and much professional photography is dominated by compact 35mm cameras like the

Above: the first compact precision camera, the 35mm Leica. This was the forerunner of the sophisticated 35mm single lens reflex cameras of today. The 36 frame roll of film produced excellent photographs at low cost.

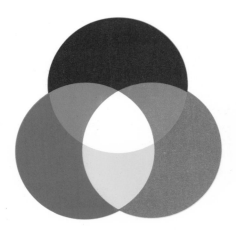

Above: diagram demonstrates the trichromatic theory. The three primary colors can be combined in varying individual proportions to produce other colors – so achieving a full-color image.

Left: the father of photography, William Fox Talbot with his wife and daughter in an early calotype print. Early photography was a cumbersome business.

Leica. Most models have a range of features allowing beginners to take good photographs, even under difficult light conditions. Single lens reflex (SLR) viewfinders, for example, enable the photographer to see exactly what his camera lens sees. Small prisms are arranged so as to direct light from the lens into the viewfinder. A linked system ensures that while the photographer composes his shot the lens is open to its widest aperture, providing a bright field of view. When he presses the shutter release, the same system instantly closes the lens diaphragm to the aperture predetermined by the photographer to give the correct film exposure.

Accurately determining exposure is the key to a good photograph. The photographer has to find the right combination of shutter speed and lens aperture to allow the correct amount of light into the camera. This difficult problem is often solved automatically by built-in light meter systems. Some of these not only measure light intensities, but actually set lens apertures or shutter speeds to their optimum values. Such systems use a group of photocells, often positioned behind the lens to give the most accurate measurement of the light entering the camera. Depending on the amount of light falling on the cells, they generate or control a tiny electric current. Until recently the selenium cell, which actually produces a current, was the most common type in use. But it is increasingly being replaced by the more sensitive cadmium sulphide cell in which electrical resistance varies with the light intensity. Such cells need a current passing through them, and this is normally provided by a miniature battery housed in the camera.

In fully automatic systems, the photographer usually presets the lens aperture. The exposure-control cells measure the light entering the camera lens and then adjust the shutter speed to ensure the optimum exposure. Such systems were made possible by the development of the electronic shutter. Unlike traditional mechanical shutters restricted to a small number of set speeds, electronic shutters can offer any speed. This means that the electronic exposure control can fire the shutter at exactly the right speed, rather than just select a setting closest to the optimum from a very limited range.

Just as modern cameras are becoming increasingly automated so they are also growing more versatile. The system camera is here to stay. Top manufacturers offer a range of camera bodies into which various types of lens can be fitted. Interchangeable lenses have, in fact, become so important that many companies concentrate on lens manufacture to the exclusion of all else. Different types of viewfinders, flash-gun equipment for indoor photography and, an increasingly popular innovation, external motors to power the film-winding mechanism, can also be fitted to the bodies. Motor-drive units clip onto the base of the camera body and enable a rapid sequence of photographs to be taken without the usual interruptions caused by winding on the film manually. Several

frames a second can be exposed to show in perfect detail the elements of a complex action sequence such as a bird rising into flight or an athlete accelerating from his starting blocks. Such equipment is invaluable to the photo-journalist for whom speed is of the essence.

Some of the best camera lenses are complex optical systems, often designed with the aid of computers. A variety of available focal lengths enable the photographer to create a range of effects. Shorter focal lengths such as 28 and 24 mm provide a wide-angle effect, ideal when a large panorama has to be included in a single shot. Taken to its extreme, the "fish-eye lens" creates an angle of viewing so wide that even the scene behind the photographer can be included. Longer focal lengths narrow the angle of viewing but produce a magnification that draws distant scenes closer. These "telephoto lenses" range from compacts with focal lengths of around 135mm to huge 1200mm lenses, so massive that special tripods are needed to support their weight.

One of the most exciting recent developments in lenses has been the introduction of the zoom, offering a variable focal length with a single lens. Zooms make it possible to change smoothly from low to extreme telephoto by shifting a simple slide grip on the lens barrel. The most recent zooms offer a range of focal lengths from wide angle through to medium tele-

Above: a man in an early airplane photographed through a fish-eye lens. Fish-eye lenses produce startling images of subjects, achieving angles of view of up to 210°.

photo. A further interesting innovation is the macro lens. This offers the hitherto unprecedented combination of a telephoto magnification with ultra-close focusing. At a distance of two or three inches the magnification produces stunning closeups of subjects such as insects and flowers.

125

The pace of change in camera technology has not been matched by developments in film. Although they are more versatile today than ever before, the basic working principle of film has remained unchanged for nearly a century. Originally only black and white prints could be produced and this is still the most widely used process. But color photography now renders near perfect color in the form of prints or transparencies.

Present-day color photography is based on the way in which the eye sees color. The accepted theory states that the eye responds to only three color stimuli: red, green, and blue. These are the primary colors. Man's perception of color is produced by combinations of these. Color film is built up of three layers, called a tripack. Each layer responds to either red, green or blue light. The film therefore reproduces natural colors in the same way as the human eye.

A major breakthrough in film technology occurred in 1947 when American inventor Edwin Land showed it was possible to take a photograph and have the print processed inside the camera itself within about a minute. Land adopted a process already used in office photocopying machines called the Polaroid system. Results from Land's early polaroid cameras suffered from poor contrast and yellowing of the prints, but the more sophisticated later models produced excellent photographs. With improved models appearing regularly, instant photography is now one of the biggest growth areas in the photographic industry. Added impetus was provided in 1963 when the first polaroid color film for prints was devised. Polaroid film for color transparencies and negatives is under development and should soon be available.

Specialized applications of photography have become increasingly important both in industry and science. Many depend on orthodox photographic techniques using other than visible light. For example, infrared photography enables images to be formed in complete darkness, the film being exposed purely by the invisible heat emissions of the surroundings. This is a useful technique for photographing nocturnal animals that would be startled by the flashlights needed for visible-light photography.

Left: with the Polaroid SX-70 Sonar Auto-Focus camera sound waves focus the lens. The Polaroid system was developed by Edwin Land. Developing fluid is stored in the camera. Color pictures are developed in 60 seconds, black and white 10.

Right: a color coded X-ray of a tooth. Pictures like this are playing an increasing role in modern dentistry. Dentine and enamel are both crystalline in structure, and show flaws typical of crystal. The core of the tooth is just visible.

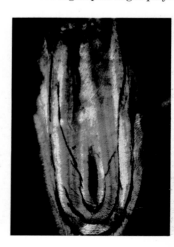

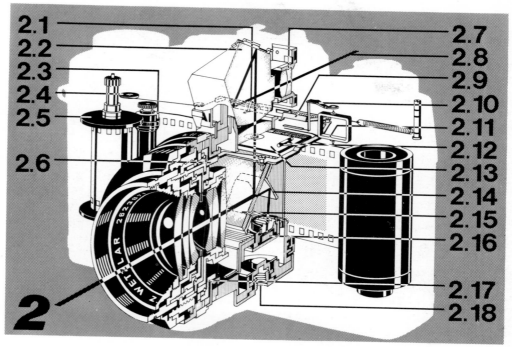

Right: internal workings of a Leicaflex SL2: **2.1** shutter speed display pointer in the viewfinder; **2.2** Pentaprism; **2.3** film transport drum; **2.4** setting disk for shutter speed display; **2.5** film take-up spool; **2.6** upper cam for lens stop display; **2.7** viewfinder eye-lens; **2.8** optical path of viewfinder; **2.9** lens stop display pointer in viewfinder; **2.10** exposure meter follow pointer; **2.11** spool with exposure meter measuring pointer; **2.12** Illuminator for measuring pointer; **2.13** viewfinder plate; **2.14** cylindrical deflecting mirror for the photoresistor; **2.15** beam splitting mirror; **2.16** photoresistor **2.17** lower cam for exposure meter setting; **2.18** ⅓in tripod bush.

Artificial satellites are often equipped to take infrared photographs to provide information for weather experts about temperature variations in the atmosphere, or to warn agriculturalists about crops and vegetation where the onset of disease can be detected by changes in heat output.

At the other end of the spectrum, X-rays also provide a valuable form of photographic record. Their role in medicine has been recognized for over 80 years and X-ray photographs continue to play a vital part in diagnosis. Similar techniques have only recently been applied in industry but are rapidly becoming an established method of checking for flaws in structures and mechanisms. The major advantage of X-ray photography is that it is nondestructive. It can reveal a fault deep within an aircraft wing for example, that could only otherwise be detected by virtually destroying the wing.

In the field of astronomy, photography has almost supplanted the direct observation of the heavens by astronomers. Long exposures of up to several hours collect enough light to form images of stars and galaxies many times too faint to be visible with the naked eye. Also, the relative positions of images on a photograph can be measured with great accuracy enabling astronomers to detect even minute displacements among stars and planets. In 1930, American astronomer Clyde Tombaugh discovered Pluto, the outermost planet of the solar system, by spotting its movement against a stationary background of stars revealed by a sequence of photographs.

The technology of photography is developing almost daily. New applications are merging and established ones are being further refined providing scientists and engineers with an increasingly valuable tool. Whether in the hands of an amateur recording a simple family scene or housed in a giant telescope probing the heavens, the camera in one of its many forms is playing a major role in the modern world.

Below: an infra-red photograph of part of Southern California. Infrared photography is useful for survey work. It eliminates haziness. Vegetation shows up red because it emits infra-red radiation quite strongly. Such satellite photography is now commonplace.

Artificial Lighting

Until Joseph Swan and Thomas Edison invented the electric light in the 1870s, all man-made lighting depended on burning some kind of fossil fuel. The most primitive light sources were blazing wood brands, which provided plenty of light but were dangerously hard to control, especially in confined spaces. Burning vegetable or animal oils and fats either as candles or in simple lamps was a major advance which provided a reasonably effective and practical form of artificial light widely used for thousands of years. Even the gas lighting used in the 1800s could not supplant the more versatile and manageable oil-burning devices.

But modern lighting is based firmly on the use of electricity with candles and oil lamps restricted to occasions when a romantic atmosphere is more important than lighting efficiency. The principle of the simplest electric light has not changed since the English chemist Joseph Swan built the first carbon filament lamp in about 1870. Developed by American inventor Thomas Edison, the modern light bulb soon emerged and revolutionized the lighting industry. Its principle relies on electricity passing through a filament with a high resistance to the current. Part of the electricity is converted into heat, and as its temperature soars the filament becomes incandescent. In his early prototypes Edison tried various substances for filaments including carbonized bamboo, charred vegetable tissue and even whiskers from his own moustache! Today, light bulb filaments are made of tungsten, a metal with a melting point of about 3400°C. The first bulbs had the air pumped out of them to prevent their filaments combining with oxygen and burning up. But in practice, the air could only be reduced to a low pressure, not eliminated altogether, so early filaments still tended to have a short life. Modern bulbs are filled with an inert gas so that no chemical reaction can take place between the filament and its atmosphere. The gas pressure is only slightly

Above: in industry, a wide variety of artificial lighting is employed to cater for many different production processes. A tungsten halogen lamp provides the operator of this grinding machine with even illumination of the grinding wheel and workpiece. Complicated work can thus be carried out with speed and accuracy.

Left: Thomas Edison holding one of his light bulbs. Henry Ford is on the left. Edison experimented with thousands of filaments. Tungsten is used today.

reduced and so is sufficiently high to prevent atoms of tungsten "boiling off" from the filament, weakening it and blackening the bulb.

The most recent development in filament lamps is the tungsten-halogen-filament lamp. Instead of an inert gas, a small quantity of iodine is used. This enables the tungsten filament to operate at a very high temperature and to provide a brilliant light. Because iodine has a chemical effect on glass, the filament has to be placed in a quartz bulb rather than the cheaper glass type used in ordinary domestic lighting. Although expensive, such lamps are becoming important for use in car headlamps, photography, and all situations where maximum brightness is needed from bulbs of minimum size.

The use of gas as a fuel for lamps was a relatively short-lived development, but just as gas lamps were being made obsolete by electricity a totally new kind of gas lighting was being developed. In the late 1800s, English physicist William Crookes showed how a sealed glass tube containing traces of a gas at low pressure could be made to fluoresce when an electric current was passed through it. Within a few years tubes filled with neon gas giving a deep red glow were being used for decoration and advertising throughout the developed world.

Since those early days lamps based on other vapors have been used to produce light in a variety of colors. The sodium lamp with its characteristic yellow light is perhaps best known today because of its use in street lighting. But less common gases are popular in

Below: the lights of Piccadilly Circus, London. Gas filled strip lighting are widely used for advertising and display purposes, as well as information signs. Neon and argon are the gases most commonly used.

multicolored advertising signs. Krypton, for example, produces a deep blue color while a mixture of neon and argon in a yellow tube gives a vivid green light.

A major breakthrough in this type of lighting took place in the 1930s when the modern fluorescent tube was invented. As before, a long, glass tube was used, filled with mercury vapor which produces a white light. The vital innovation was to coat the inner surface of the tube with a substance that fluoresced when exposed to the mercury light. This multiplied the light output about a hundred times, and at last produced a practical form of lighting wherever a vivid, functional illumination was needed. Such "strip" lights are now widely used in kitchens, offices, laboratories, and factories.

Lighting technology is still developing rapidly. Improvements have been made to the traditional mercury and sodium vapor street lamps by increasing their vapor pressures. Electric arc lamps producing massive outputs have been made for searchlights and theatrical spotlights. Microscopic lights with tiny filaments have been developed to enable surgeons to illuminate the interior of a patient's body. Lighting designers now know enough about the physics of the electric light to invent one to meet almost any need.

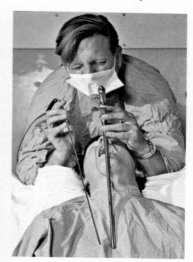

Above: a simulated bronchoscopy. The bronchoscope houses a torch which illuminates the bronchi. Secretions are removed by the sucker in the surgeon's right hand. **Below:** the surgeon's eye view.

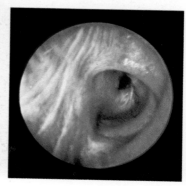

Spectroscopy

Spectroscopy is a method of detecting the presence of different chemical elements in an object by analyzing the light substances given out when they are heated. This idea is not new and the basic technique was worked out as long ago as 1861 when German physicist Gustav Kirchhoff analyzed sunlight. He identified the presence in the Sun of numerous common elements.

The simplest spectroscope is a box with a narrow slit through which light can pass. The light falls onto a prism made of high quality polished glass. Light usually consists of a number of different wavelengths, each refracted a slightly different amount by the prism. The result is numerous distinct images of the original slit each representing a particular wavelength in the original light. These slit images, usually recorded photographically, are called spectral lines. Each line or group of lines is an unmistakable "signature" of one or more chemical elements. If the light originates from a substance containing sodium, for example, and is heated so strongly that it glows, the spectroscope will reveal a bright yellow line in a particular position in the overall spectrum, indicating the presence of atoms of sodium in the light source. Most of the known elements have a characteristic

Above: a scientist producing a spectrogram of an alleged 16th century portrait in Boston Museum. Using a laser, he revealed the picture to be a forgery. The presence of zinc was discovered. Zinc was not used in paint pigments until 1820. The painting had fooled many experts.

Left: the first spectroscope was built in 1814 by its inventor Joseph von Fraunhofer (*center*). His spectroscope consisted of a lens, a prism, and an observation telescope. Fraunhofer detected about 700 dark lines crossing the spectrum of sunlight. These are now named after him.

spectral signature. It is even possible to identify compounds (combinations of atoms of different elements) by the groups of spectral lines they produce.

Spectroscopy is an important tool for studying the internal structure of atoms and molecules. In the early 1900s, it was largely on the basis of spectroscopic evidence that the first modern theories of atomic structure were verified. Today, the mechanisms by which atoms emit light of characteristic wavelengths are fairly well understood. Detailed analysis of the changes in a spectrum, caused perhaps by powerful

in modern times several important forgeries have been revealed that had passed innumerable visual inspections by experts.

In the field of fighting crime, Dutch forensic scientist Anton Witte has just pioneered a similar technique, but one requiring a breathtaking degree of accuracy in aiming the laser beam. Witte has succeeded in matching individual paint layers in minute chips of paint by vaporizing microscopic amounts within a single layer with a laser beam so accurately focused that the adjoining paint layers are unaffected.

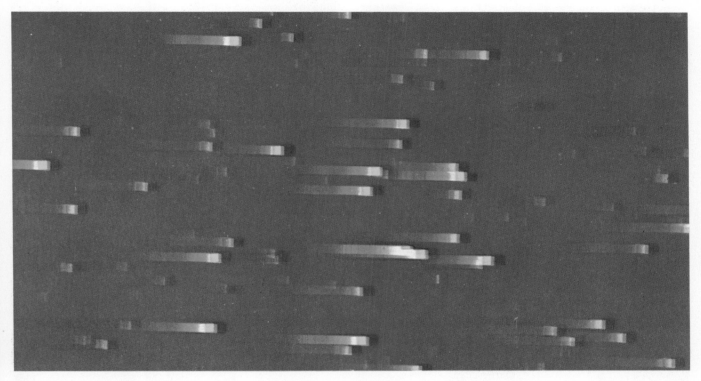

electric or magnetic fields, can provide researchers with important insights to corresponding physical changes within atoms and molecules.

The most widespread use of spectroscopy is as an aid to analyzing unknown substances and determining their composition. In most cases the telltale spectrum is produced from the light generated either by heating the unknown substance, or by vaporizing a small sample of it in a powerful flame or electric arc. Modern techniques are so accurate that traces of an element in minute quantities of a few parts per million can be detected.

One of the most recent developments in high-precision spectroscopy has been made possible by the invention of the laser in 1960. Because it can be focused on minute targets with sufficient intensity to vaporize even minerals and metals, it is being used to vaporize microscopic samples of substances so that the light produced can be spectroscopically analyzed. In authenticating works of art, for example, this enables the composition of the paint used on a supposed "old master" painting to be analyzed from a sample so small it can be removed from the canvas without visibly damaging the work. By detecting pigments only used

Above: a spectroscopic view of starlight. Through spectroscopy astronomers can learn valuable information about stars billions of miles from Earth. The stars shown here are from the Taurus constellation.

The spectroscopic analysis of such minute light impulses as are produced by this technique is perhaps even more delicate a job than the handling of the laser itself.

Light analyzed by spectroscopes need not be restricted to earthbound laboratories. The same scientific principles apply even if the light originates in distant stars and galaxies. Modern astronomers are now able to use spectroscopic analysis of starlight to determine not only the chemical composition of stars, but their temperatures and degrees of stability as well. Some exciting discoveries have resulted from astronomic observations of this kind. In the late 1960s light from some of the huge dust clouds drifting in interstellar space revealed the presence of organic molecules thought to be a preliminary stage in the chemical evolution of living tissue. Evidence of this kind may one day confirm beyond doubt that mankind is not alone in the universe, and pinpoint the existence of primitive extraterrestrial life.

Power from Light

Nineteenth century scientists knew that certain metals such as zinc acquire a positive electrical charge when exposed to sunlight. German physicist Philipp Lenard believed that the light was somehow "knocking" electrons out of the surface of the metal, and that with the loss of negatively charged particles an overall positive charge remained. Confirmation and explanation came in 1905 when Albert Einstein worked out that the mechanics of photoelectricity – as the effect became known – depended on light behaving as a flux of tiny particles, later called photons. The photon impacts, Einstein demonstrated, physically dislodged electrons from the surface of the metal. The freed electrons constituted an electric current, the strength of which depended upon the intensity of the light producing it.

The photoelectric effect remains the basis of photocells, although a number of types exist with slightly different working principles. One of their first applications was in detecting very faint light sources. Connected to a suitable amplifier, a photocell would generate a pulse of electricity when struck by even a single photon. The modern version of this device is called a photomultiplier. It uses a series of photocells – electrons from which trigger one another in an increasing number of reactions. Eventually a single photon can give rise to an avalanche of electrons and an easily measurable current. Such ultrasensitive detectors play an important role in atomic physics – spotting the minute scintillations of light subatomic particles cause when they collide with certain materials. Used in this way photomultipliers not only warn of the presence of a particle, but can actually count the number of particles by the pulses of electricity they produce, and give an accurate measure of atomic radiation levels.

Industry has also found a range of uses for photocells. One of these is the control of large-scale manufacturing processes when an automatic counting system is needed. The completed goods are usually carried by conveyor belt through a light beam targeted onto a photocell. As long as the beam is connected to the photocell, an electric current flows. As each

Above: a photomultiplier cascade for an image storage tube. Astronomers use numbers of photoelectric cells to measure light from distant stars. Faint light is built up into a cascade of electrons by selenium plates in a photomultiplier tube.

Left: a lot of research into magnetism and electricity was carried out before the nature of light waves was understood by scientists. This 18th-century print demonstrates the attractive effect of an electric charge. The boy is being charged with friction between his slippers and the revolving sphere.

object on the conveyor belt breaks the beam, the current is interrupted. The drops in current are recorded by a digital counter. Therefore, a running tally of the objects passing in front of the photocell is provided.

A similar principle is used in certain types of security systems. Here a beam – usually infrared rather than visible light to avoid alerting an intruder – maintains a current in a photocell which in turn prevents an alarm sounding. If the beam is broken, the current is interrupted, and the alarm triggered.

One of the most common uses of photocells is in photography. The simple relationship between the intensity of light and the strength of current it produces makes a photocell an ideal means of measuring lighting conditions and calculating photographic exposures. Sophisticated metering systems are available in cameras, linked to microelectronic systems which instantly adjust shutter speeds or apertures to provide near-perfect photographs. All the photographer has to do is frame his subject in the viewfinder and press the trigger. In the best systems the photocells are arranged behind the camera lens so that the light measured is exactly the illumination the lens, and therefore the film, receives.

In the world of television and movies, photoelectricity plays a vital role. Almost all television cameras use a mosaic of photoelectric cells behind their main lens in order to change the initial optical image of a subject into an electron image. The electrons are emitted from the mosaic in numbers proportional to the intensity of the light falling on it. This means that the bright areas of the subject will generate more electrons than the darker areas. The information in this electron image is transformed in turn into a pattern of positive charges on a glass plate situated behind the photocells. When a secondary beam of electrons is used to scan the plate, it is partially absorbed by the most strongly positively charged areas, and partially reflected by the weakest. The

Above: selenium cells installed in the photoelectric shutter mechanism of a modern single lens reflex 35mm camera. Shutter speed and lens aperture are controlled automatically by the selenium cells' response to varying light intensities.

modified scanning beam returns to an amplifying system which produces an electrical signal containing all the information from which a television picture of the original subject can be built up.

While photoelectricity produces the image in television, it is responsible for the sound in the movies. A common audio system in use today consists of a lamp focused on the "sound track" on the moving film. The light passes through it onto a photocell creating a varying photoelectric current, which is a replica of the current produced in the recording microphone when the film was made. Amplified and fed into speakers, the signal is converted into the voices, sounds, and music which are essential to today's movies.

Photocell systems for large-scale power generations are still in their infancy. Most photocells are very inefficient, many converting no more than one percent of the light energy falling on them into electricity. While this scarcely matters in the devices described above, it is a major problem in the power generation industry where high conversion efficiency is essential to economic viability. But small scale use is becoming increasingly common. Panels of photocells are nearly always used to power satellites and space probes. On a smaller scale still, the mid-1970s saw the advent of the first solar powered wristwatch. A similar development in 1978 was the appearance of a miniature electronic calculator aptly called "photon." Powered by a small panel of photocells, the manufacturers claimed that if there was enough light to see the calculator, there was enough to power it.

Below: the Teal Photon – the world's first solar powered calculator. Power is converted by solar cells wherever there is reasonable light. Even the light of a single candle is sufficient.

Interference and Diffraction

Scientists now accept that there are two distinct ways of describing light: as a wave of electromagnetic energy or as minute particles or energy "packets" called photons. In some circumstances only the particle theory can explain the behavior of light, but in others, the wave theory is needed. Interference and diffraction are two important phenomena which can only be explained by regarding light as a wave, and when first discovered were in fact used as a proof of the wave theory against all other contending ideas about the nature of light.

Interference was first discovered in 1801 by English physicist Thomas Young. In an experiment he passed light from a single source through two pinholes in a card so that two beams were cast onto a screen behind the card. It seemed that in the dark areas, in a pattern of light and dark bands, the light from the two beams together somehow produced darkness.

of light. They were astonished to discover just how small they were. The wavelength of red light, for example, was less than a third of a thousandth of an inch.

More recently the accuracy of interference patterns in measuring even such minute distances, has been used to fix the fundamental standard of length against which all scientific measurements are now made. The wavelength measurement in the spectrum of krypton gas would give more than the 1,500,000 wavelengths in a single yard.

Interference is widely used in the optical glass industry where it can reveal minute imperfections in lens surfaces. It is particularly valuable in testing so-called optical "flats," glass plates designed to have absolutely flat surfaces. When placed against a standard flat surface and illuminated, uneven parts of the glass will show up as irregularities in what should be a parallel, equally spaced set of straight bands of light and darkness.

The shortness of its wavelengths is important in explaining the behavior of light. The reason light waves seem to travel in straight lines, casting sharp shadows is that they are incomparably smaller than ordinary objects. But, if we carefully examine the way light behaves when it passes through a small hole or past a straight edge, we find that it bends to some extent around the hole or edge. In bending, it also undergoes a kind of interference with itself creating a

Left: this picture shows how clearly a diffraction grating separates colors. The degree of diffraction depends on the wavelength of the light and the distance between rulings on the diffraction grating.

Right: Newton's rings on the interface of glass sheets. They are caused by interference between the reflected waves from the glass and those from the lens.

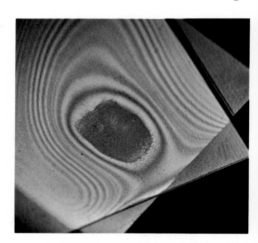

Strange as Young's discovery appeared, the wave theory explained it quite simply. The bright bands were regions where waves from one light beam were reinforcing the other, both of their peaks coinciding. This became known as constructive interference. The dark bands, meanwhile, were regions where the peak of one wave met the trough of another. Here the waves were interfering destructively, cancelling each other out and creating darkness.

The distance between the peak and the next trough of a wave is called its wavelength. The width of the light and dark bands in an interference pattern is therefore directly linked to the size of the wavelength of light involved. Interference patterns for the first time enabled scientists to work out the wavelengths

characteristic pattern of light and dark fringes. This phenomenon is called diffraction.

The effects of diffraction are most obvious when light is passed through a specially designed diffraction grating. Such gratings consist of glass or optical plastic scored with thousands of parallel lines very closely spaced. Modern gratings can have as many as 20,000 lines to the inch. As with ordinary interference patterns, the diffraction pattern enables scientists to measure wavelengths of light to a remarkable degree of accuracy. More important, light waves of different wavelengths are diffracted by different amounts. This means that light containing a mixture of wavelengths will be split by a diffraction grating into a colored spectrum.

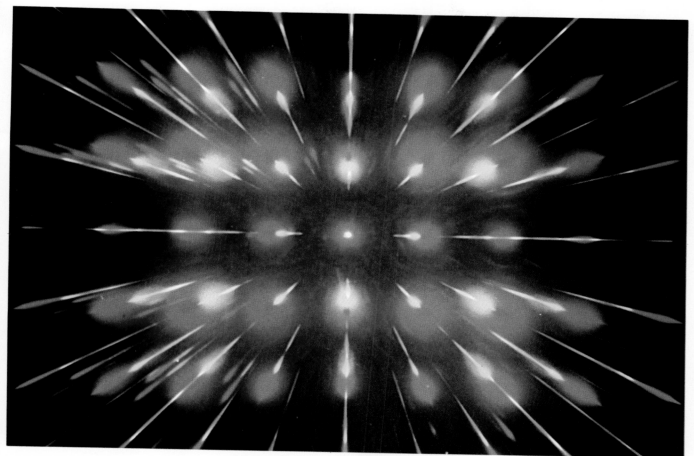

Developments in accurately ruling gratings and in manufacturing them in specialized shapes and sizes have led to their playing an important role in spectroscopy. Traditionally, a glass prism is used in a spectroscope to break light into a spectrum. The nature of the spectrum reveals the presence of certain atoms and molecules in the source of the light. Diffraction gratings are replacing prisms. The ease and accuracy with which they can now be made is bringing a fresh versatility to spectroscopes in fields as diverse as forensic science, chemical analysis and astronomy.

Above: a wind tunnel experiment. The picture shows grid patterns formed by laser beams in an interferometer used to make visible the pattern of air flows in the wind tunnel.

Below: two diagrams of a model breakwater. On the left along **OC** and the other heavy lines, reinforcement by waves from **P** and **Q** produces high crests and low troughs. Nearly still water is produced by cancellation along the lighter lines. The righthand diagram illustrates that at the points **A**, **C**, and **E** on the sea wall inside the breakwater, the waves rise high and fall low. At points **B** and **D**, the water is almost calm.

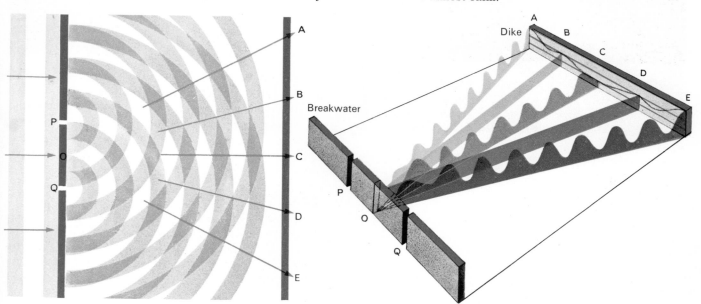

What is a Laser?

Toward the end of 1960, American physicist Theodore H. Maiman made a dramatic breakthrough in the technology of light. Applying the research and design suggestions of fellow-scientist Charles H. Townes, Maiman built the first laser – an instrument capable of producing a thin pencil of ultra-intense light.

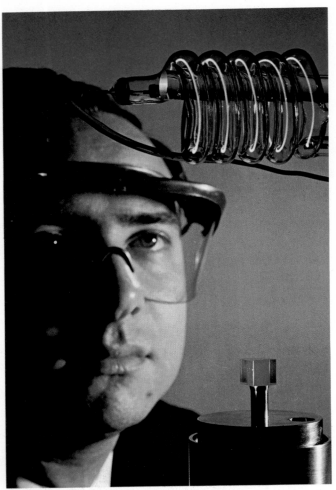

Above: the inventor of the ruby laser, Theodore H. Maiman with the essential parts of his instrument. The glass helix at the top is a flashtube supplying the energy.

Right: the vital part of a ruby laser is the cylindrical ruby crystal. A glass helical flashtube is fitted round this. Mirrors at each end of the crystal reflect light back and forth. The laser beam emerges through the partly silvered righthand mirror. The crystal is cooled by a flowing water system.

Maiman and Townes had placed in the hands of scientists and engineers one of the most versatile tools in the history of technology. Powerful enough to punch holes in the toughest materials, the laser beam can be used even in delicate eye surgery. In the years since Maiman's first prototype, a variety of different kinds of laser have been developed, and an ever-increasing range of applications for them is emerging.

The word "laser" is taken from a phrase describing how a laser works (Light Amplification by the Stimulated Emission of Radiation). Although the technical problems of building lasers are complex their basic working principle is not difficult to understand. The idea originally conceived by Townes was based on the way in which atoms can store different amounts of energy. Atoms containing comparatively little energy can be "excited" into high-energy states, usually by heating the substance of which they are a part. Excited atoms are usually unstable and quickly return to their lower energy state, emitting their surplus energy as they do so. Often the energy is released as visible light. This is why, for example, metal glows when sufficiently heated. Normally the different atoms emit light of many different wavelengths, but Townes realized that there might be a way of triggering excited atoms to release all their unwanted energy simultaneously as a light beam of great intensity and of a single wavelength.

Towne's first experiments were with extremely high frequency (UHF) radio waves called microwaves. Using a crystal of artificial ruby he "pumped up" the ruby atoms into high-energy states by feeding microwaves into the crystal. When the atoms were sufficiently excited he triggered them with a further weak pulse of microwaves. Instantly the atoms released their store of energy as a powerful beam of microwaves of a single wavelength. Townes called his invention a maser, but went on to suggest the possibility of applying the same principle to visible light and so producing the first laser.

Maiman followed Towne's suggestion and after several setbacks eventually succeeded. Like Townes, Maiman also used an artificial ruby, exciting its atoms by a process now called optical pumping. A powerful electronic flashtube, helical in shape, was wound around the cylindrical crystal. The ends of the crystal were silvered with an area of one only slightly covered so as to be semitransparent to light. The pumping process relied on firing the flashtube at a certain in-

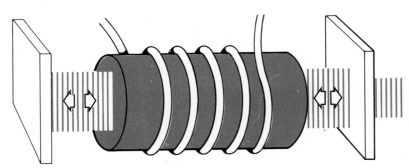

136

intensity had built up for it to penetrate the semi-transparent region. What emerged was a short-lived flash of intense red light, the first laser beam.

Since Maiman's pioneering work, laser technology has advanced rapidly. While the first laser could manage no more than a single pulse of light, today's instruments use a nonstop pumping process to keep atoms excited, so producing a continuous beam. New materials are now used, replacing ruby crystals and other solids which tend to break down because of the heat generated by the pumping process. Liquid and gas lasers have become increasingly important. In the liquid laser, the solid crystal is replaced by a cell containing a special fluid. Of greater practical importance was the invention in the mid-1960s of the first high-efficiency gas laser, destined to become a workhorse in the many industrial, military and scientific applications of lasers. The earliest gas lasers using neon and helium were inefficient, converting little of the pumped-in energy into laser light. But the emergence of the modern carbon dioxide laser in 1965

Left: a maser employed by the British Post Office to amplify the signal at Goonhilly Downs. Maser stands for Microwave Amplification by Stimulated Emission of Radiation. It was the maser that gave the idea for the laser, which amplifies light waves.

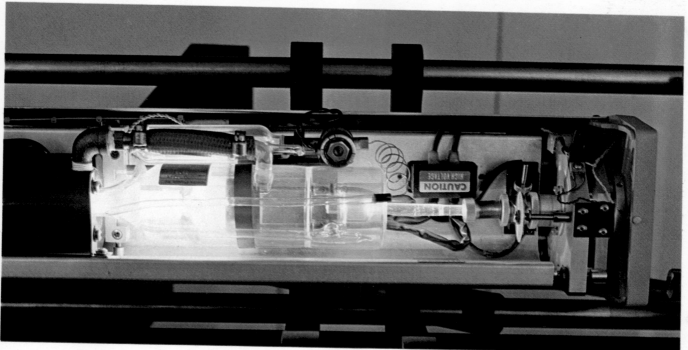

Above: the inside of a modern ion argon laser. Mercury ion lasers failed to prove as effective as first hoped. Other gases were experimented with, and argon was eventually widely adopted.

tensity, raising the atoms in the crystal to a highly excited state. Once a single atom had released its surplus energy, the tiny pulse of light produced triggered other atoms, setting off a kind of energy cascade. The resulting light was reflected to and fro against the silvered ends of the crystal until sufficient

provided an easy to use, powerful and efficient device, ideal for a wide range of purposes.

Potentially more efficient even than the gas laser, is a new generation of devices based on semiconducting substances, of the kind used in transistors. Although their power output is limited, semiconductor lasers are compact and cheap to make. Current research suggests they may also become the most efficient kind of laser, converting almost 100 percent of the pumped-in energy into its light beam.

Lasers in Medicine

Practical uses of lasers have probably been carried further in medicine than in any other field. The narrow but intense laser beam is ideal for performing operations on small, well-defined areas of the body and is now used in eye surgery, dentistry, and the treatment of certain kinds of cancer.

Laser surgery is furthest developed in the treatment of the eye, especially in curing a detached or damaged retina, the light-sensitive tissue coating the back wall of the eye. Retinal damage is often an after effect of illness, or a sudden blow, and can cause partial or total blindness. Once a serious condition, difficult to treat

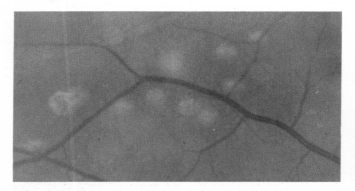

Above: part of the retina of a human eye. The light spots are welds made by a laser during surgery. Many delicate eye operations were impossible before the laser was introduced into the operating theater. **Right:** surgeon using a laser opthalmoscope.

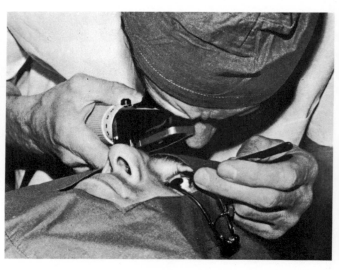

successfully, the laser has made surgery comparatively simple. The technique used is to fire a carefully controlled pulse of laser energy directly into the patient's eye. Precisely targeted onto the site of the damage, it creates extremely localized heating which induces a welding together of the tissue. Welds can be made in a fraction of a second. The time taken is so short that the patient does not blink so the surgeon need not clamp the eye. Also, very little heat is generated, so the patient suffers no serious discomfort and does not need anesthetic.

The use of lasers in dentistry has been pioneered by American dental surgeon, Dr Leon Goldman, who has used a laser focused onto cavities to destroy decayed areas. The laser beam is absorbed by the darker decayed parts of the tooth while the white healthy areas reflect most of the energy and remain undamaged. Pulses of about a thousandth of a second are enough and because the heating is minute and local-

ized, the patient suffers none of the discomforts of the dentist's drill.

Current research is trying to establish a preventive role for laser dentistry. Research workers in the United States believe that decay often starts in minute cracks in tooth enamel. Laser beams are being used to weld them together, preventing decay-forming bacteria from entering and reaching the vulnerable inner regions of the tooth.

Perhaps the most exciting use of laser surgery is in fighting cancer, one of the most feared diseases of modern times. The laser is used as a kind of "light scalpel" capable of simultaneously destroyed cancerous growths and cauterizing the surrounding healthy tissues. The most common kind of tumor treated is skin cancer.

The effect of laser energy on skin varies widely between individuals. But dark skins generally are more susceptible than light. Even the blackest skins, however, require a fairly high energy level to tackle serious skin conditions. A major problem facing surgeons, therefore, is avoiding the use of a level of energy that would seriously damage healthy tissues

along with the diseased. But some of the results of laser treatments seem almost miraculous. In one of the earliest operations a highly malignant tumor over an inch across was treated with nine bursts from a laser. Within 10 days the tumor had disintegrated and after six weeks the affected area was clear and completely healed. Unfortunately, not all kinds of skin cancer are affected by laser beams. Of the 20 known kinds only seven can yet be treated with any hope of success. But experiments with lasers of different wavelengths hold the promise of increasing the range of cancers that can be healed.

In recent years, the laser has also been used to treat internal cancers. In these cases the laser is used like a scalpel but has the advantage of being far more thorough in removing diseased tissue, a key factor in preventing the recurrence of tumors. One of the latest developments has been the use of special optical fibers. These flexible synthetic fibers only allow light

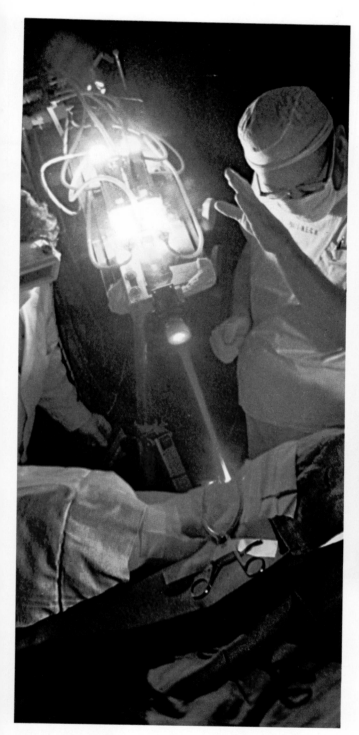

the site of the tumor by conventional, surgical means.

Research on further applications of lasers in medicine continues. A particularly interesting possibility being explored is the use of a laser for surgery at a genetic level. Using ultra-thin and very accurately targeted laser pulses, it is hoped to modify the nuclei of individual cells or even parts of their chromosomes.

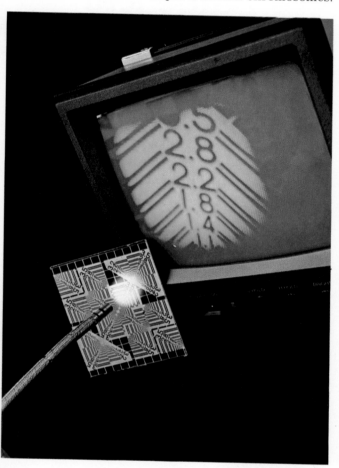

Above: image transmission in fiber optics being demonstrated by the test card being shown on the monitor screen. Fiber optics enable doctors to view internal parts of their patient, avoiding the need for exploratory operations.

Above: skin cancer being treated with the help os a laser. Without a laser, such treatment would be uncertain and distressing for the patient. A laser beam can destroy diseased cells quickly and precisely.

Right: fiber optics explained. As long as a light ray strikes the interface of the tube at a particular glancing angle, light cannot escape.

to pass along their length, none can escape through their walls. This means that light can be guided through bends and curves without loss of intensity. Using such fiber-optic guides, surgeons can reach growths deep inside a patient's body without opening

In this way the characteristics of cells may be radically altered with possible beneficial effects on other parts of the body. Such genetic engineering may hold the key to curing some of the most terrible diseases of our time.

Lasers on the Battlefield

When the laser was invented in 1960 it seemed to the popular imagination the epitomy of the "death ray," a much beloved weapon of science fiction writers. Even in the armed forces some weapons designers believed that laser cannons might soon become a reality. In fact, the power levels currently available from lasers are far below those required for effective weapons, and it is unlikely that developments in the

Right: contrary to the "death ray" images of science fiction, military lasers are usually used to aid range finding and aiming. This 13-pound laser designator is being used to pinpoint targets for homing missiles in an experimental test.

Below: a soldier using a second generation ground illuminator. Lasers can provide excellent lighting over territory where ordinary arc lamps would be too cumbersome. TV stations use laser lighting too.

foreseeable future will alter this. But the invention of the carbon dioxide laser, far more efficient than the early solid crystal lasers, has made possible a weapon that could at least set fire to the clothing and equip-

ment of enemy troops at great distances. So, while not efficient weapons for killing people, gas lasers may one day be used in war for their psychological effects, wreaking terror among a bewildered enemy by setting their clothes ablaze with a silent and near-invisible beam.

Most practical military applications of lasers are based on using them to measure distance with great accuracy. The simplest method for doing this is to measure the time taken for a laser pulse to be reflected from a target. Although light travels at great speed (about 186,000 miles per second) electronic timers can easily measure the minute time intervals and instantly calculate the distance implied. But the most precise technique involves mixing the light "echo" from a target with light leaving the laser. This creates an interference pattern which provides information on distance to a phenomenal degree of accuracy. Scientists recently used a laser beam in this way, bounced from a reflector on the surface of the moon, and measured the moon's distance from the earth to an accuracy of about one foot!

Tests are already under way on laser detecting sys-

Above: a lidar (laser radar) device. The narrow beam width of the laser and its high frequency provides much higher definition than microwave radar.

Above: an attacking aircraft flying close to the ground to dodge enemy radar along path shown by yellow line. Piloted automatically, radar (white) feeds back information on the terrain ahead. The laser beam (red) provides the accuracy that allows such low flying.

tems capable of pinpointing enemy troops and installations. Accurate location of such targets is especially difficult in jungles where conventional radar provides only a hopeless confusion of echoes. But a laser system can pick out a single target among a number of others. One idea currently being evaluated relies on ground troops aiming a laser beam at the target which would reflect and scatter the light. A laser detector and video display mounted in an attacking aircraft gives the pilot exact data on the range and position of the enemy. It is even possible to feed the data directly into a fire control system aboard the aircraft, so that missiles would fire automatically at the target.

The first operational military equipment using lasers will be the battlefield rangefinder. The laser is so accurate in measuring distances, that it will be possible to estimate the range of a target several miles away to an accuracy of a few inches. Although rangefinders using microwaves – similar to conventional radar systems – have been used for some years, they suffer from the effects of the fairly large beam width characteristic of this kind of radiation. Using microwaves it is often impossible to measure the distance of an enemy installation because of echos from obstacles that, although not directly in line with the target, are close enough to interfere with a fairly broad and diverging beam of radiation. But lasers provide an extremely narrow beam that hardly diverges even over huge distances, enabling a gun or missile crew to pinpoint the range of even a small target hidden among the complex of buildings or trees.

Laser radar, often called LIDAR, is really just a more sophisticated version of the rangefinder. Already in restricted use, it works in much the same way as ordinary microwave radar, except that the transmitter

is a laser and the receiver is a special optical system. Because of the narrowness of the laser beam, LIDAR is able to resolve much finer detail than ordinary radar, enabling pilots of attack aircraft, for example, to fly faster at even the lowest altitudes with complete confidence.

Laser communications systems are also being investigated. Just as with radio waves, the basic principle of laser communication is to modify the the radiation used so that it carries the information of a message from the transmitter in a way that can be decoded by a receiver. This process is known as modulation. Ordinary light is useless for this purpose because it invariably consists of a wide range of different wavelengths, making modulation that can subsequently be decoded quite impossible. It is also unlikely to be intense enough to carry information any real distance. A laser beam has none of these disadvantages, and has the added feature of being highly directional. As a result, a modulated laser beam can be aimed at receivers huge distances away, and messages can be decoded with virtually no loss of signal strength. The system suffers from one major drawback. Unless a complicated system of relays is used, the only people who could receive the message are those in the line of sight of the transmitter. Even a minor hill would be enough to block the signal completely. But military experts are happy to trade this limitation against the vastly increased security of a laser message, even over huge distances. Unlike radio transmissions, laser beams would be almost impossible to tap without revealing that some interference had taken place.

In war as in peace, lasers are beginning a technological revolution. Their first military applications now emerging will almost certainly be followed by more sophisticated developments until the armed forces of the future have as much to fear from silent, powerful beams of light as from high explosive, and even nuclear blasts.

Lasers in Industry

Limited was making over 250,000 suits each year and over 1,000,000 yards of cloth were having to be cut to pattern by laborious hand and semimechanical techniques. The firm's research department soon realized, however, that a laser beam could cut through a double thickness of synthetic cloth leaving behind it neatly melted edges. The first laser system used gave the Company a cutting rate of nine feet a minute, saving time, manpower, and providing a perfectly accurate cut pattern for every suit. Unfortunately, despite their

One of the laser's most dramatic effects is the ease with which it punches holes through the toughest materials. This does not mean that the overall power output of a laser is necessarily high – often it is no greater than a soldering iron – but simply that the power is extremely concentrated in a fine beam. It is therefore unlikely that lasers will ever be used in industry in large-scale cutting or blasting operations, but rather where the work is both limited and of a particularly precise nature.

One of the earliest industrial uses of laser-beam drilling took place in the 1960s when the General Electric Company in the United States successfully bored tiny holes in diamond, the hardest substance known to man. Subsequently, another company, Western Electric, put such laser drills to work on its production line, piercing holes in diamonds used as wiredrawing dies (for making wire by drawing metal through them). A laser takes about two minutes to drill each diamond, compared with two or three days required to grind out the hole in the conventional way using diamond dust.

The only semi-large-scale cutting application to be developed was conceived by a British firm of ready-to-wear tailors. In its best years, Montague Burton

Above: a hole being burned by laser beam in a diamond, the hardest known material. Diamond drilling was the first industrial process in which lasers were employed in the early 1960s.

far-sightedness in the technology of their business, Burton's failed to keep up with the changing fashions in men's clothes and by the mid-1970s were suffering a major financial crisis.

The fact that energy from a laser beam can be controlled much more precisely than an electrical arc or a flame makes it useful for high-precision welding. Because lasers deliver energy to a joint very quickly, welds are completed rapidly with little heat lost by conduction in the metal being welded. It is even possible to weld through glass using a laser. This makes welding in an inert atmosphere – often important in preventing a metal undergoing chemical changes at high temperature – a fairly simple matter. Metal to be welded is placed in a sealed glass container with an atmosphere of, say, argon gas. The laser beam is then directed from outside the container onto the metal joint which is then welded in a totally nonreactive atmosphere.

Left: a laser cloth cutting machine put to work by the British clothing chain Montague Burton. By using a laser, men's suits can be produced with a cut of perfect accuracy.

Some exciting applications for lasers are also emerging in the field of pure science. One of these with enormous longterm promise for mankind is laser-induced nuclear fusion. A nuclear fusion reaction is a fusing together of the nuclei of simple atoms such as hydrogen. Such reactions are accompanied by the release of huge amounts of energy which may one day provide man with an almost inexhaustible supply of power. The drawback is that fusion only takes place at about 1,000,000 degrees Centigrade. Lasers are at last providing the means of reaching these incredible temperatures. The world's biggest fusion laser, called Shiva, is situated in the Lawrence Livemore Laboratory, California. Its research program began in 1978 when its 20 beams of high-powered laser light blasted a target of an isotope of hydrogen with 26 million million watts of light energy. The pulse, lasting about a 10,000 millionth of a second, successfully created an entire series of fusion reactions in the target. Although the problems of controlling and safely tapping fusion energy will probably not be solved before the next century, lasers are at least making possible the fundamental research on which the breakthrough will be founded.

The laser's capacity to measure distance extremely accurately is widely used in geodesy, the science which concentrates on the size and shape of the earth. Using artificial statellites in special triangulation techniques, laser-based optical methods are beginning to supplant the more traditional radio systems that often suffer from distortions caused by the earth's atmosphere.

Above: a side view of the world's most powerful fusion laser, Shiva. Situated at Lawrence Livemore Laboratory, California, its 20 beams are together capable of 26 million million watts of light energy.

Left: lasers have been put to work in meteorology. This picture shows a laser used to measure the height of clouds, and thereby help estimate the likelihood of rain.

Lasers are also helping scientists learn more about the way light interacts with matter and – one of the greatest mysteries – the nature of light itself. One developing area of research concerns the so-called nonlinear behavior of light. Just as sound can be made to produce harmonics (extra sound frequencies mixed with the main frequency of a particular note), it is also possible at extremely high beam intensities to produce similar harmonics in light. Lasers have made it possible for scientists to observe nonlinear behavior in detail for the first time. A particularly dramatic illustration of the effect is produced when the original beam from a crystal laser is near the infrared part of the spectrum and is therefore invisible. The second harmonic of such a beam falls within the visible part of the spectrum and appears to an observer as a beam of blue or green light apparently originating in the middle of the laser crystal. Studies of these nonlinear effects are in their infancy but already promise important new insights into the nature of light.

Holography

Holography is the science of reproducing objects as truly three-dimensional images. The idea behind the process is not new. As long ago as 1948 Hungarian-born physicist David Gabor produced the first holograms using mercury vapor light, the strongest illumination then available. His results were dim, blurry images, serving only to prove his remarkable theory. But lacking a sufficiently intense light source to bring the images into vivid focus, scientists regarded

Above: how to project a hologram. The rectangular box in the center of the picture is the laser source, with the beam splitter in the background. In the foreground is the photographic plateholder.

Right: the holographic image reconstructed. A laser beam is directed at a developed hologram. This produces a visual image that is three-dimensional when viewed from different angles.

Gabor's work as no more than an interesting exercise. But the advent of the laser in 1960 transformed the situation. Suddenly a sufficiently powerful light source was available, and holography could be developed into a promising new technology.

The technical aspects of making holograms is complex but the underlying idea is comparatively simple. A laser beam is first split into two. One component, called the objective beam, is spread out by a lens so that it can bathe the subject. The light is then reflected onto a special photographic plate, usually made of glass. The intensity of the reflected light varies with the shape and texture of the surface of the subject, and in this way the light carries information about the subject which eventually forms the basis of a holographic image. The second component of the beam, called the reference beam, is directed at a particular angle on to the photographic plate. Where the two beams meet, the light waves overlap forming an interference pattern which is recorded on the plate.

Holographic plates are coated with a layer of gelatin in which tiny crystals of silver bromide are dissolved. The exposed plate is developed in the same way as ordinary film. It is then "fixed" and bleached so that the precipitated silver is washed away leaving the bromide crystals to form a complex diffraction grating. When illuminated by reconstructing light directed on the plate at the same angle as the original reference beam, a fully three-dimensional image of the subject appears behind the plate, the same size, in the same position and at the same distance from the plate as the subject when it was recorded. By using a more complex and costly process, it is even possible to reconstruct the image *in front* of the plate. When this is done the illusion of reality is so strong that the person viewing the hologram can scarcely resist reaching out to "touch" the image.

A unique property of holograms is that more than one image can be stored in a single plate. Simply varying the angle of the reference beam enables other images to be recorded on the plate. In theory, there is no limit to the number that can be stored and subsequently recreated by a reconstructing beam aimed at the appropriate angles.

Although holography is still a young technology, new manufacturing processes are bringing the cost of holograms within the reach of many branches of

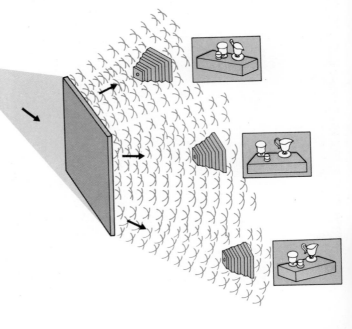

commerce and industry. In television, plans are already afoot for using holograms as backdrops to studio transmissions. At present, conventional two-dimensional backdrops mean that directors must keep their cameras stationary. By contrast, holographic backdrops enable panning and tracking to be freely used, since the three-dimensional backgrounds will change perspective with the movement of the camera just as if they were real. In the theater it will be possible to project three-dimensional objects into scenes,

appeared in Harrods, the famous London department store. Plans are already under way for widespread use of holograms on billboards. It is now even possible to dispense with lasers entirely and produce holograms with much less expensive lighting equipment, making their use still more attractive.

Holography is still in its infancy but as its technology is further refined, even more applications will appear until holograms are a natural and important part of everyday life.

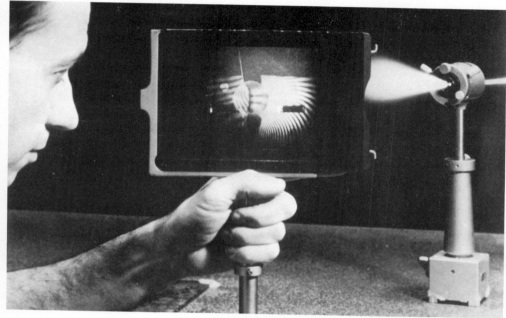

Left: Dr LeRoy D Dickson of IBM examines a hologram in part of the company's General Systems Division laboratory. He is trying to find out what caused similar parts of an experimental machine to fail. The finger-print-like patterns on the image show precise areas of stress in a part as they occur while the part is actually in operation. Faults can be quickly spotted.

Below: a hologram of a car. Holograms make it possible to view images of objects in three dimensions. They are now being used in developing new designs for automobiles.

even superimposing them over real objects or people. Holographic techniques of this kind would make possible the most convincing ghost scenes ever staged while three-dimensional backdrops could lend a startling sense of reality to almost any show.

In industry in general, holography is already being used for testing engineering structures. Researchers at the University of Michigan have developed practical techniques for measuring vibrations and small deformations. Vibrations of aircraft components, and deformations in metals under stress, are two examples. Their method, now widely used, involves superimposing a holographic image of the object under test on the real object itself. Even the most minute differences in size or shape between object and image produce an interference pattern, and changes in the object's shape as it undergoes vibrations or stress are immediately apparent.

Holography stands on the threshold of many new and exciting developments. Three-dimensional television is a possibility, though the formidable technical problems involved suggest that it will be many years before it becomes a reality. In Moscow, Professor J. Komar has developed what he calls multiple-frame pulsed holographic film which is providing the first genuinely three-dimensional movies. Holograms will soon revolutionize advertising and marketing techniques. In 1978 the first holographic sales display

CHAPTER 6

SOUND AND ITS USES

Sound is of vital importance to man's survival and advance. Man has exploited sound to achieve remarkable progress in industry and science. This chapter fully describes the theory of sound, and explains what sound is. Sonar and echosounding equipment make it possible to "see" with sound. Useful in war to detect the presence of enemy submarines, sonar has been put to use in peacetime to locate shoals of fish. The spoken word distinguishes man from other animals, and speech is fully explained in this chapter. Twentieth-century technology has achieved enormous advance in reproducing and recording sound. Recordings are becoming closer to the "real thing" all the time. Architects now give serious attention to designing their buildings acoustically, so that the sound within is distributed to the best possible advantage. Ultrasonics is not yet fully understood, though it has important uses in medicine, industry, and pure science. Infrasonics is another aspect of sound of which more must be learned, for infrasound is a potential menace.

Opposite: the interior of the Sydney Opera House. The design and construction of the opera house was carried out with maximum acoustical efficiency in mind. Many public buildings are now designed according to established acoustical principles.

What is Sound?

"If a tree falls in a forest and no one is there to hear it, will there be a sound?" This question has puzzled many would-be philosophers. But the scientist's answer is simple: yes and no – it depends what is meant by sound. If sound is regarded as a sensation transmitted to the brain by the eardrums, the tree certainly falls in silence. But if sound is taken to be the physical vibration of the air causing the sensation, then there is sound whether or not it is heard, or indeed hearable. There are, therefore, two distinct

Below: Sound made visible. These sound waves have been converted into a stream of electrical signals – here displayed on an oscilloscope screen. The oscilloscope was developed to visually observe the oscillations of sound waves caused by varying disturbances.

interpretations of the word "sound" and each is important in its own field of influence.

The physical effect of sound is basically to transfer energy from one place to another according to particular scientific laws. By contrast, the sensation of sound is more difficult to pin down to anything so exact as scientific laws. Sound adds a dimension to man's experience of the world, just as the sensation of sight does. It can express complex ideas and feelings in language or music. And in the form of noise, it has the power to destroy concentration and even health.

Sound waves are generated whenever an object vibrates in air or indeed in any medium capable of transmitting a vibration. During the object's forward motion, the air in front of it is compressed, and during its backward motion the air is rarefied. So, in the course of its back and forward motion, the object generates a series of alternate rarefied and compressed layers of air which move away at the speed of sound. These moving air layers make up sound waves. One wave is a single compression and rarefication.

The speed of sound varies with the kind of medium through which it passes. As a rule it travels more quickly through solids and liquids than through air. The speed of sound through air is just over 1000 feet per second, while in water it is about 4700 feet per

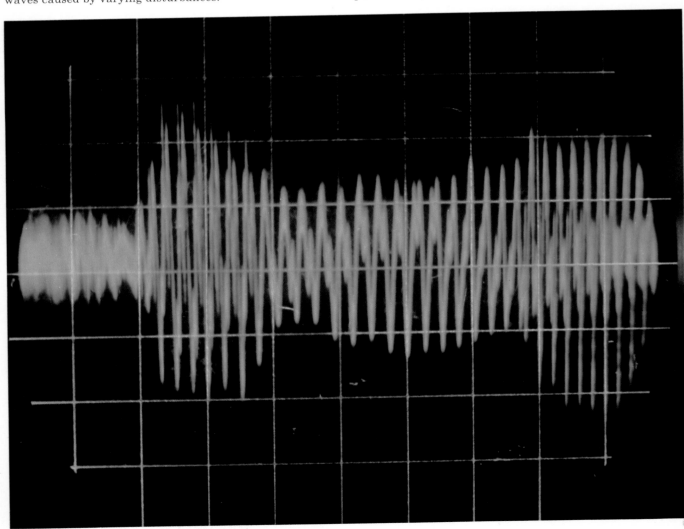

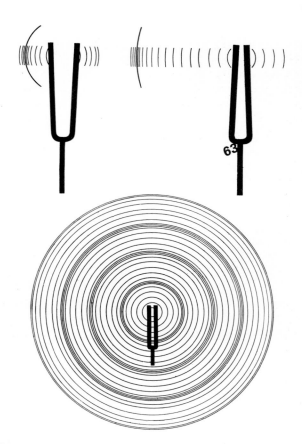

Above: a vibrating tuning fork. The prongs of the fork move outwards (*left*), pushing and compressing the air nearest them. Then the prongs move inward again (*right*), allowing the air nearest them to expand. The diagram shows the effect produced by the lefthand prong only. Tuning forks have been an aid in tuning musical instruments for many years.

of 10 miles, and instruments almost 100 miles away could detect it.

Because sound is basically a mechanical vibration it can cause dramatic effects in objects and structures that are themselves free to vibrate. Most important of these is a phenomenon called "resonance." Most objects and structures have a natural frequency determined by their shape and the materials contained in them. If struck they will vibrate at their natural frequency. For example, a tuning fork is designed so that when struck it will vibrate at a particular frequency, passing this onto the air around it as a pure musical note. Resonance occurs when a sound wave strikes an object with the same frequency as the object's natural frequency. The object will begin to vibrate strongly as the incoming compressions and rarefications reinforce its natural motion. Opera singers are often able to shatter a wine glass by hitting a particularly high note. When the pitch of the note matches the natural frequency of the glass, it will cause the glass to vibrate hard enough to shatter.

Below: audible sound waves can cause remarkable patterns to be formed in various substances. The top picture shows the effect of sound oscillation on a viscous substance. Heated kaolin (**bottom**) turns plastic under oscillation, and round forms circulate within themselves.

second. In a solid such as steel it reaches 16,000 feet per second. This means that if a powerful explosion occurred 50 miles away, the detonation would take four minutes to reach the ear. The vibration traveling along a nearby railroad track would cover the same distance in only 15 seconds.

The number of sound waves striking the ear each second produces the sensation of pitch. Since the speed of sound is usually constant, high pitched sounds are those having short wavelengths (the distance between consecutive rarefications and compressions) while low pitched sounds have long wavelengths. Scientists call pitch the "frequency" of a sound wave and measure it in a unit called the "hertz" (Hz). Each hertz represents one cycle or wave per second.

The range of frequencies audible to the human ear is from about 20 to 17,000 Hz. Sound with frequencies below the lower limit is called "infrasound." Ultrahigh frequency sound above the upper limit is known as "ultrasound."

The frequency of sound waves affects how easily they are absorbed in the atmosphere. Generally, the lower the sound the less it is absorbed. This is why fog horns are extremely low pitched. The horn used by the famous passenger liner *Queen Mary* sounded at 27 Hz. It could be heard by the human ear at a distance

Acoustics

As an audience listens to a symphony orchestra playing in a huge concert hall or an actor on a theater stage speaking the lines of his part, its enjoyment and response owes as much to the quality of the sound as to the content. An expertly handled musical instrument or actor's voice can achieve a great deal but the auditorium has a major effect on the quality of the sound. It determines whether it is loud and distinct, or muffled and echoing. The way physical surroundings modifies the sound heard is known as architectural acoustics. It is playing an increasing role in the design of modern buildings as more is known about it.

American physicist Wallace Clement Sabine founded the modern science of architectural acoustics in the 1890s. As a professor of physics at Harvard he was asked by Harvard's president to try to solve the acoustical problems of a recently opened lecture hall in the university. Having no previous research to guide him, Sabine started from first principles. With the aid of a few organ pipes, a stopwatch, an acute sense of hearing, and a good supply of cushions, he converted the lecture hall into the world's first acoustical

Above: the Philharmonic Hall at the Lincoln Center, New York, before its refurbishment in 1976. It was renamed the Avery Fisher Hall at the same time. Acoustic baffles are visible, as are the walls, which have been broken up into numerous small areas to minimize echo and distortion.

laboratory. The good listening qualities of lecture halls, concert theater auditoriums, and even schoolrooms built since, depend on the discoveries Sabine made.

His main finding was that the listening quality of any room depends on three scientific factors: reverberation, interference, and resonance. Sabine's im-

Left: the Greek theater at Epidaurus. Most Greek and Roman theaters were semicircular. This meant that in the open air tiers of seats acted as reflectors that focused the sound from the stage back to the stage. Sound reflected from seatbacks may also suffer serious distortion in frequency. Seatbacks are nearly always padded, and sometimes absent.

Left: an acoustical engineer experiments with a model auditorium. He can study the behavior of the sound and alter the model until the acoustical qualities are just right. The design of the model can then be reproduced in the finished building. This procedure obviously avoids costly mistakes. Architects now regard good acoustics as one of the essentials of a modern building.

mediate problem in the Harvard lecture hall was reverberation. Any sound in an enclosed space tends to bounce off the containing walls, falling away to silence as it is gradually absorbed. In the Harvard lecture hall a word spoken in an ordinary tone of voice could be heard for over five seconds, reverberating around the room and becoming mixed up with subsequently spoken words so that speech soon deteriorated into a hopeless confusion of sounds. Using the stopwatch and organ pipes he measured the reverberation

Below: when London's Royal Albert Hall was opened in the late 19th century it was an acoustical disaster. In 1969 more than 100 glass fiber saucer, from six to 12 feet across, were suspended about 80 feet above the floor. A 60 foot long glass fiber reflector was fitted behind the orchestra. The result was spectacular, with the hall's appalling echo eliminated. The picture below shows engineers testing the glass fiber saucers before installation. **Bottom:** the Albert Hall after the conversion.

time and began experimenting with the cushions as sound absorbers. After laboriously filling and unfilling the hall with cushions, Sabine was able to work out a scientific formula linking sound absorption and reverberation time. His formula has been used ever since in building design.

The other two acoustical factors, interference and resonance, can be controlled quite easily. Interference occurs where sound waves mix and, in certain parts of a room, reinforce each other. This makes the sound appear louder than it really is. The acoustical engineer attempts to diffuse all sound reflections so that throughout the room a multitude of sound waves will be meeting from random directions. This effectively rules out any real interference problems. To achieve this, the engineer covers the walls with nooks and crannies that break up the sound and reflect it in different directions.

The best known example of resonance is the singer who can hit a note of the same natural frequency as a wine glass. But even rooms and buildings have natural frequencies and if sound of sufficient intensity hits those frequencies, an unpleasant resonance builds up. In rooms and auditoriums, resonance can usually be damped out by using sound absorbing structures fitted to walls and ceilings.

One of the great successes of modern acoustical engineering was the improvement made in 1969 to the auditorium of the Royal Albert Hall, the internationally famous London concert center. Built in the late 1800s, the auditorium was an acoustical disaster. Although its reverberation time was excessive, its biggest problem was the related one of echo. The shape of the hall's domed roof actually focused some of the echoes making it almost painful to sit in certain parts of the auditorium. Early in 1969 over 100 fiberglass saucers from six to 12 feet across were suspended about 80 feet above the floor. A few months later a 60 foot long fiberglass reflector was fitted behind the orchestra to project a kind of acoustical image of the orchestra so that the sound was clearly audible even at the back of the hall. The results were a spectacular success, transforming the listening qualities of the auditorium.

In the 1970s computers were used for the first time to work out how to achieve required acoustical effects. The problems are complex. There is no "best" approach. In some cases, for example, too little reverberation can be as bad as too much, making a room acoustically "dead." Success depends on tailoring the design of a room to its particular needs. Computer programs are being used to provide theoretical models of a room or hall upon which simulated experiments can be performed to determine the best acoustical characteristics even before building work begins. Such new techniques are playing an increasingly important role in the design of new buildings helping architectural acoustics to make a real contribution to the quality of everyday life. It enhances the audibility of the spoken word and the full-scale concert.

Speech

In building up a vast and complex civilization man has been immensely aided by the power of speech. Language has enabled him to develop a culture and a degree of social organization that far outstrips any other creature. Speech and thought are closely interlinked — one is virtually impossible without the other. The "words" in which we think are mental images of the

or off and at what temperature to cook a meal.

It is not difficult to build a speech recognition device to cope with a few words from one voice. As long ago as 1950, Bell Telephone Laboratories built one that could recognize the 10 digits spoken over the telephone by a particular speaker. But when another speaker tried to use the system, the device became unreliable. This emphasizes the basic problem of speech recognition: to create a machine usable by a wide range of voices, each with their own distinct accents and inflections. In 1956, a similar but improved device was invented by the Radio Corporation of America and used to operate a typewriter. The machine could recognize 10 monosyllabic words from a particular

Below: a British experimental machine for recognising vowels in the English language. It is being operated by Brian Pay of the Computer Science Division of the National Physical Laboratory.

Above: Dr Harry F. Olson with his phonetic typewriter. Developed for RCA the machine was unveiled in 1956. It had a vocabulary of 10 monosyllabic words, recognised from a single speaker to an accuracy of 98 per cent. Machines now exist with a vocabulary of thousands of words.

enormously complicated pressure patterns in the air that make up speech signals. Such signals can carry a wealth of information beyond basic facts. The nuances of a person's voice, which give a simple sentence a range of subtle meanings, provides the essential richness and variety of human communication.

Modern speech technology concentrates on two areas, the development of speech recognition machines and the mechanical simulation of speech. In both areas of research the computer plays a central role and has already won limited success which may soon lead to major breakthroughs.

One of the dreams of computer engineers is to be able to talk directly with their machines. But the development of speech recognition devices would extend far beyond orthodox computers. For example, it might become possible to dictate directly to a typewriter, or even to "tell" an oven when to turn itself on

speaker. The basis of the device was a computer memory designed to recognize frequency patterns in the various syllables. By 1961, RCA's phonetic typewriter was handling 100 syllables, and by 1970 this had been further increased to around 2000.

RCA has also developed a speech recognition technique called Analog Threshold Logic (ATL). This depends on an assembly of transistor circuits each designed to work very like a human nerve cell. The ATL circuits are linked to a frequency analyzer that breaks down speech into frequency bands. By the mid-1960s, ATL could recognize a wide range of words from as many as six different speakers. So successful has ATL been that a special experimental system has been tried by the United States Post Office Department for sorting mail. RCA has also been working on an astronaut maneuvering unit that can understand up to 14 basic commands spoken into the microphone in the astronaut's helmet.

The greatest challenge in speech recognition is coping with running speech. As yet, no machine can break up a phrase into component words or store the necessary knowledge of syntax and grammar to comprehend what it hears. Only when this problem is overcome will man at last genuinely talk to his machines.

Speech synthesis machines are quite different from the well established speaking clocks and answering machines available on telephones. These devices merely use recordings of human speech. The true speaking machine generates its own speech like sounds. One of the earliest was built by the British scientist Sir Richard Paget in the 1920s. It used vibrating reeds as resonance chambers made of clay. Modern speech synthesizers use electronic techniques to produce speech. There are two basic approaches. One attempts to convert the physical process of producing human speech – the movement of vocal chords, tongue, and lips – into electronic terms. The other concentrates on how the ear perceives speech and reproduces signals of particular frequencies and intensities to simulate spoken words.

Before making his classic film *2001 – A Space Odyssey*, the director Stanley Kubrick asked computer experts from IBM to describe the kind of computer that would be technically feasible by the turn of the century. They suggested HAL (named after the three letters preceding their own company's initials). HAL could synthesize speech perfectly. Such a machine would be able to convert all the basic parameters of speech together with the rules which give speech its intonation, stress patterns and rhythm, into electronic signals controlling a synthesizer.

Experts in many parts of the world are already tackling the basic research that could make machines similar to HAL a reality. Before the end of this century Kubrick's fantasy may be part reality, with computers that can listen to instructions and reply in polite well-modulated voices.

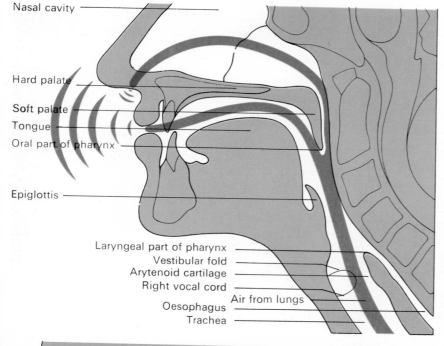

Nasal cavity

Hard palate

Soft palate

Tongue

Oral part of pharynx

Epiglottis

Laryngeal part of pharynx
Vestibular fold
Arytenoid cartilage
Right vocal cord
Air from lungs
Oesophagus
Trachea

Left: speech synthesis machines' greatest application is in the field of computers. A computer that has understood a spoken instruction will respond by speaking the answer via its speech synthesis apparatus. From the "speaking" computer it is only a small step to developing a machine that could read aloud from a printed text. Machines that can translate from various languages are also a possibility. The *top* diagram is of the human vocal tract. The *bottom* diagram is an electrical circuit designed to represent a working analogue of the vocal tract. Component values in the various sections of transmission line represent the dimensions of corresponding parts of the vocal tract. The two diagrams illustrate the articulatory approach to speech synthesis.

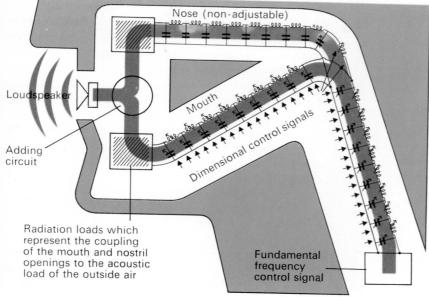

Nose (non-adjustable)

Loudspeaker

Adding circuit

Mouth

Dimensional control signals

Radiation loads which represent the coupling of the mouth and nostril openings to the acoustic load of the outside air

Fundamental frequency control signal

153

Music

Speech conveys mainly ideas and information. But music works on man's imagination, evoking impressions and feelings, less precise than the communication of speech but equally powerful in influencing the cultural growth of civilization. Despite the strong emotional element in all music, its effects are based on firm scientific principles. The way musical instruments work is a practical demonstration of the behavior of sound, and the quality of the music depends on the exact shape and structure of the sound waves produced by them.

The Greek philosopher and mathematician Pythagoras began the scientific study of music 2500 years ago. But his rudimentary investigations did little to modify the design of musical instruments. For cen-

Below: Pythagoras experimenting with music. These illustrations are taken from a book published in Italy. They depict Jubal, the biblical father of music, and Pythogoras with Philolaus testing the tones of bells, hammers, glass, string, and pipes of differing sizes and weights and shapes.

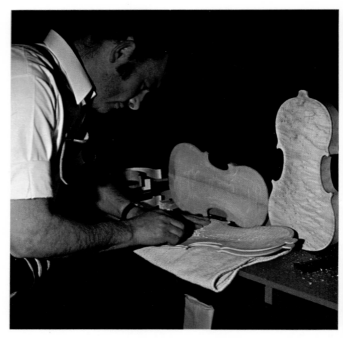

Above: a craftsman at work on a violin. It is now possible for technically perfect violins to be manufactured by machine. As long ago as 1958, eight instruments based on the violin family were produced to provide maximum tonal quality. But many musicians prefer to trust traditional handmade craftsmanship.

turies, the making of instruments remained a craft, depending on practical and instinctive skills. Modifications to a traditional design were made purely on the basis of trial and error. But even today, instruments produced by the famous craftsmen of the past are often regarded as the finest examples of their kind. The violins of the 17th-century Italian Antonio Stradivari, for example, are prized above all others, fetching vast sums of money in auctions.

It is only in the last 50 years that technology has begun to directly influence the design of musical instruments. Today, instruments can be made according to strict scientific standards which equal if not surpass the quality of their traditionally made predecessors. One of the most radical developments has been the invention of a whole family of new stringed instruments. In 1958 American acoustical scientists Frederick A. Saunders and Carleen Maley Hutchins began work on a series of eight instruments based on the violin family, but in a new series of tunings and sizes covering almost the entire range of pitch required by classical music. The instruments – treble, soprano, mezzo, alto, tenor, baritone, small bass, and contrabass – were designed according to the principles of acoustical science to produce maximum tonal quality. The first concert using the instruments, held in 1966, was a great success. But further developments have been slow. Most classical musicians resist efforts to bring technology into the world of serious music, preferring the older traditional skills and values.

The greatest revolution in recent years has been the

rise of electronic music. Although originally a reflection of the musician's search for new modes of expression, electronic music has its practical origins in simple human inventiveness. As long ago as 1895, Thaddeus Cahill built a remarkable device, the teleharmonium. Largely made up of electrical generators and telephone receivers, it was the first instrument to convert electrical signals into a kind of music. But Cahill was at least 50 years ahead of his time, and without the knowledge to build adequate amplifiers or to simplify the circuits to make them more reliable, the teleharmonium never became more than a passing novelty.

By the 1930s the first modern electronic instruments emerged. The most successful were the early electric organs, patented by Laurens Hammond in 1934. These instruments reproduced the tones of conventional wind-powered organs by converting electrical signals to audible ones. By touching keyboard controls, the performer activated an electronic wave generator. This produced electrical signals converted into sound by a device called a transducer. The initial signals could be made to produce sound waves with all the complexity of structure needed to recreate the full range of pitch and tone of the traditional organ.

At around the same time the first instruments were being built producing electronic sound that did not attempt to imitate the music of any orthodox instrument. Some new musical compositions were also performed using no instruments at all. The electronic sounds were introduced directly onto magnetic tape by using electric circuits to modify the input of an ordinary tape recorder. A whole new movement in music emerged led by composers such as the German Karlheinz Stockhausen. Outlandish though the new compositions still seem to people bred on the traditional rules of the classics, they have gradually become an accepted if controversial part of the world of serious music.

The biggest impact of electronic instruments has been in the popular music field. In the 1950s the first electric guitar became widely used, and helped to create an entirely new kind of sound that became known as "rock and roll." Linked to powerful amplifiers, the electric guitar can produce a wide range of guitarlike sounds together with other unique audio effects. The influence of electronics became increasingly sophisticated in the 1960s with the invention of the first full-scale music synthesizers. Consisting of numerous signal generators, electrical modulators, oscillators, and filters interlinked with a complex system of circuits, synthesizers can produce the sound of any instrument – brass, woodwind, string, or percussion – together with a seemingly infinite variety of purely electronic sounds.

American engineer Robert Moog was the first to market compact, moderately priced synthesizers, and their availability rapidly transformed popular music. Today the "Moog" synthesizer is used to produce both partly instrumental and purely electronic compositions. Because of its enormous range and flexibility, the synthesizer has become an accepted and almost essential part of modern popular music.

Right: Roy Emerson of the rock group Emerson, Lake, and Palmer checks his synthesizer. Synthesizers can widen the range of musical styles available to performers.

Below: electric guitar of the 1950s. Introduced in the 1930s, it is now almost universal in popular music.

Recording and Reproducing Sound

The record player or cassette tape recorder is almost as much a part of the basic equipment of a home as a television or radio. The goal of recording technology is quite simply to make the reconstituted sound seem as near as indistinguishable from the original as possible. Everyone with a modern hi-fi stereo system will appreciate how far audio technology has progressed in recent years, both in recording quality and standards of reproduction. In the early 20th century, recordings were limited to opera or music hall singers. The tone was harsh, quality poor, and the frequency

reversed to achieve reproduction. Another needle was set vibrating by running it through the grooves in the disk. Its vibrations were then converted to sound waves amplified mechanically to an audible level. Modern systems use a microphone to first pick up the full range and quality of the sound and convert the sound waves into electrical signals. These signals, containing all the information in the original sound, are converted into a mechanical vibration in a cutting needle or stylus made of diamond. The stylus engraves a spiral groove into a thin layer of acetate lacquer coated on a flat metal disk. The original sound is therefore "written" into the acetate as a long wavy groove.

The record produced in this way is called the master. From it is produced the intermediary masters or "stampers" used to make the thousands of duplicates that eventually reach the public. Modern fine-groove masters have grooves less than three thousandths of an inch wide and about one thousandth of an inch deep. Electroforming techniques by which duplicates are struck give a reproduction accuracy better than one millionth of an inch.

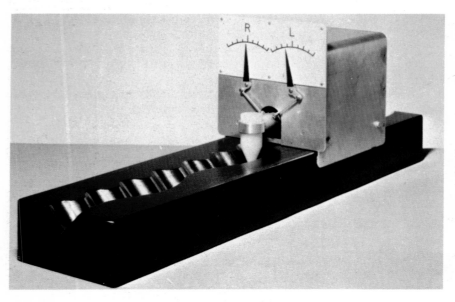

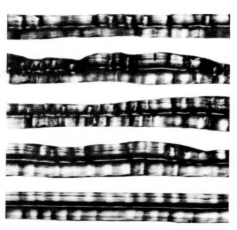

Left: a model of the groove of a stereophonic record with magnified stylus. **Above:** stereophonic grooves at × 250.

range severely limited. Most people nowadays cannot distinguish between the high quality stereophonic reproduction of music and the real thing.

To record any sound, the mechanical energy of its waves must first be converted into electrical and magnetic energy and then stored in some suitable manner. Reproducing the sound relies on the opposite process. But although the general principle is simple enough, the technology required to produce high quality audio systems is complex and sophisticated. The two most common recording mediums are the wax disk and magnetic tape.

Wax disks were first used in 1888 by Emile Berliner but until 1925 sound was recorded and reproduced by mechanical means, without the aid of electricity. Sound waves were amplified, converted into the vibrations of a needle which were recorded by making it cut grooves in a soft wax disk. The process was

A potential problem of fine-groove recordings is that the grooves are normally so close together that a sudden loudening can cause grooves to overlap. The first solution to the difficulty was simply to reduce the recording level for loud sections of a particular work. But this impaired the quality of the recording. A better answer developed recently is variable groove spacing. This technique employs an electronic means to "warn" the cutting stylus that a loud passage is imminent. The stylus immediately begins cutting more widely spaced grooves. When the loud passage is passed the stylus returns to close spacing.

Reproducing the sound from the grooves requires a delicate and sensitive pickup system. The wavy grooves set the pickup stylus vibrating. This vibration is converted into an electrical signal, amplified and fed to a system of speakers. The fluctuations of the electricity are converted into a mechanical oscilla-

Above: modern hi-fi equipment is so sophisticated that the sound reproduced is virtually indistinguishable from the original performance.

Right: a moulding press for pressing 12-inch records.

tion in the diaphragms of the speakers, and this produces sound waves almost identical to those originally recorded. To insure minimum wear to the groove surfaces, modern record pickups weigh less than an ounce. They practically float in the record grooves. The stylus is normally made of diamond, and is spherical at its end with a radius of about one thousandth of an inch.

Stereophonic records became a commercial reality in 1958. In a live music performance sound reaches the human ear from different directions creating a kind of auditory perspective. By recreating this perspec-

ually amplified and each played back through a speaker. The two speakers are placed a few feet apart. The ideal listening position is midway between them and a few feet in front of them, so as to form the apex of an equilateral triangle.

Realism has been carried still further by four-channel recording and playback. Such quadraphonic systems use four speakers to create not merely a lateral sense of perspective but also an impression of depth. Instruments can be distinguished not only from left to right but also behind one another. Impressive though the achievement is, few listeners can detect the improvement above stereo reproduction, and quadraphonic systems with their considerable extra expense have not yet become popular.

Tape recording has undergone a revolution which began in the 1970s with the introduction of high quality cassette tapes at low prices. Today, for the first time, prerecorded tapes are challenging records for popularity. Recording tape itself consists of a plastic base coated on one side with a very thin layer of iron oxide particles, each of which is only about one fifty thousandth of an inch long. To record sound onto the tape, the electric current from a microphone is fed into small coils of wire inside the recording head of the tape recorder. As the current rises and falls, a magnetic field is generated in the head, and tape passing by it is magnetized more or less depending on the electrical fluctuations which control the magnetic field. When the tape is played back, the reverse takes place. The magnetized iron oxide particles moving past the playback head generate tiny

Above: a modern hi-fi system built by Sony. It consists of a turntable, receiver, cassette deck system and two speakers. Speaker stands are also provided.

Left: Frank Sinatra recording a song in the 1950s. The recording will be monaural (mono). Virtually all recordings made today are stereophonic. Quadrophonic sound (where a signal is received through four channels) has been introduced, but has not been popularly accepted. The improvement over stereo is not thought big enough to merit the extra expense. Unlike Sinatra in the 1950s, a singer today would almost certainly record the vocal track separately from the backing track.

tive, stereo systems produce a far greater sense of realism than monaural (mono) reproduction. Stereo requires two distinct sound channels with separate speakers for playback. In cutting the wax disk, the stylus is made to perform a complex vibrating motion which cuts each wall of the groove with one sound channel. The pickup stylus has to be even finer than for monaural playback. The two signals are individ-

electrical currents in the wire coils inside the head. This current is identical to the one that originally magnetized the recording head. When fed back through an amplifier to a loudspeaker, it reproduces the sound recorded on the tape.

Stereo tape recording relies on complex recording and playback heads capable of recording up to eight tracks and reproducing from them four two-track stereo recordings. This means that a single tape can contain a vast amount of recorded material, a distinct advantage over disks. Most commercial cassettes have four tracks providing two stereo recordings, like the two sides of a disk. The cassette itself consists simply of a feed reel and a take-up reel together with tape, enclosed in a rectangular plastic package. The tape begins on the feed reel and passes through guides, traveling across the face of the cassette where it comes into contact with the heads of the playback unit. After the play is completed, the cassette can be inverted and played in the opposite direction, reproducing the other two stereo channels.

Tape recording is more flexible than using wax disks. On a disk, sound can only be recorded once, just as the microphone hears it. But on tape, it is possible to record over existing recordings, superimposing new sounds on the old. Similarly, different tracks can be recorded independently but played together. In the popular music industry, in particular, these techniques are widely used. Called "dubbing" and "multitracking," they enable an instrumentalist or vocalist to simulate several instruments and voices participating together. Groups can produce even more complex patterns of sound and harmony. Most modern pop records are, in fact, made from a prerecorded master tape which has on it all the audio gymnastics intended for a particular work. It is often impossible to reproduce such music on stage.

Above: British technicians learning Italian in a language laboratory. Their tutor can listen and speak to them individually.

Below: the mixing console of a modern recording studio, equipped to record stereophonic sound. The producer can precisely control the kind of sound he wishes to create.

Ultrasonics

Although the human ear is a remarkable device, it is limited in what it can detect to a particular range of sound frequencies. The lower limit of hearing is about 20 Hz while most people cannot hear sounds shriller than 17,000 Hz. Sound with a frequency above this upper threshold is called ultrasonic.

Scientists have known about ultrasound for over a century. As long ago as 1883 English inventor F. Galton developed the first ultrasonic whistle, used mainly as a "silent" signal for dogs, which can detect higher frequencies than man. Since then various kinds of ultrasonic whistles and sirens have been designed and are widely used in the many applications of ultrasound. The real breakthrough in generating ultrasound came with the first piezoelectric transducer. This device converts an oscillation in an electric current into a mechanical vibration in a special kind of crystal. Ultrasonic frequencies can be produced at will by simply adjusting the frequency of the electrical oscillation.

Ultrasonics has become the most fruitful area of development in the technology of sound. Ironically, it is the one kind of sound that we know least about. Despite its important uses in industry, pure science, and medicine, scientists are at a loss to explain the theory behind many of ultrasound's applications. Much work is still to be done on the theory of ultrasound, and a growing understanding of it is sure to mean an ever increasing range of uses for ultrasonics in the future.

One of the earliest practical applications was in detecting flaws in metal. The first systems were developed in the 1930s. In Germany in 1933, O. Muhlhauser worked out the basic principle, amounting to

Right: Mulhauser's diagram demonstrating how ultrasound can be used to test materials. A receiving transducer (**A**), when moved in relation to a transmitting transducer (**B**) can discriminate between energy levels to indicate whether the material is flawed.

Below: a piezoelectric transducer used in ultrasonics.

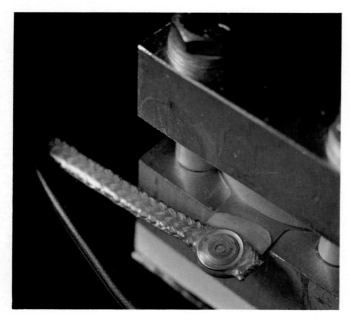

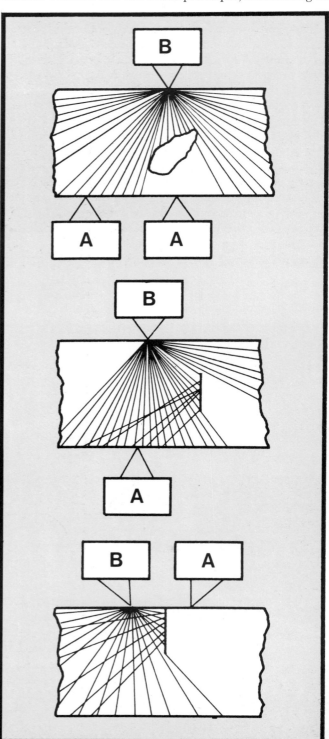

160

a kind of ultrasonic radar. A transducer placed against the metal under test transmitted a continuous ultrasonic signal through the metal. Other transducers, acting as receivers, picked up the signal after it had been scattered by internal flaws. The scattering pattern revealed gave enough information for an engineer to work out the location and size of the flaws. In 1935, Russian physicist S. J. Sokolov improved the system

Below: ultrasonics employed in testing a pipe for flaws. A sending transducer emits a pulse of ultrasound into the material under test. Any reflections (echoes) are picked up by the same transducer or a receiving one.

by submerging it in a liquid. This increased the efficiency with which the signals were transmitted and received. Sokolov's approach has been adopted ever since.

Sokolov went on to propose the first ultrasonic "camera." His idea was to build a receiving transducer for the flaw detector which would retain a pattern of electrical charges corresponding to the ultrasonic signals it received. The charges could then be scanned by an electron beam. This scanning would modify the beam so that it effectively carried the information about the charge pattern to an image

converter where a "picture" of the flaws could be built up. Sokolov's remarkable idea was soon put into practice and his ultrasonic camera is still in use and undergoing further development.

In the early 1940s, the defining power of the flaw detector was improved still further by using pulsed rather than continuous ultrasound. The reflectoscope, as the new device was called, is now the most common detector in use, and finds a wide range of applications in industry and agriculture – from locating hairline cracks in railroad lines and testing vital welds in pipelines and boilers to measuring the relative amounts of fat and lean tissue in livestock.

Another early application of ultrasound was in developing ultrasonic cleaning machines. Early efforts merely involved submerging the materials to be cleaned in a liquid and passing an ultrasonic signal through it at as high a frequency as possible. Researchers thought that the agitation of the water caused by the signal would be sufficient to dislodge any particles of dirt or grease. Their ideas were too simple and few of the prototypes worked at all successfully. It was not until the 1950s that the key to ultrasonic cleaning was discovered. Instead of using high ultrasonic frequencies it was necessary only to use a frequency that would create a phenomenon in the liquid known as cavitation. Hydraulic engineers had known of cavitation for years as a complicated erosion process that could quickly destroy fast moving metal surfaces in water. The classic victim was the ship's screw which could be quickly eaten away by the process.

Nobody really understands cavitation fully but it appears to be a process in which strongly agitated water ruptures into tiny cavities, each containing gas at a high pressure. When the cavities burst they produce an intense shock wave. At sea this can slowly erode metal ship parts, but in an ultrasonic cleaner it is sufficient to strip away dirt particles and disperse any oil or grease. Today, a wide range of materials are cleaned ultrasonically: tiny electrical circuits, watch components, engine parts, hypodermics and other

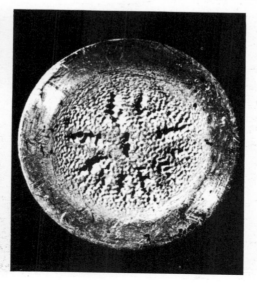

Left: these two pictures show the effect of cavitation on two objects, the thread of a screw (*far left*) and a stainless steel probe. Cavitation produces extensive erosion and results in loss of efficiency. An ultrasonic cleaner can strip away dirt and disperse any oil or grease.

Above: the first commercial ultrasonic homogenizer appeared in 1951. Research had been going on for many years previously. The main problem was to design a suitable transducer. The liquid whistle was developed, and is now used in most ultrasonic homogenizers.

Above: ultrasonic homogenizers are today used extensively in industry. Products ranging from emulsion paints to baby foods receive their attentions. This picture shows medicated ointment being produced with the aid of a multiple-transfer homogenizer.

surgical equipment, sheet metal, and even fruit and vegetables.

The same process can also be used to mix substances. Many industrial and consumer products have to be mixed more thoroughly than normal mechanical stirring can achieve. The problem is often made more difficult by mixtures being required of nonmixing substances such as oil and water. Cavitation produced by ultrasound has proved the answer, and has made it possible to homogenize substances containing a wide range of ingredients. Oil and water, for example, are converted into an emulsion – a fine mixture of the two – and as such forms a basis for many modern paints. Ultrasonic mixers also play important roles in manufacturing cosmetics, polishes, and pharmaceutical preparations.

Ultrasound is also used for welding. Although scientists are baffled by the physical principles involved, the technique is a great practical success. The two metal surfaces to be welded are pressed together with gentle pressure and then made to vibrate against each other at an ultrasonic frequency. No heat and very little preparation of the metal is required, and the two surfaces bond perfectly. The technique is now used for most of the standard kinds of welding and has been used on jobs as diverse as huge aluminum heat exchangers for industrial plants, and microelectronics where tiny circuit elements have to be spot welded.

Equally baffling to scientists is the ultrasonic drill. The equipment required is straightforward. A drill bit of an appropriate cross-section is made to vibrate at an ultrasonic frequency. It is not applied directly to the material to be drilled but acts through a kind of abrasive sludge. By agitating the sludge, delicate holes and even a variety of intricate shapes can be drilled and cut in the hardest material. Ultrasonic drills are of particular importance in machining the tough borides and carbides used as dies for such purposes as drawing metal wire. Industrial diamonds and refractory materials for kilns and furnaces are also drilled and cut with ultrasound.

Microelectronics is also feeling the benefits of ultrasonic technology. Increasingly, ultrasonic "delays" are replacing their more bulky electrical counterparts. A typical delay consists of a vibrator set in motion by an electrical signal in the circuit. The vibrations travel as an ultrasonic wave through a strip of material to a converter which turns the vibration back into an electrical signal and reintroduces it into the circuit. In practice, the whole device only measures a fraction of an inch and fits easily into a miniaturized circuit. Because ultrasonic waves move very slowly compared with electrical impulses, the time taken by the ultrasonic signal builds a significant delay into the movement of the electricity around the circuit. To produce the same delay electrically, yards of extra wiring would be required. Creating delays is particularly important in devices such as radar and television, and ultrasonic circuitry is making possible major advances in miniaturization.

The field in which ultrasonics makes a contribution most directly affecting life is medicine. Many of the techniques have been derived from the industrial applications of ultrasound. Among the earliest developments was a kind of ultrasonic X-ray. The industrial pulsed flaw detector was the basis for the idea. The apparatus is a simple hand-held ultrasonic probe, enabling rapid examination and diagnosis. It is of particular value for examining head injuries where a conventional X-ray treatment can take up to an hour and can even risk causing brain damage by radiation. An ultrasonic examination can be made safely in about one minute.

Developments of the basic probe now enable a doctor to "listen" to a wide range of body processes by detecting ultrasonic echoes. In heart transplant surgery, the technique is used to monitor a patient's new heart. Any slight swelling of the heart, indicating that its new body is beginning to reject it, will be instantly revealed by changes in the echos. More sophisticated techniques are enabling doctors to make thorough examinations of women during pregnancy to insure that the unborn child is well and progressing normally. It is also possible for them to discover whether a mother is carrying more than one baby.

A recent and still experimental technique uses the potentially disruptive power of ultrasound. Researchers have succeeded in using ultrasonic waves applied to a patient's body to disperse gallstones – hard nodules of a substance called cholesterol which build up in a person's gall bladder. This painful condition could hitherto only be relieved by a fairly major surgical operation, which could be risky for elderly patients or those with weak hearts. But ultrasonics offers an alternative that dispenses with the need for surgery altogether.

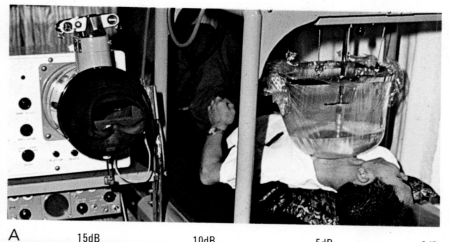

Left: a thyroid ultrasonic tomography in progress. The probe scans the subject through water bag. Resulting pulse reflection echoes are displayed on an oscilloscope screen. A tomogram is a polaroid photograph of this image.

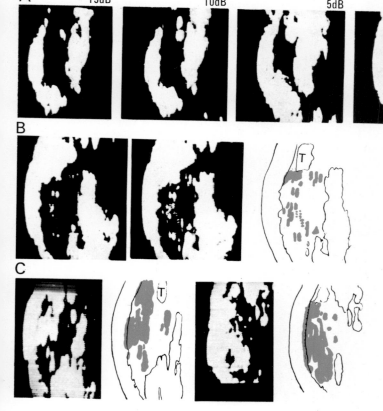

Left below: row **A** is a series of thyroid tomograms produced at four different ultrasound intensities. Those made at 10 and 5dB are the easiest to read and diagnose. Row **B** shows tomograms from a patient with a benign (non malignant) thyroid growth (interpreted in blue). Tomograms and interpretations from two patients with malignant thyroid growth are shown in row **C**. "T" marks the echo from the trachea in every case.

Seeing with Sound

Seeing depends on either receiving a signal from an object or sending out a signal and picking up the reflection or "echo." In everyday life, light is the medium man uses to see. If an object does not give out light itself, it reflects light from another source – the Sun or an electric light. But light, while ideal for seeing in most situations, is very limited underwater. In the seas and oceans both light and radio waves have a range of only yards before they are absorbed by the

water. But sound waves can penetrate thousands of yards, even miles, depending on their frequency and intensity. In all undersea technology, therefore, man depends on sound for his long distance "vision."

One of the earliest uses of underwater sound was in a military detection system called *sonar* (*so*und, *na*vigation, and *r*anging). Developed during World War I and refined in World War II, sonar depends on sending out from a transmitter on a ship's hull a short pulse or "ping" of high frequency sound energy. A submerged object such as a submarine sends back a characteristic echo. From the time taken for the echo to return, the range of the submarine can be easily calculated.

One of the drawbacks of the first sonars was their inability to indicate the direction from which the echo had come. The crew of a destroyer hunting a submarine would know the range of their target but not its direction. If it were confined to a narrow beam, like a searchlight, both range and direction could be determined. The answer was to build more complex transmitting systems. Instead of a single sound source, several are grouped together on the hull so that the sound waves produced interfere with each other, reinforcing in some places, and canceling each other out in others. By carefully arranging the transmitters the reinforcement can be made to predominantly occur in one direction. This produces the effect of a narrow, directional beam of sound. Most sonar systems now use this technique to combine accurate range and position fixing.

One of the most important nonmilitary uses of sonar is in the fishing industry. Introduced in the 1930s, echo sounders coupled to recording devices have revolutionized the industry. Initially, sonar was

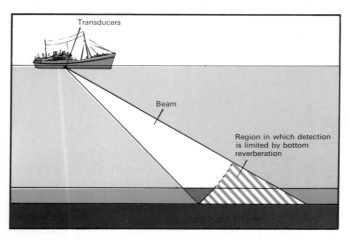

Transducers

Beam

Region in which detection is limited by bottom reverberation

Above left: this diagram shows the limitation of a nonvertical long-range sonar beam. Its ability to detect shoals of fish close to the bottom is restricted. Moreover, echoes in the area shown red are received simultaneously with back-scattering from the sea bottom. "Bottom reverberation" is at a high level and usually blacks out echoes emanating from shoals of fish. Therefore, many a valuable fish catch escapes detection. High resolution sonar can detect the presence of fish much closer to the bottom. With controlled sound and frequency of the sound pulse individual fish can be detected and measured. Even species can be identified. This is useful when looking for increasingly lucrative catches like cod.

Left: three Norwegian trawlers combine to handle a huge catch located by sonar.

used in detecting shoaling fish such as herring and pilchard. But while shoals are still a prime target, high resolution sonar can detect fish that live close to the seabed. By modifying the frequency and time of the sound pulse and making the sound beam very narrow, it is possible to identify individual fish, measure their size, and sometimes even identify species.

In most sonars the information obtained is displayed visually, often on a paper recorder. As each pulse is sent out a pen moves across the paper. The shape of the trace shows whether the ship is approaching or moving away from the fish.

One of the latest sonar developments provides a remarkable cathode ray display that actually shows up underwater movements of individual fish or shoals. The system uses an electronically controlled scanning technique. Instead of combining the transmitter and receiver in a single device as with conventional sonar, the two are separate. The transmitter has a very wide

Above: a United States Navy frogman equipped with portable sonar equipment. His lightweight handheld sonar system has been developed to locate possible underwater obstructions. Powered by flashlight batteries, it is fitted with earphones so that the diver can pick up an aural signal of objects detected by the sonar beam.

beam. But the receiver itself sends out a narrow-beam signal, which rapidly scans over the area through which the transmitter signal is passing. Echoes are recorded only when the receiver beam falls on an object simultaneously receiving a signal from the transmitter. The fast scanning combines high resolution with a very wide area of search – two important advantages that are usually mutually exclusive in orthodox sonar.

Echo sounders are also used for undersea surveying, building up a picture of the seabed. For deep water use, such sounders employ several pulses traveling through the water simultaneously to make better use

of the long delay between sending and receiving a signal. In the late 1940s the British Royal Navy developed a "sideways-looking" sonar. This produced a profile of the seabed rather than the plan view to which ordinary sounders are restricted. The system's range is usually restricted to about a mile. But versions with ranges of nearly 15 miles have been built. These are bulky devices, consuming huge amounts of power, and usually housed in their own miniature submersibles towed behind the survey ship.

Although seeing with sound is largely confined to underwater technology, there is one exciting application emerging on land. While for those people with normal eyesight light is the best medium of vision, for the blind or partially sighted sound may offer a practical alternative. Borrowing an idea evolved in nature by bats, creatures that use an ultrasonic echo-sounding system instead of conventional vision, a kind of ultrasonic device has been designed for the blind which looks like a pair of conventional eye-

Above: a United States Navy helicopter recovers a submarine detecting sonar device from the sea. Sonar was first put to use in naval warfare in World War II, and now meets a variety of civil needs as well.

glasses. It is in fact a sophisticated sound transmitter and receiver. Changes in an audible signal tell the user when other objects are near. After using them for a while, a blind person can build up a sensitive awareness of his surroundings, enabling him to carry out a wide range of activities without guidance or previous practice. The device holds the promise of restoring at least a type of vision to the blind, enabling them literally to "see" with sound.

Infrasound

Sound waves below the lowest limit of human hearing are called infrasonic. Unlike ultrasonics with its many promising industrial and medical application, infrasound abounds with potential menace. At extremely low frequencies the sheer power of the waves of compression and rarefication can be extremely dangerous, damaging both building structures and biological tissues.

Although scientists have been aware of infrasound since the turn of the century, little is really known about it. Modern research in the field has been pioneered by the French physicist Vladimir Gavreau, who has a very personal reason for his interest in the subject. While working in his laboratory he was suddenly exposed to an intense infrasonic signal. Despite the fact he could hear no sound, he felt a painful build-up

ordinary whistles – although, of course, its force would be felt rather than heard. Gavreau has also worked out plans for a monster whistle over 20 feet in diameter, with an output equivalent to over 170,000 whistles.

He has also achieved extreme infrasonic frequencies using very little power, so damage was avoided. Using organ pipes nearly 80 feet long he has produced waves with a frequency of around 3 Hz. His research has enabled him to design specialized microphones capable of detecting infrasonic waves.

Gavreau is also trying to devise techniques for blocking infrasound. Such techniques could become vitally important if infrasonic weapons were ever developed. There is no doubt about their terrifying potential. A directional beam of high-power infrasound would be virtually unstoppable. It could literally tear buildings and people apart.

In nature, infrasound with frequencies of less than 1 Hz can be generated by earthquakes and volcanic activity. The main shock of an earthquake, in particular, is usually preceded by smaller infrasonic shocks. Early warning systems for earthquakes are being developed using detectors capable of picking up the preliminary signals. Similar techniques are being considered to forewarn against volcanic eruptions.

Infrasonic shock waves in the earth are put to

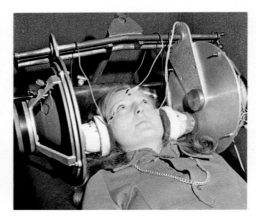

of pressure in his ears. Objects around him began to vibrate strongly. He tracked down the source of the infrasound to a faulty ventilator in a nearby building rotating at a low speed and generating infrasonic waves.

Fascinated by the phenomenon, Gavreau and his colleagues designed an infrasound generator rather like an ordinary whistle. His experiments very nearly proved fatal. Imbedded in concrete, the infrasonic whistle produced powerful signals that soon set up painful vibrations in Gavreau's body. There is no doubt that had he persisted with the test he and his colleagues would have suffered serious hemorrhages as their internal organs rubbed against each other under the force of the vibrations.

Undeterred, Gavreau went on to design a larger whistle, nearly five feet in diameter. If it were used at full power, the blast would be equivalent to about 2000

Above left: Dr Vladimir Gavreau's first infrasonic gun. The signal from the whistle imbedded in concrete produced painful vibrations in Gavreau's body. **Above:** this woman is wearing earphones, carrying infrasonic waves, and skin electrodes to test the effect of infrasound on involuntary eye movements.

practical use by geologists and mining engineers. By triggering explosions on or just below the surface of the ground, they artificially produce low frequency shock waves. Information collected by infrasonic detectors at ground level enables experts to work out the nature and structure of the underlying rock strata. The technique is extensively used in mineral and oil surveys and has recently been employed in detecting breaches in the underground nuclear test ban treaty.

Infrasound from both natural and artificial sources can have a marked physical effect on people, even if the signals are of very low power. In Great Britain in the mid-1960s an investigation was started when draftsmen working in the drawing offices of British Airways became disturbed and agitated, refusing to work in the offices but unable to explain why. A research team led by Dr Michael Bryan found that the offices were close to the test bed for the engines of the supersonic airliner Concorde, then under development. They discovered that when the engines were running, the offices were bathed with infrasound. Clearly, the inaudible sound waves were having a subtle but real physiological effect on the men.

Further research has shown that even fairly low levels of infrasound can cause symptoms such as nausea, headaches, and a general feeling of unease and discomfort. More sinister still is the possibility that infrasound may have subtle effects on the way we behave and on our powers of reasoning. Experiments on pilots and astronauts, for example, have shown that infrasound affects the ability to perform logical tasks. Bearing in mind the vast number of po-

tential sources of infrasound – from waves on the seashore to industrial or even domestic equipment – it is possible that a large part of human life is ruled by these powerful and unheard low frequency sound waves.

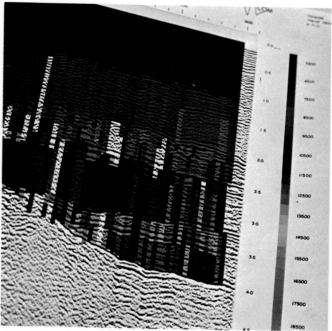

Right: infrasonics has proved a valuable tool for geologists and mining engineers. This is part of an infrasonic Seis-Com of a section of the North Sea oil shelf.

Below: one of the Anglo-French supersonic airliners Concorde being given infrasonic testing at Toulouse to assess the soundness of the fuselage.

Living with Noise

One of the prices man has had to pay for an increasingly technologically advanced society is a steady rise in everyday noise levels. While technology provides great material benefits, the quality of life is being brought under a mounting attack by chaotic, uncontrolled sound – the rumble of heavy traffic on highways, the scream of jet engines at airports, the roar of heavy machinery in factories and workshops. People's leisure time is also constantly disrupted by the unwanted sound of televisions, radios, and stereo or quadraphonic hi-fi systems.

There is still little known of the long-term effects of noise on physical and mental health. But those people who live close to airports or major roads understand only too well the misery of disturbed sleep and the total lack of tranquility. There is direct evidence that excessive noise over long periods can damage hearing, while exceptionally loud blasts can cause perforated ear drums and even internal bleeding. More subtle effects include the impairment of concentration and reasoning power, leading to industrial accidents.

In order to combat the noise menace, technologists have had to work out ways to measure noise and to investigate its sources. Noise measurement is a problem complicated by the subjective response of different people to different kinds of noise. A dripping tap, for example, may not be loud but can be more disturbing for some than even the constant roar of road traffic. In 1959, American researcher Karl Kryter introduced a system now used internationally which, although not eliminating the subjective element entirely, at least allows for the different responses of the human ear to various frequencies. His unit of noise is called the *perceived noise decibel* (PNdB). A

busily working typing pool in an office would produce about 80 PNdB, and a jet airliner taking off, 129 PNdB at 150 yards; the threshold of physical pain is about 130 PNdB.

Road traffic is by far the greatest source of noise in towns and cities. The vast increase in the number of privately owned automobiles over the last 20 years has been matched only by the increase in highway and general road construction. Concrete and asphalt highways have cut swathes through countryside and towns alike, turning once peaceful environments into a nightmare of noise and fumes. During the 1970s, roads were more widely used for transporting freight than ever before, and the introduction of huge freight vehicles has increased noise levels substantially. In some parts of Europe, giant vehicles of up to 56 tons laden weight are permitted. Noise emissions from them are subject to internationally agreed control, but the maximum level of about 100 PNdB applies only to a new vehicle leaving the manufacturer. No control is exerted on its noise emission when it is no longer brand new. The problem is made worse by the way in which heavy freight vehicles have to be driven. Unlike automobiles, they are at full throttle for long periods of time, producing maximum noise.

Left: an automatic noise indicator installed in a Tokyo street by the Capital Construction Company of the Metropolitan Government. Noise here is calculated in "phons", which measure apparent loudness. The pitch of a sound is as important as volume in noise nuisance.

Above: a 650-ton load of two enormous cylinders being transported by road from Cammel Laird's shipyard. Loads like this on the public highways cause considerable noise nuisance in town and country. Such loads also break up the surface of many highways.

Taken on its own, an automobile is comparatively quiet. Some manufacturers such as Rolls Royce actually base their appeal on silent luxury. But automobiles usually travel in continuous, often congested streams of traffic. Constant braking and acceleration and the blasts of car horns make manufacturers' noise standards for new vehicles little more than academic.

Ironically, technology has answers to traffic noise which society will not allow it to use. The British Road Research Laboratory has shown how a heavy freight vehicle may be made more quiet than an automobile. The principal modifications are simple and obvious. The engine is enclosed in a sound-proof medium and the exhaust fully silenced with baffles. But there is little chance of such modifications being made by manufacturers. The increased cost would mean higher selling prices and, because quieter running is not a persuasive sales point, customers will not pay the extra. Until public demand for less noisy vehicles grows, or national government intervention takes the choice away from manufacturers, little progress will be made. It is possible, however, that decisions will be forced by the dwindling reserves of oil. Within the next century, man may be compelled to turn to battery or gas powered engines which are far quieter than the internal combustion engine.

Aircraft noise, although often more dramatic than the sound of road traffic, actually affects far fewer people. The main reason is that planes spend most of their flight time at altitudes of 30,000 feet and above. At any height above 15,000 feet the noise nuisance caused at ground level is negligible. The only exception to this general rule is the supersonic airliner Concorde, and a number of supersonic military jets.

Left: sound insulation materials are essential if noise in a modern city is to be coped with. This is heat and sound insulating material made of foam plastic clad with aluminum. It is commonly used in office ceilings.

Right: a modern sound insulation material. This absorption system relies on thousands of tiny glass beads sandwiched between perforated end plates. The absorption prevents noise from carrying from room to room and eliminates echo.

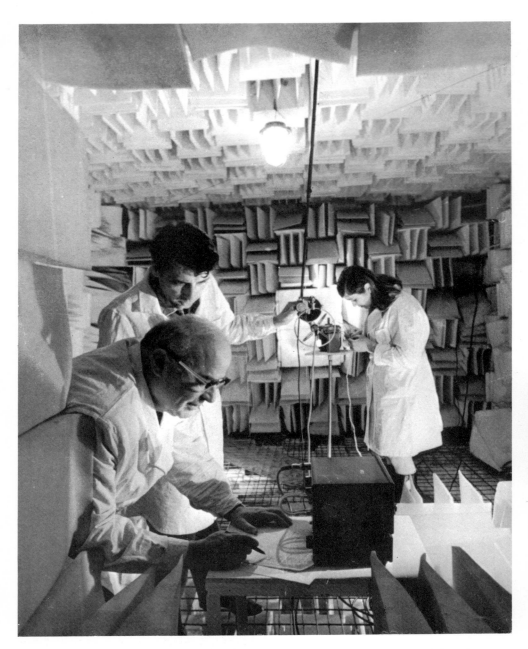

Any aircraft flying faster than sound generates a roughly cone-shaped shock wave stretching behind it. On the ground the shock wave is heard as a double "sonic boom." Fears about the effects of sonic booms meant that Concorde faced fierce public opposition before beginning scheduled services in the mid-1970s. By restricting the supersonic phase of its flights to periods when it is over sea or sparsely populated land areas, the nuisance has been minimized. Research is, however, underway on techniques to reduce the intensity of sonic booms. One early suggestion was to fit structures to the wings and bodies of planes to make the shock wave generated in one part of the structure cancel the shock wave produced in another. The idea foundered when it was discovered that the required structures would prevent a plane from achieving sufficient lift to get off the ground. A more realistic proposal was made in 1968 by the Northrop Corporation of America. Their idea was to electrify the air layers

in front of the plane to spread out and so reduce the abruptness of the shock wave. The electrified air would effectively create a "phantom structure" increasing the apparent length of the plane. Northrop estimated that a 50 percent increase in apparent length could reduce the impact of the boom by half. A year later the idea was attacked by engineers from Boeing who produced a counterproposal for improving aerodynamic styling to reduce the boom by 20 percent.

The real nuisance of aircraft noise is concentrated around takeoff and landing. Not only are aero engines running at nearly full power at these times, but also the airplane itself is closest to the ground. Noise is therefore a major source of annoyance to the community living in the vicinity of airports. Most national authorities have introduced restrictive legislation compelling pilots to use noise minimizing flight procedures such as less rapid climbs after crossing airport boundaries, and steeper descents. On the

Above: noise measurements being taken in a boiler shop in an engineering factory. An acoustical engineer is taking readings from a sound-level meter (*left*), and a sound frequency analyzer (*right*). Heavy engineering is one of the industries known to cause hearing damage.

Above right: a major area of research is aimed at the reduction of aircraft engine noise. This is a Rolls-Royce RB 211 jet engine under test. Its three-shaft design allows noise from fan, turbine, and jet to be juggled to create minimum noise whatever the thrust.

ground buffer zones of forest or agricultural land divide airports from the surrounding community, and government help is often available for insulating houses and offices against noise.

Noise abatement procedures tend to restrict airport capacity and are, therefore, extremely costly. But aeronautics engineers are making important progress in designing less noisy engines.

One of the most important advances has been the development of the turbofan principle, the most efficient example of which is the famous Rolls Royce RB211 engine. The basic idea is to blanket the fastest and, therefore, noisiest area of the engine's jetstream with layers of slower, quieter gas. In the United States the Pratt and Whitney Corporation has used the concept in its engines built for the Boeing 747 "Jumbo" Jet. Their efficiency is such that on takeoff the huge airplane is less noisy than most other jet aircraft despite the vastly greater size and power output. In

fact, so much progress is being made in aero engines that researchers at Rolls Royce have been encouraged to announce that within 40 years they will produce a jet engine no noisier than an automobile.

While the obvious problems of road traffic and aircraft noise receive much public attention, the more insidious dangers of occupational noise have only recently been investigated. There is nothing new in deafness being caused by a noisy working environment. In 1782, for example, Admiral George Rodney commanding the British warship *Formidable* was struck deaf for 14 days after firing 80 consecutive broadsides from his ship. Modern working environments are rarely as noisy as the gundecks of old warships, but many jobs do involve sound levels that eventually cause deafness. Hearing deterioration occurs at a rate determined by the frequency of the sound, its intensity, and the length of exposure. It can be cumulative and often imperceptibly slow. The first sign of trouble may be no more than a dullness or slight ringing in the ears.

Factory work, especially in the engineering and construction industries, is most likely to involve noisy working conditions. The usual means of cutting noise levels is fitting sound absorbing materials to ceilings and walls and firmly securing machinery that might otherwise vibrate. Many industries have been slow to acknowledge the damage workers can suffer to their hearing. In Britain, deafness is not even classified as a possible occupational disability, and sufferers cannot claim compensation. But medical research is now producing irrefutable evidence to the link between the noise of the workplace and deafness. Public pressure is beginning to insist on action beyond remedying problems in the factories themselves. Increasingly, people are demanding that machinery is designed to be less noisy in the first place.

171

CHAPTER 7

ENERGY AT WORK

For the first time, man is confronted with the fact that the energy resources of his planet are finite and, often, unrenewable. Today the technology of power is a technology in crisis. But crisis has imposed a welcome discipline both on man's use of energy and on his search for new resources. Increasingly alternative power technologies are being studied. Nuclear energy, its early promise now beginning to mature, is coming under close scrutiny. What are its dangers? Is it really an answer to the long-term problem of man's soaring demand for energy?

Ultimately there is no single answer to the energy crisis. But if terrestrial technology fails to keep pace, there may yet be new and hitherto unexplored possibilities beyond the Earth. Man is already developing a more economical form of spaceflight, and many energy scientists believe the power generation industry is about to take its first major step into space where it can tap the energies of the Sun.

Opposite: the Dez Dam in Iran. Water is the most economic motive power for the generation of electricity. Engineers control water flow through the turbines of a hydro-electric power plant. The modern waterwheel works on the same principle as those of 2000 years ago.

Energy, Work and Power

Energy is the capacity to do work. Because it is only through work that change can take place in the universe, energy is the motive fuel of nature. The civilizations of mankind have been built up on the conversion of different kinds of energy into useful work. Early man survived and prospered by learning to kill animals that were stronger and faster than himself. Using a bow and arrow he could convert his own muscular energy – by drawing the bowstring – into the energy of the arrow's movement toward its target. Just in warming himself by a wood fire, man unconsciously released its chemical energy as heat. With growing technological insight, man built the first machines, converting energy from one form to another and doing useful work in the process. The early steam locomotives, the forerunners of the modern technological revolution, used the conversion of heat into mechanical energy to haul passengers and goods over the first railways.

Energy – from the Greek word meaning "containing work" – is of two basic kinds: potential and kinetic. A weight suspended in the air, a stretched spring, or a piece of coal – all these possess energy. The energy exists because of the weight's position, the spring's stretch, and the coal's chemical structure. This kind of stored energy is called *potential* energy. It is energy that has to be released in order to do work. If the weight is dropped its potential energy will change into the energy of its motion, called *kinetic* energy. Similarly, if the spring snaps back it produces movement in the surrounding air – a kind of kinetic energy – and coal when burned heats and agitates the molecules of the air also producing kinetic energy. In order to store energy therefore, it must be turned into potential energy. To make it do work it has to be converted from potential into kinetic energy.

Whenever an object is moved against a force, work is done and energy is expended. The hunter with his bow and arrow, uses muscular energy to do the work necessary to draw back the bowstring. Momentarily the energy is stored in the taut string and bent wood of the bow. When he releases the string, the potential energy instantly turns to kinetic energy that does work, propelling the arrow through the air.

The efficiency of any process depends on how much energy is actually converted into *useful* work. The aim of all engineers is to make machines that give out nearly as much energy – as useful work – as is fed into them in the first place. In an automobile, for example, only a small part of the energy locked up in the fuel is turned into the energy that eventually drives the car. Most of it is "lost" as heat caused by

Below: an archer drawing a bow to its fullest extent. His muscular energy is converted into the energy of the flying arrow. When the archer releases the bowstring potential energy is converted into kinetic energy.

Below right: the constructive use of energy demonstrated. Whenever the man and the weight is moved against a force (gravity) work is done and energy is expended.

friction between moving parts or heat in the exhaust gases. But energy can never simply disappear. It is a fundamental law of nature that energy can neither be created nor destroyed. It can only change its form. The heat of the burned fuel is only "lost" therefore in the sense that it does not help to drive the car, the prime purpose of burning the fuel. In most automobile engines only 25 percent of the fuel's chemical energy powers the vehicle. The rest is wasted.

Man's technological achievements are based on his skill in converting as much energy as he can into useful work, wasting as little as possible. He is not only interested in the efficiency but also in the speed with which he can make the conversion. The rate at which work is done is known as power and it depends on how quickly energy can be transformed into work. For example, a man carrying a suitcase up a flight of stairs has to do an amount of work determined by the height of the stairs and the weight of the suitcase. If he walks up he uses relatively little power. But if he

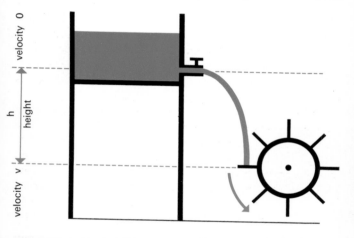

Above: the principle of the waterwheel applied to a water driven turbine. The falling water from the stopcock drives the turbine, thus producing kinetic energy. All matter in motion has the capacity to do work. Water falling under gravity drives the turbine.

Right: the device that provided the motive power for the British Industrial Revolution – the steam engine governor developed by James Watt. It adapted the power output of a steam engine automatically to the load by means of in-built mechanical feedback.

has the strength to run, although he does the same amount of work, he does it in a shorter time and therefore expends much more power.

The generation of power by any of a variety of means is really the production of energy in a form in which it can quickly be changed into work. Technological progress depends on ready supplies of such energy and today's power-generation industry has to work at full stretch to keep up with demand. The most usual form of energy the industry provides is electricity. Many techniques are used to generate it but all rely basically on the conversion of the energy of motion into electricity by a phenomenon called electromagnetic induction. But the motion itself needs some fueling

energy and as demand soars so the fuels on which power generation depends come under increasing pressure.

Fossil fuels such as coal and oil still dominate the energy-supply market. But the technological revolution that has transformed the world over the last century has consumed in less than 100 years about a third as much power as was used in the whole of the previous 2000 years of human history. Now, under the pressure of the overwhelming demand for yet more power, man is confronted with the limitations of his planet's resources. For the first time, man is compelled to seek new kinds of fuels and new ways to generate power from them.

175

Fossil Fuels

Coal, oil, and natural gas are fossil fuels, relics of plant and animal life that flourished millions of years ago. Once used, these fuels cannot be replaced. Yet man's consumption of them is rising rapidly as the pace of technological progress quickens.

Coal was first mentioned by the Greek philosopher and scientist Theophrastus as long ago as the 3rd century BC and the first records of the organized mining of coal go back to the 12th century. The great underground seams exploited by the mining industry were formed about 250 million years ago. The climate then was tropical over most of the Northern Hemisphere and the land was covered with dense forests of primitive plants, mainly mosses and giant ferns. These plants grew, reproduced, and died forming

Right: the Spindletop oilfield on the Gulf of Mexico in Texas, pictured in 1902. Derricks proliferated and fortunes were made and lost after oil was found in 1901.

Below: an English coal mine of the 1790s depicted by an unknown artist. It is clear from the picture that the British Industrial Revolution had barely begun. Within a decade coal mining dominated the landscape.

thick layers of decaying vegetation. Mud and sand washed down by rivers covered the vegetation. More plants grew on the mud and sand, died, and were covered in the same way. Over millions of years this process was repeated. Then the climate altered and great earthquakes changed the face of the land, covering the alternating layers of mud, sand, and vegetation with rock often thousands of feet thick. The tremendous weight of rock compressed the decayed vegetation, heating it until eventually it changed into coal.

Coal, which is essentially a type of carbon mixed with organic chemicals formed from plant remains, is of two main types: bituminous coal, which is hard and black, and lignite, probably formed later and under less pressure, which is soft and brown. For centuries coal was used as a domestic fuel but in more recent

176

times it has become one of the most important fuels used by the power generation industry. Constant demand is being offset by new mining techniques allowing deeper and more inaccessible seams to be reached. But reserves are being rapidly depleted. Nonetheless, coal remains the richest of our fossil fuel resources with an estimated 7000 million tons still remaining in the earth. This should last even our present coal-hungry society another 200 years or more.

Petroleum oil has a less certain future. Because it is an especially convenient fuel, the demand for it outstrips even coal and despite newly discovered oil fields our resources are not expected to last much into the next century.

No one is certain how oil was formed. The most likely theory is that it was produced millions of years ago by the decay of tiny sea creatures in much the same way as coal was formed from dead vegetation. The creatures formed layers on the sea bed which were buried in mud. Over millions of years these deposits were trapped by upheavals in the earth's crust. Heated and compressed, they underwent chemical changes and slowly turned into underground reservoirs of petroleum.

Oil from surface seepages has been used by people for thousands of years. Traces of oil in spring water in Texas led prospectors to make the first drillings for underground oil and so started the great American oil rush of the late 1800s. When no telltale seepages

Below: an inland natural gas drilling rig in the United States. Natural gas often forms a pocket above an underground deposit of oil. This is why the two resources are often drilled for in the same area.

reveal oil's presence, geologists have to scrutinize the structure of the earth's crust in the vicinity of possible oil fields to find the signs that usually indicate oil-bearing rocks. But the geologist can never be certain of his findings. Even today, the only sure way to locate oil is by drilling.

In power stations and internal combustion engines, the world is now burning over a million barrels of oil every hour. No one is sure how much oil remains. Recent new finds beneath the sea bed have given the oil industry a new lease of life. But offshore drilling is extremely costly, at times only barely economic. Even with these additional supplies most fuel scientists believe the world will run dry of petroleum within the next 100 years.

Natural gas, used mainly for industrial and domestic heating, is found together with oil deposits, often in huge subterranean pockets. The gas is tapped by drilling in much the same way as for oil. The largest source of such pockets has for long been the oil-rich states of America. But demand is fierce, rising on a world scale by nearly eight times between 1950 and 1980, and American resources are now seriously depleted. In Europe a great deal of the interest in natural gas has arisen from recent finds in the North Sea. The nearness of the deposits has made it economic for Great Britain to convert its domestic and industrial users of gas from the old "town gas" – produced from coal or oil – to natural gas. But as with all fossil fuels, natural stocks of gas cannot be renewed. It is difficult to predict just how much gas remains. Most experts estimate it will outlast oil but will be used up long before the last coal is removed from the earth.

Below: natural gas being burned off on a rig in the North Sea. Some gas in a deposit is always burned off. But natural gas, which is largely methane, could be used to make polymers instead.

Electricity

The power generation industry is in the business of producing and supplying electricity on a large scale. The basic scientific principle upon which it depends – electromagnetic induction – was discovered in the 1820s by the English physicist Michael Faraday.

Before Faraday's time, electricity could be generated only by converting the chemical energy of the first crude batteries into a fairly low level of electrical energy. While adequate for laboratory use, the early chemical techniques held no promise of large-scale supplies of electrical power.

Faraday's vital discovery must rank as one of the great breakthroughs of science, for without it the technological progress of the past century would have been impossible. Faraday found that when an electrical conductor such as a loop of wire was moved through a magnetic field, an electric current suddenly sprang up in the wire. The faster the movement through the field, the higher was the voltage produced. He also discovered that if the wire was kept stationary and the magnet moved, current was still induced in the wire. In other words, the key to the electromagnetic induction of electricity lay in producing *relative*

Below: the world's first electric lighting power station, opened in Pearl Street, New York, 1882. Designed by Thomas Edison, it was considered a wonder of its age. Edison founded the electric light and power industry in 1880. By 1900, electricity had sales of $100 million.

Above: electrical windings in the stator of a modern generator. The flow of electricity through the windings creates a magnetic field which generates electric current in the rotor. Smaller generators are often used where it is uneconomic to lay on a main supply.

motion between the conductor and the magnet. All generators in power stations use this basic principle, employing steam or water-driven turbines to produce the relative motion needed.

The age of electricity really began in earnest on September 4, 1882 in New York when Thomas Alva Edison opened the first commercial power station. Although for some years previously factories and workshops has used their own small generating plants, it was Edison who took the bold step of establishing a central station for the express purpose of supplying current to the public. Initially, Edison probably only intended to encourage the wider use of the electric lamp that he had invented three years earlier. But demand outstripped all expectations. When the service opened, 400 lamps were connected each using about 80 watts. Within nine months, 10,000 lamps were being supplied with over 800,000 watts and customers were beginning to buy current for machines and appliances. The years following saw demands for electricity rocketing throughout the developed world. Power station followed power station until, by the 1970s, the total generating capacity of the world's stations exceeded one million megawatts.

Most modern power stations burn coal or oil to boil water, producing vast quantities of high pressure steam. As the steam passes through the blades of massive turbines, it expands and cools, transferring much of its energy to the blades that rotate a central drive shaft. Attached to the rotating shaft is a powerful magnet, on each side of which are fixed coils of wire. As the magnet spins on the shaft an electric current is induced in the coils. The electricity is then fed into a complex network of cables that make up the national power grid.

In some parts of the world hydroelectric power is as important as generation based on fossil fuels. This is particularly true of areas where such fuels are hard to come by and where instead rivers can be dammed or mountain lakes utilized as reservoirs. The working principle of hydroelectric power stations is to make use of the potential energy of stored water, converting it to kinetic energy by releasing it – usually through sluice gates in a dam – and allowing the tremendous rush of water to strike the blades of a turbogenerator.

Distributing electricity efficiently to consumers can be as complex a problem as generating it in the first place. The most common technique is to use a system of overhead cables held aloft by tall pylons. Although unsightly, overhead power lines are usually the only economic way of conveying electricity over long distances. Their efficiency depends mainly on the fact that all large-scale generators produce alternating

Above: electricity pylons stand in line across the countryside of every advanced nation. They may be unsightly but they bring vital power from the generating stations to every factory, farm, and home.

Below: Battersea Power Station in London was one of the largest in Europe when it was opened in the 1930s. It is powered by coal, which is now likely to gain new popularity.

Above: an enormously powerful 275,000-volt transformer for stepping up the voltage of the current supplied by a power station. The current is then carried by high voltage cables, slung between pylons. Current provided for domestic use is stepped down to a safe voltage.

Above: the upper reservoir of the Ffestiniog pumped storage power station, Wales. Four pumps rated at 75 milowatts apiece are used for about seven hours during the night to fill the upper reservoir. The plant is used as a standby for extreme peak demand.

current. This is a form of electricity in which the current reverses direction many times a second. One of its principle advantages is that its voltage can be changed easily and cheaply using a simple device known as a transformer. This is vitally important when current has to be carried many miles in a cable, enabling the loss of power from the cable to be kept to a minimum.

Power losses are inevitable when electricity travels through a conductor. A power line always has a resistance to the current flowing through it and this resistance turns some of the current into heat. Fortunately, the amount of power wasted in this way depends on the voltage of the current. For example, using voltages as low as 200 volts more than 10 percent of the transmitted power can be lost as heat. But by stepping up the voltage to 200,000 volts the loss can be cut to a fraction of one percent. In practice, power stations produce electricity at between 11,000 and 33,000 volts. Before being fed into the national grid, it is stepped up by transformers to about 400,000 volts. When it reaches its destination, transformers at substations step down the current to voltages safe for distribution along street power cables. A final step-down transformer reduces the voltage once more to a standard level, which varies from country to country but is usually about 200 volts.

Although clean, convenient, and relatively safe to use, electricity suffers one major drawback. It is very difficult to store. Chemical accumulators are only useful for smallscale storage and power stations cannot be turned on or off at will to meet fluctuating demands. In fact, to be most economic they must work at more or less peak capacity at all times. What can engineers do with the unwanted electricity that builds up each night as industrial and domestic demand slumps to minimal levels?

The latest answer to the dilemma has been the use of pumped storage schemes. One of the first of these was built in the 1960s in the mountains of North Wales near the town of Ffestiniog. Here a natural high-mountain lake was dammed and connected by pipelines to a lower-level man-made reservoir. The system pioneered at Ffestiniog relies on converting surplus electrical energy into potential energy and reconverting this to hydroelectricity when required. In practice, engineers use night-time electricity to power giant pumps that raise water from the low-lying reservoir to the higher one. As it is raised, the water acquires potential energy roughly equal to the electrical energy expended in pumping it. When electricity is required to meet daytime demands, sluice gates in the upper reservoir open allowing water to pour down to the lower one. On its way it turns strategically placed turbine blades and much of the water's kinetic energy is converted into electricity.

Pumped storage systems have great flexibility. All or part of the water can be released as required to supplement peak period electricity generation. Alternatively, it can be stored for long periods as an emergency standby in the event of a major generator failure in the main power station.

But although pumped storage works well in regions with hills or mountains where water supplies are plentiful, it is impractical in dry or flat countries. One possible alternative being studied at present uses the principle of pumped storage but with gas replacing water. The idea is to use overnight electricity to pump air into large underground cavities – possibly disused mines – until considerable pressure has been built up. By day, the pressure could be released, driving air-powered turbogenerators and producing the electricity needed to meet peak demands.

Both these pumping techniques store electricity by first converting it into something else. So far no satisfactory technique has been developed for the large-scale storage of electricity itself. One theoretical possibility is to use a special type of material called a *superconductor*. At temperatures close to absolute zero, superconductors lose all resistance to electric current. In principle, therefore, a current could circulate in a semiconductor coil indefinitely without losing energy. As long as they could be kept cold enough to retain their superconducting properties, such coils would be ideal for storing current. Unfortunately, at present, keeping a coil cold enough consumes so much power that the cost of storing electricity in this way would be wildly uneconomic. For the present, therefore, superconducting storage systems must await a breakthrough in refrigeration technology or the development of materials that remain superconductors even at normal temperatures.

Left: the Dinorwic pumped storage power station under construction. With the increasing demand for electrical power, engineers have been concentrating on practical means of storing power on a large scale. Water is playing a major role, chiefly in the development of pumped water storage. Conventional generating plant pumps water to reservoirs to supply power when needed.

Below: a giant electro-magnet coil rests on its cradle, ready for a journey from southern England to the Max Planck Institut fuer Plasma Physik in Munich, West Germany. The coil is for research into controlled thermonuclear fusion. Nuclear fusion can use sea water as a virtually inexhaustible supply of basic fuel. It is also "cleaner" than nuclear fission.

Batteries and Fuel Cells

Batteries are compact, portable electrochemical generators. Their working principle relies on a complicated reaction between two dissimilar metals immersed in a suitable liquid or paste, known as electrolyte. The two metals act as electrodes and when connected to an electrical circuit, provide a low voltage current.

The most widely used batteries today are dry-cell batteries. They are used for powering appliances such as flashlights, portable radios, electronic cameras, and even some wristwatches, and are a modern development of one of the earliest types of battery, designed by Georges Leclanché in 1865. The commonest dry cell consists of a central carbon rod acting as a positive electrode, surrounded by a core of compressed manganese dioxide. This is enclosed by a cylindrical shell of zinc – the negative electrode – and the inner space between core and shell is filled with the electrolyte, a gell of ammonium chloride. A single dry cell generates a steady 1.5 volts and in practice batteries are groups of cells linked to produce various multiples of this voltage depending on the require-

ments of the appliance they are to power.

A dry cell is a primary battery, one whose power is exhausted once its internal electrochemical reactions have run their course. A secondary or storage battery – such as an automobile accumulator – has to be charged before current can be drawn from it. Charging involves forcing mains current through the battery and causing certain chemical changes in it so that some of the electrical energy fed in is stored as chemical energy. After charging for a few hours a voltage of 2.2 volts can be drawn out from each cell. An automobile battery consists of plates of lead and lead oxide immersed in a solution of sulfuric acid. Usually six cells are linked to provide the 12 volts commonly required by an automobile electrical system.

Secondary batteries have an obvious advantage over the primary battery – they can be used repeatedly. However, they are limited by that very advantage because without the ready availability of another source of electricity for recharging, their lifetime is short. Traditionally storage batteries have always been heavier and bulkier than dry cells with the added danger of possibly spilling acid. Recently however, a new generation of storage batteries have emerged. Using nickel-cadmium or nickel-iron combinations for their electrodes, they are compact, extremely light and far more robust than lead-acid accumulators. Increasingly, dry cells are being displaced by small rechargable power units in appliances such as cordless electric shavers and toothbrushes.

Although useful as a portable source of electricity, both types of battery are extremely expensive ways of producing current. It is purely their convenience for

Right: a possible answer to gasoline shortages and air pollution is the electrically driven car. But engineers are still striving to develop a fuel cell of practical size and weight.

Below: a hydrogen-oxygen (hydrox) fuel cell. The electrodes are made of porous nickel, and the electrolyte is a potassium hydroxide solution. The two metals in the electrolyte act as electrodes and provide a low voltage current when connected to a supply.

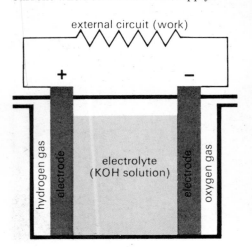

external circuit (work)

+   −

hydrogen gas — electrode — electrolyte (KOH solution) — electrode — oxygen gas

supplying modest currents that makes them attractive. For large-scale generation they would be hopelessly uneconomic.

The fuel cell is a much cheaper kind of electrochemical generator. The earliest and still most widely used is known as a hydrox cell. The first working prototype dates back to 1829 but the modern versions have only emerged in the last 30 years. The working principle remains unchanged. Oxygen and hydrogen gases are made to combine under pressure in the presence of an electrolyte to form water. In the process a current is generated at two strategically placed electrodes.

The most important application of hydrox cells so far has been in powering many of the on-board systems of the American Apollo spacecraft during the manned lunar expeditions of the 1960s and early 1970s. Their importance was dramatically emphasized when an explosion in the oxygen supply tanks destroyed the ill-fated Apollo 13's hydrox cells. For several days the three crewmen fought for survival in their dangerously under-powered spacecraft until they could be brought safely back to earth.

On the whole hydrox cells are reliable, despite the Apollo 13 disaster. They are also more compact and lighter than the equivalent in primary or secondary batteries. An added advantage is that their only waste product is pure, drinkable water.

Valuable though they are in specialized situations such as spaceflight, hydrox cells are still too expensive for large-scale generation on earth. Most current research in fuel cell technology is focused on developing new cells running on cheaply available fuels such as methane gas. Some liquid fuels such as methanol and paraffin are also being investigated. The chief problem to be overcome is the relatively slow rates at which these cheap fuels react to produce current and major efforts are being made to discover suitable catalysts to speed them up.

If the technical problems can be solved, fuel cells will offer an economic and portable means of generating quite high levels of electric power. Compact packs of cells generating about 10,000 watts would be attractive to campers and explorers and a useful emergency stand-by for hospitals and remotely situated houses or farms. Large fuel-cell complexes producing up to a megawatt could be used as independent generating units for apartment buildings and factories.

Left: a hydrox fuel cell of the type used in the Apollo 13 spacecraft. An explosion in Apollo 13's oxygen supply tanks put these cells out of action.

Below: nickel cadmium batteries being checked. They provide reliable emergency power in places as diverse as hospitals and oil rigs. The basic raw materials are nickel sulphate and cadmium oxide, which are converted into nickel hydroxide and cadmium hydroxide.

183

Solar Power

For thousands of years man's life has been dominated by the sun. Primitive peoples worshiping the forces of nature, treat it as the most powerful and benevolent of the gods. Sun-worshipers argue that theirs is the most logical of all religions since it pays homage to the source of all energy and life on earth.

The sun is the powerhouse of the solar system, radiating energy each second equivalent to burning 10,000 million million tons of coal. Only a small fraction of this radiation reaches the earth and a good deal is reflected back into space by the earth's atmosphere. But on a clear day the amount of solar energy reaching the ground is the equivalent of well over one horse-power over each square yard. This energy is divided almost equally between visible light and longer-wavelength infrared radiation with a small proportion of ultraviolet.

Efforts to tap this virtually inexhaustible supply of energy are not new. In 1615, a Frenchman, Salomon de Caus, developed one of the first solar-activated machines. Using the expansion of air caused by the heat of the sun, his device could be made to pump small quantities of water. But later interest in solar energy dwindled with the growth in easily available fossil fuels such as coal and oil. It took until the 1950s for interest in solar energy to be rekindled when it was realized that stocks of such fuels were falling in the face of a vast and ever-growing worldwide demand. By the 1970s, the developed nations of the world were faced with a real energy crisis and urgent attention was being paid to the technology needed to harness

Right: this 17th century engraving shows a design for using the Sun's heat to activate a fountain. Lenses focus the Sun's rays onto water containers. The heated water expands, rises up the pedestal of the fountain, and spurts into the pool. The arrangement of the lenses looks similar to that of a panel of solar cells.

Top: solar panels on a house in the United States. Solar panels are used to heat water which is conveyed to radiators and hot water tanks. Another way of using solar power is by converting sunlight directly into electricity by means of power solar cells. **Above:** a solar collector of the type used in the panels of the house above. It features an arched acrylic absorber of aluminium extrusions locked around $\frac{3}{8}$-inch copper tubes. Some houses rely on solar power.

solar energy. The United States and Japan are now both budgeting thousands of millions of dollars for research and development while France, Australia, and the Soviet Union are also developing national solar energy projects.

The two main approaches currently being explored are the conversion of solar energy into heat and electricity. The sun's radiation can easily be converted into heat by using a black surface to absorb it. The most important device currently available, known as a flat-plate collector, is commonly used in producing power for buildings. It consists of a blacked plate in which a continuous system of tubes is fitted carrying a fluid to pick up the collector's heat. The sides and back of the plate are carefully insulated to reduce heat loss and the device is mounted in a fixed position exposed to maximum sunlight.

The flat plate collector is likely to offer the first large-scale applications of solar energy. Already some houses have been built with panels of flat-plate collectors in their roofs and the first commercial organizations to offer domestic solar heating are beginning to advertise the benefits of "free" energy from the sun. The first house totally heated by the sun was built in 1954 in Arizona. A small bungalow, it was fitted with 315 square feet of roof-mounted collector plates. By the early 1970s, over 11,000 square feet of collector

plates were being used in the world's largest solar-heated building, the Shenandoah Solar Recreation Center in Georgia, USA.

In cold climates or those with a good deal of cloudy weather, it is impossible to depend exclusively on solar heating. In practice, most modern solar houses also have a conventional heating system to fall back on when the store of solar energy become depleted. Although fully solar-heated houses are rare, perhaps a few hundred have been built worldwide, solar-powered water heaters are being used in their thousands. Despite the high costs of installing them, householders are increasingly taking the view that even if solar heaters are barely competitive with fossil-fuel systems at present, they are likely to be well worth their initial expense over a period of years, especially with the continuing rise in the costs of coal, oil, and gas.

In addition to space and water heating, solar energy is also being used for distillation. The most important solar stills have been built in Spain, Greece, and Australia where they provide a drinkable water supply from salt water. Solar stills are the simplest of solar-energy devices. Consisting of a black basin covered with sloping transparent covers, the still is kept filled with brine that is steadily evaporated. The water vapor rises, condenses on the inner surface of the

185

Above: a tiny piece of silicon magnified several times. It forms part of a solar cell. Developed by two Scottish universities, the silicon is grown on stainless steel plates, and sandwiched between nickel to form a cell.

covers and runs down their sloping surface. At their edges the distilled water is collected in troughs and channeled into storage tanks. Although obviously limited to coastal areas or regions close to inland salt lakes and dependent on plenty of strong sunlight, solar stills are providing a valuable supplement to many local water supplies. Indeed, the conditions under which stills have to operate usually prevail where extra supplies of fresh water are most urgently needed.

A solar heating device now used in certain kinds of research into the behavior of materials is the solar furnace. Instead of utilizing a flat-plate collector, a solar furnace relies on a special focusing collector – usually a parabolic mirror – to concentrate the sun's energy onto a small area. The curved reflectors are linked to tracking devices that keep them turned toward the sun as it passes across the sky. Solar furnaces can produce sudden localized temperatures of over 5000°F and are therefore ideal for studying the effect of violent thermal shock on various kinds of metals. Although most furnaces are quite small, a huge installation has recently been built near Odeillo in the mountains of the Pyrenees in France. The furnace has a total energy-collecting area of around 20,000 square feet and it uses over 9000 curved reflectors to bring the sun's energy to a single focus.

By contrast with the relative ease with which solar energy can be transformed into heat, its conversion into electricity is extremely difficult to achieve economically. The means of producing electricity from light has been known since the last century when light-sensitive metals were identified, notably the oxides of copper and selenium. In 1954, a breakthrough in solar cell technology was made when it was discovered how to produce a fairly efficient cell from a semiconductor material known as silicon. A typical modern solar cell consists of a thin slice cut from a single crystal of silicon to which traces of the elements boron and phosphorus are added for increased efficiency. When sunlight strikes the cell's surface, electrons are knocked free of their parent atoms. This disturbs the normally well-balanced distribution of positive and negative electrical charges in the silicon

Above: Odeillo Solar Power Station in the French Pyrenees. Solar power is nothing new. Thousands of solar water heaters were in use in Los Angeles by 1910.

and causes an overall movement of the free electrons. It is this movement that constitutes a small electric current. Because it takes a certain minimum level of energy to knock an electron free, the cells only work with radiation having at least this much energy. This means they respond best to visible and ultraviolet light.

The current from a single silicon cell might be a fifth of an ampere with a power output of about 0.1 of a watt. Although far more effective than the old-style light-sensitive oxide cells, it still takes about 10 silicon cells just to light a small torch bulb. Unfortunately the cost of producing them in sufficiently large quantities to generate power on a major scale is far too great to make them economic. Their principal use therefore is confined to situations where their high cost is outweighed by practical considerations. This is particularly true in spaceflight where they can provide an indefinite supply of power. A few small-scale terrestrial uses are also emerging, mainly as novelties rather than for sound practical reasons. Among these are their use as a back-up power source for wristwatches and electronic calculators.

It is in space, however, that the solar cell is most

extensively used. Usually panels of cells directed at the sun are sufficient to power the various subsystems of artificial satellites and deep-space probes. The most spectacular use of solar panels was the Skylab Workshop program in the course of which three teams of three American astronauts spent a total of 171 days aboard their roomy orbiting space station. Launched in May 1973, Skylab is a large, roughly cylindrical structure with winglike panels of solar cells. Disaster almost struck the program even before the first astronauts boarded Skylab. During the launch, the solar panels had been damaged. In orbit, one failed to deploy while the other deployed only partially. Over 27 feet long, the panels were the largest ever used in space and were intended to provide most of the Workshop's power needs throughout its long stay in space. Eventually astronauts Charles Conrad and Joseph Kerwin, members of the first team to board Skylab, performed a hazardous spacewalk and made running repairs to the partly deployed panel that at least secured the future of the space station.

The Skylab Workshop represents probably the largest-scale use of solar cells yet attempted. On earth, where other power sources are readily available, the cost and physical inconvenience of such large panels of cells make similar applications both uneconomic and impractical. Until technologists can both make the cells far more efficient – so that fewer are needed – and cheaper to produce in quantity, the development of solar energy will continue to be centered on its conversion into heat rather than electricity.

Right: around the world sailor Robin Knox-Johnson on his catamaran *British Challenger*. He used batteries for powering the navigation equipment. Current for recharging them was provided by photoelectric cells mounted on an outrigger. Research spending on solar energy is bound to increase.

Left: extra-vehicular activity being carried out to repair damaged solar panels on *Skylab*. It was damaged during the launch. Two boarding parties, each of three men, carried out running repairs. A sunshade was mounted over *Skylab* to reduce the temperature. *Skylab's* occupants climbed out to release the jammed solar wing. In the end *Skylab* overcame its problems and was flown for three missions of 28, 56, and 84 days respectively. It came down in pieces over Australia in July 1979.

Energy from Wind, Sea and Earth

Although solar energy appears to hold the greatest promise of freely available power in the immediate future, technologists are also looking at ways of harnessing the energy of the winds, tides, and the hot interior of the earth itself.

Man has made use of the wind for centuries. With sailing ships he controlled the forces of the wind using them to carry him across the oceans. On land, windmills have been used for grinding corn or pumping water and in some parts of the world continue to be used in this way. Modern technologists are now trying to develop economic ways of using the windmill principle to produce electricity.

It is not difficult to build a simple wind-powered generator although modern designs are radically different to traditional windmills. The emphasis today is on designing blades that will swing freely in light breezes without becoming uncontrollable in stronger winds. Such blades tend to be long and narrow, pitched rather like aircraft propellors. They are linked to a rotating shaft that drives a conventional power

Above: a wind powered generator in the United States. This generator, like its contemporaries, has been designed to function even when there is very little wind. Its advantage is that the supply of wind is limitless.

Left: Saxtead Green Post Mill is still in operation. Even the slightest breeze has enormous energy potential, needing only to be exploited by appropriate design.

generator, producing a current that fluctuates with the prevailing wind speed.

In the United States, in particular, considerable advances have been made in designing practical generators. On the windswept plains near the town of Clayton, New Mexico, the American Department of Energy has built an experimental installation. Whenever the wind speed exceeds 8 miles per hour, the machine's two 63-foot blades rotate, feeding enough electricity into the local grid to supply 60 of the town's 1300 homes. By the end of 1978 similar generators were in action near New York and on Culebra Island in Puerto Rico. In Boone, North Carolina, a giant installation with 100-foot blades is producing 2,000,000 watts. Encouraged by their early success, some experts are claiming that 10 percent of America's total energy need could be supplied from wind power before the end of the century. In Europe the possibility of off-shore windmills is being studied. Proposals have been made for clusters of up to 400 windmills sited on platforms a few miles out to sea, close to the centers of on-shore demand such as major coastal cities. Researchers estimate that windmill installations of this kind could produce an average output of nearly 500,000,000 watts.

Although the wind itself is free, the cost of building, installing, and maintaining windmill systems for large-scale generation is extremely high. Additional problems of storing an output which fluctuates with wind conditions or trying to match it to demand, make wind-powered generators one of the least promising new sources of large-scale power supply. For the forseeable future they remain valuable supplements to more conventional generating systems.

Although technically more challenging, harnessing the energy of the tides appears to offer a more economic source of power than wind generators. Already two major tidal power stations are in operation. The first was completed in 1967 on the north coast of Brittany, near St. Malo, France and produces about 240,000,000 watts – enough to power a large town. A smaller experimental station was opened in 1969 at Kislayaguba in North Murmansk in the Soviet Union.

Tidal power stations make use of the tidal rise in river estuaries to fill large reservoirs. When the tide falls, water is released from the reservoir and electricity is produced by conventional hydroelectric generators. The siting of such power stations therefore depends on finding coastal locations where very high variations occur between high and low water levels.

Above: any country with a long coastline can potentially supply a significant proportion of its electrical generation through wave-power – particularly if there are high tides. Britain, for example, could supply half its needs from a 600-mile stretch of ocean. These diagrams show a rectifier in which waves can create a hydraulic "head" to drive a turbine. Water flows between reservoirs through two sets of non-return valves.

Left: a laboratory model being used for research into tidal power. Mr Stephen Salter of Edinburgh University is conducting experiments into the "nodding duck" system. This relies upon a free-floating concrete sea station, from which a series of vanes attached to a "backbone" are strung out. The vanes measure about 600 yards long, 12 yards wide, and 30 feet high.

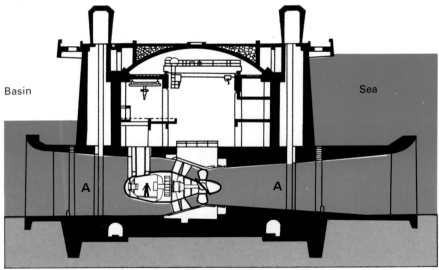

Above: an aerial view of the Rance barrage in France. Visible from left to right is the navigation lock, generating station, fixed dyke, and sluice gates. At high tide, seawater flows into the basin and the sluice gates are closed. At low tide, the water flows through the turbines in the generating station back into the sea. Power is generated by the constant flow of water through the turbines.

Left: diagrammatic cross section of the Rance tidal power barrage. One of the 24 bulb turbines is shown. Each of these contains a turbo-alternator of 10 milowatt capacity. The sluice gates (**A**) are opened when the difference in water levels between the sea and the reservoir is sufficient for the turbines to operate.

The French scheme consists of a 2500-foot-long dam across the estuary of the Rance river where a maximum tidal variation of over 40 feet is common. The power station itself is built into a hollow section of the dam and turbine generators produce electricity both as the tide comes in and when it goes out.

The major limitation on tidal power stations is finding locations that have sufficiently large tidal variations and are also close to centers of population that will consume the power produced. Although only two stations are currently in operation, plans are being studied for new sites near the head of the Bay of Fundy in Canada and on the British coast at the estuary of the Severn river. The Soviet Union has announced firm plans to follow its experimental station with a much larger installation on the White Sea Coast near Mezenskaya. Russian engineers are

hoping for a peak output of up to 6000 million watts from their new station.

Because it is difficult to find economic locations for tidal power stations, the energy of the sea is less likely to make a major contribution to world power supplies than the geothermal energy locked up in the earth itself. Wherever we are in the world, we are standing on a virtually limitless supply of heat energy. If we drill into the earth's crust, the temperature rises by about 1.5°F for every 100 feet we go down. At a depth of about 10 miles therefore, rock temperatures are over 800°F.

If a sufficiently large borehole – perhaps two or three feet in diameter – could be drilled this deep, we would have the basis of a geothermal energy well. Cold water could be poured down the borehole to absorb the heat of the underlying rocks. Because pressure near the bottom of the well would be over 16,000 pounds per square inch, the water could become superheated to temperatures of around 700°F without turning to steam. An insulated inner pipe in the well would allow the hot water to rise to the surface, the density difference between the cold and hot water being sufficient to drive the hot water to the surface without the need for a pump. Heat from the emerging water would be collected by a system of heat exchangers to produce a plentiful supply of steam which, in turn, would power a conventional turbine generator.

The spent water, still under pressure and at a temperature of nearly 400°F, could generate yet more power in hydroelectric plants and also provide plentiful supplies of industrial and domestic hot water.

Unfortunately the technical problems of drilling widebore shafts to great depths are formidable. As yet no economic techniques have been worked out. But the possibility of being able to literally mine the inexhaustible wealth of geothermal energy is so attractive that engineers throughout the world are undertaking serious programs of research into new kinds of drilling technology.

Below: the Rotorua geyser in New Zealand with the steam escaping into the atmosphere. A number of countries are exploiting the power provided by their geothermal wells. It is not yet certain how much energy the world might be able to glean from the steam and hot water that lie just beneath the Earth's crust.

Above: the geothermal power stations at Larderello in northwest Italy. Larderello's source of energy, steam, gushes from specially drilled wells and is used as the motive power for turbo-generators. These power stations generate enough electrical power to operate most of Italy's railroad system.

Although deep-drilling techniques to tap geothermal energy must await some future engineering breakthrough, large numbers of small geothermal power stations are already in action in regions of natural volcanic activity. In such areas the high temperatures otherwise encountered deep below the earth's crust, occur within a few thousand feet of the surface or perhaps even at the surface itself. Most existing geothermal installations consist of wells drilled to a sufficient depth to tap the vast underground reservoirs of superheated steam which characterize volcanic regions. Once brought to the surface, the steam runs conventional turbine generators producing anything from a few hundred to several megawatts of electrical energy.

Geothermal wells of this kind exist in many parts of the world. Their drawback is that they are limited by their nature to volcanic regions and these are often the least populated regions and so the lowest in energy demand. However, there are many exceptions and in Lardarello, Italy, over 400 generating sites have already been developed. By the 1960s the American Department of Energy had established about 600 sites in the area of California known as The Geysers, with firm plans for over 1000 more. Some energy planners estimate that with further development these natural geothermal power stations could meet up to 25 percent of the Western world's current energy needs.

191

Nuclear Power

On December 2, 1942, the era of nuclear power began with the cryptic message, "The Italian navigator has entered the new world." These words from scientists at the University of Chicago were a prearranged signal marking the successful opening of the world's first nuclear reactor. The Italian "navigator" heading the team of researchers responsible for the breakthrough, was the physicist Enrico Fermi. Only four years earlier, Fermi had been awarded a Nobel prize for his remarkable discoveries about nuclear fission, the working principle of the first reactor.

In the 1930s, Fermi and others had found out that atoms of an isotope of uranium, known as uranium 235 (or U235) would split into two lighter atoms when bombarded with neutrons. The splitting into two became known as nuclear fission. In the process, more free-moving neutrons are produced together with a flash of energy. Fermi realized that the newly produced neutrons could go on to split further atoms of U235 and this effect could rapidly "snowball" leading to a runaway nuclear chain reaction. The energy released from such a process would be huge and catastrophic.

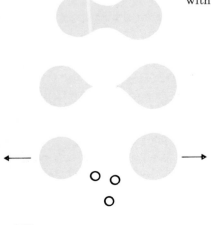

Left: this diagram demonstrates the process of nuclear fission. A neutron enters a uranium nucleus, and the compound nucleus splits into two fragments of similar size. Two or three neutrons are also emitted. Nuclear fission is considerably more efficient than burning fossil fuels.

Below: the world's first nuclear powered submarine USS *Nautilus* was equipped with a water cooled reactor.

Uncontrolled, the chain reaction would produce a devastating explosion. But restrained and controlled, the energy release could mean a vast new store of power for mankind.

To build the first reactor, Fermi had to solve two important problems. First he had to produce a means of actually making a chain reaction possible. Having done this, he had to work out how to control the reaction, slowing it down so that the energy produced could be tapped from the reactor's core before temperatures became too high and the device turned into an impromptu atomic bomb. By solving these problems, Fermi laid the foundations for all modern nuclear reactor technology.

In nature, uranium occurs usually as an isotope called U238. Atoms of U238 rarely undergo fission. About one of the uranium atoms in 140 is the more unstable isotope U235 needed for the chain reaction. Neutrons produced by the splitting of a U235 atom, move very fast. If they happen to strike another U235 atom further fission take place. But if, as is more likely, they strike U238 atoms, they are simply absorbed. Slow-moving neutrons, however, still split U235 atoms but are not absorbed by U238. Fermi therefore created the conditions for a chain reaction by surrounding the reactor core with graphite, a form of carbon that slows down the neutrons as they bounce off it. The graphite was called a moderator and the neutrons richochetting from it back into the mass of fuel were safe from being absorbed and free to sustain a U235 chain reaction.

Fermi controlled the rate of reaction by adjusting the number of free neutrons in the core. He did this by inserting moveable control rods of an element such as boron that readily absorbs neutrons. By raising or lowering the rods, the chain reaction could be slowed down or speeded up as required.

Nuclear fission generates heat just like burning a fossil fuel. The difference is that much less nuclear fuel produces vastly more heat. The heat generated is converted into electricity by conventional turbine generators. The heat is transferred by a kind of coolant from the reactor core to a simple water boiler. In many reactors water is itself used as a coolant and in America this type of system was used to power the world's first nuclear submarine, the United States Navy's *Nautilus*.

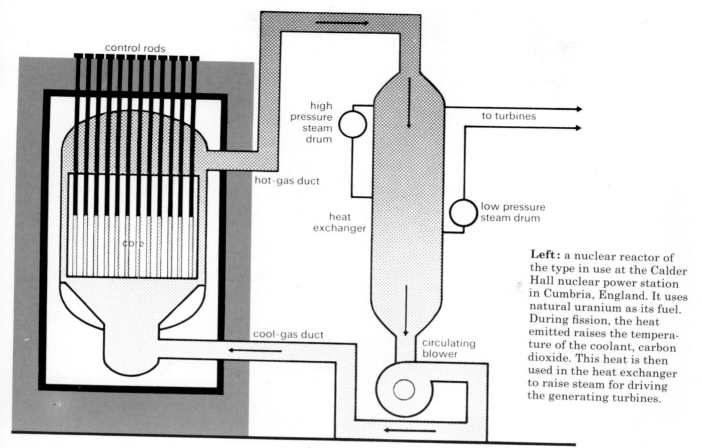

control rods

high pressure steam drum

to turbines

hot-gas duct

heat exchanger

low pressure steam drum

core

cool-gas duct

circulating blower

Left: a nuclear reactor of the type in use at the Calder Hall nuclear power station in Cumbria, England. It uses natural uranium as its fuel. During fission, the heat emitted raises the temperature of the coolant, carbon dioxide. This heat is then used in the heat exchanger to raise steam for driving the generating turbines.

Below: fast reactor research at Dounreay Experimental Reactor Establishment, Scotland. Scientists are working on providing suitable rigs for the irradiation of fuel. This is a post irradiation examination of active fuel pins through a five-feet thick zinc bromide window.

In Britain and France, most reactors are gas-cooled. Usually carbon dioxide gas is introduced into a closed circuit of piping from reactor core to the water boiler. Electrically powered blowers circulate it through pipes around the fuel elements in the reactor's core, down through the boiler and back to the reactor. As the gas circulates around the uranium fuel it not only picks up heat, it also becomes increasingly radioactive. By monitoring the level of radioactivity, engineers judge when rods need to be raised or lowered to adjust the rate of the chain reaction.

A major breakthrough in reactor technology took place in the 1950s when the first fast-breeder reactors were developed. Instead of using uranium as fuel, fast breeders use a radioactive metal called plutonium. The remarkable feature of such reactors is that in the course of their heat-producing work, they actually manufacture more fuel for their future use.

Plutonium usually exists as a single isotope called Plutonium 239. The isotope undergoes fission much like U235, each atom splitting to produce neutrons that can continue the chain reaction. Unlike uranium, however, in plutonium no other isotope is present to capture neutrons and interrupt the reaction. This means that a moderator is unnecessary. The most revolutionary aspect of the reactor is its capacity to produce more fuel than it actually consumes. When the plutonium core is surrounded with uranium, some neutrons from the plutonium fission strike the uranium case converting U238 atoms into more plutonium 239.

Above: fears about the possibility of a serious accident have resulted in stiff opposition to the use of nuclear power for generating purposes. The first accident in Britain took place in 1957 after a fire at the Windscale power station in Cumbria. Radioactive dust escaped. The wind dispersed most of it over the sea, but some settled on grassland. The milk of the grazing cows became radioactive – putting the local population at risk.

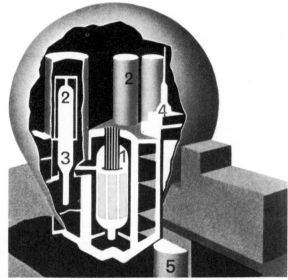

Left: a diagrammatic representation of one of the gas cooled reactors at Windscale, Cumbria. The reactor uses uranium oxide and heats the cooling gas (carbon dioxide) to a higher temperature than the Calder Hall reactors nearby. The gas enters at the bottom of the pressure vessel **1**, passes up between the fuel elements and graphite, then on to the heat exchangers **2**, and back through the circulator **3**. Fuel elements can be unloaded **4** and stored in a deep pit **5**. The whole apparatus is encased in a steel sphere for added safety.

A prototype breeder reactor was built at Idaho in 1951 by the Canadian-American physicist Walter Zinn. A few years later in 1959 a much larger reactor and operational power station was built at Dounreay in the North of Scotland.

Although the development of nuclear reactors is now taking place throughout much of the developed world, protest movements are emerging opposing their proliferation. Fears are growing about the large-scale use of nuclear power. Many people are not convinced that adequate precautions are taken against technical failures that could lead to terrifying accidents. Although the chances of a reactor going out of

control and turning into a vast bomb are very small, real possibilities do exist for radioactive waste being vented into the atmosphere. Widespread death from radiation sickness could follow such accidents. In addition, the waste products of reactors remain radioactive and potentially lethal for thousands of years and few nuclear scientists have satisfactory proposals for the safe disposal of the waste or for its secure storage over such huge periods of time.

Public concern over the rapid growth of fission reactors is unlikely to halt their development. But within the next 100 years it is likely that a far safer and more efficient reactor will emerge. Harnessing the

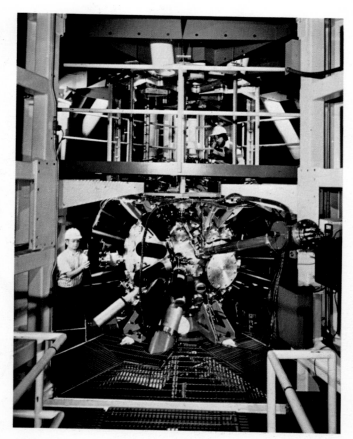

energy-generating process of the stars, nuclear fusion reactors will provide a virtually inexhaustible supply of power derived from one of the commonest substances on earth: ordinary seawater.

Fusion is the exact opposite of fission. Instead of an atom splitting into two fragments, two atoms merge together to form a single atom of a heavier element. In the process a tiny fraction of the mass of the two atoms is converted directly into energy. The most common fusion reaction in the universe involves the lightest of the elements, hydrogen. Nuclei of hydrogen atoms fuse to form an atom of the next heavier element, helium. In the process every millionth of a pound of hydrogen mass converted into energy is equivalent to burning well over one ton of coal.

The key to producing a controlled fusion reaction on earth is to find a way of generating and maintaining the very high temperatures – in the order of millions of degrees Centigrade – without which fusion cannot take place. This means heating the fuel and then containing it without loss of heat. If this can be done, even for only about a second, the fusion reaction would start and its own energy releases would sustain the necessary temperature. Once these problems are solved, fusion reactors would be much more convenient than their fission counterparts. Producing much more power and no radioactive waste, fusion reactors would also be reassuringly safe in operation. If any fault developed, the reaction would simply "go out."

Achieving the high starting temperature needed for fusion is already possible with existing technology. One technique is to use a giant multiple laser. The most recent of these, the American device called Shiva, went into operation in May 1978. Blasting a hydrogen target with a pulse of 26 million million watts of light energy, numerous fusion reactions were initiated.

The remaining problem facing fusion technology is therefore to contain the extremely hot fuel long enough for a fusion chain reaction to take place. If a material container was used, it would instantly vaporize and the fuel's temperature would drop below the fusion level. One answer seems to be the use of a magnetic field to bottle up the fuel.

When a gas reaches fusion temperatures the electrons normally surrounding its atoms have been stripped away leaving bare, positively charged particles, no matter what their temperature. Modern fusion research is therefore directed toward producing a field of the right shape and strength to hold the fusion gases without leakage. So far it has only been possible to do this for a fraction of a second before instabilities destroy the magnetic "bottle." When this has been extended to a second or more, the first practical fusion reactor will be in sight.

Despite the problems still to be overcome, controlled nuclear fusion is widely regarded as one of the greatest technological prizes awaiting man. Once operational, fusion reactors will provide as much energy as we will ever need by tapping the isotopes of hydrogen bound up in seawater. In space, fusion-based engines may well boost astronauts of the not-too-distant future beyond the solar system on the first voyages to the stars.

Below: all over the world there is growing public concern about the possible dangers of nuclear power. A leakage from a nuclear fusion power station would create havoc. This protest took place in Trafalgar Square, in 1978.

Power from Space

Although techniques are already emerging on earth for using the energy of sunlight, converting it to heat and electricity, the real future of solar power generation lies in space. Here the full intensity of the sun's radiation can be utilized. There is no atmosphere to filter its strength and no night or day to limit the hours of power generation.

Solar cells in winglike panels are already used on a small scale to power many of the onboard systems for space vehicles either in earth orbit or during interplanetary missions. But plans are now being studied for large-scale power generation in space, using microwave beams to return the power to earth for distribution to its towns and cities.

The earliest serious proposal for solar power from space was made by American engineer and businessman Peter Glaser in 1968. Working with a small design team, Glaser outlined a scheme for a giant power-generating satellite, called a *powersat*, to be placed in a geosynchronous orbit 22,300 miles up. In this orbit the powersat would always be above a fixed point on earth where a receiving station would be sited to pick up the powersat's energy beam.

Glaser's plan is to use solar cells to generate electricity. Although this in itself is hardly an original concept, the scale on which he proposed to use the cells is revolutionary. Glaser describes two large panels extending from the body of the powersat, each about three miles square. These panels would not be entirely covered by solar cells. Part of their area would be taken up with banks of mirrors which would concentrate the sun's rays on to the cells, enhancing the electrical output. The electricity would be fed into a central transmitting antenna, converted to microwave radiation and beamed to the waiting ground on the earth's surface. The microwaves would then be converted back to electrical energy ready for

Below: an artist's impression of a solar power station satellite. It shows solar cell panels and reflector panels. The satellite should be launched in the 1980s, and will be located about 36,000 miles above the Earth's surface.

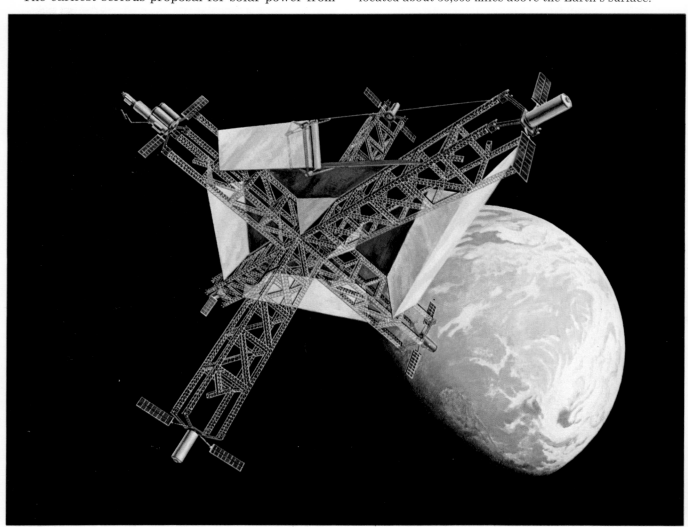

distribution in the national power grid.

The technique of using microwaves to transmit electrical energy has already been successfully demonstrated in America. The two real drawbacks of Glaser's scheme are the prohibitive cost of making solar cells on a large scale and the difficulty of placing such a huge structure as the powersat in orbit. Both these problems may well be resolved in the next few decades.

Following Glaser's proposals, another American engineer, Gordon R. Woodcock, showed how a powersat project could at least avoid the high cost of solar cells. He suggested building a huge solar furnace in space. A powersat in Woodcock's plan would be a fairly orthodox electrical generator, its turbine powered by heated helium gas. To produce maximum heating of the gas, the sun's radiation would be focused on to the generator by large aluminum-coated films of plastic floating nearby in space.

To achieve either of these proposals would stretch our present technology to its limits and our national economies to breaking point. The result, however, would be boundless supplies of energy which, once the initial investment had been recouped, would be very much cheaper than terrestially-generated power. The emergence of NASA's space shuttles offers a real possibility of ferrying into orbit the men and materials needed to build powersats. Some experts are already forecasting that within the next century a large proportion of our terrestrial power supplies will be generated in space.

Other schemes currently being evaluated include proposals to create a series of minor nighttime suns. The idea is to launch satellites consisting of huge reflectors capable of directing sunlight onto selected darkened areas of the Earth. Three possible levels of illumination are contemplated. A satellite called *Lunetta* would produce on a clear night a brightness equivalent to between 10 and 100 full moons over a land area of about 1100 square miles. Such illumina-

Below: the four spot beams of this solar power station satellite illuminate, and thus provide power, to selected areas of the Earth's surface. Solar cells are mounted on solar paddles, which always point toward the Sun.

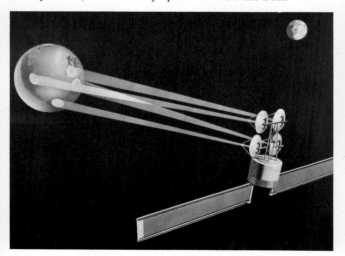

tion would make agricultural and general construction work possible throughout the night hours. An upgraded version of Lunetta could provide about 50 percent of the Sun's daytime brightness. This would be useful in enhancing the growth of selected crops. The ultimate level of illumination uses a vast number of solar reflectors – possibly covering as much as 20,000 square miles of the night sky – providing up to 70 percent of daytime sunlight. If this could be achieved, solar power generators on Earth could operate around the clock as if they themselves were in space.

All such plans, however, must await considerable technological advances. Still further into the future lies the possibility of utilizing space on a much grander scale. Some scientists suggest that taken to its logical extreme, harnessing the Sun's energy will lead man to actually dismantle one of the outer planets, probably Jupiter, and use the material to build a loosely structured sphere around the Sun. Space-borne communities living on the surface of the sphere would be able to utilize virtually all the radiated energy of the Sun instead of just the small fraction that reaches the vicinity of the Earth. Far-fetched though the idea sounds, the technology needed to accomplish the awesome scheme may well be in man's grasp within a few hundred years.

Below: an artificial sun designed at the Solar Energy Unit at University College Cardiff, South Wales. This "sun" enables scientists to reproduce sunshine intensities to be found anywhere in the world.

CHAPTER 8

MATERIALS

The key to man's progress from cave-dweller to the architect of a complex industrial civilization, has been his growing ability to use the materials around him. For most of human history, knowledge of working with glass, clays, metals, and simple composites such as wood laminates, has depended on trial and error, inspired guesswork, and occasional lucky accidents. Not until the late 1800s did the study of materials emerge as a true science in itself. Today materials science can tell us why some substances are hard, others soft, some brittle, others malleable and how their internal structures can be changed to suit them to specialized technological purposes.

Increasingly, unique new materials are being "made to measure" for specific engineering uses. Future possibilities seem almost limitless. Man's growing mastery over materials will be the most important single factor in assuring continuing technological progress in tomorrow's world.

Opposite: perhaps the most controversial building in Paris, the Pompidou Center. The building makes imaginative use of metal, plastic, and glass. Heating and plumbing systems are exposed to view with pipes of many colors and widths.

Materials for a Changing World

Some time more than 20,000 years ago – the exact moment is lost in prehistory – men realized that rocks and stones could be useful weapons with which to attack enemies or hunt small animals. Once the basic potential of stone was appreciated, man shaped and honed it into tools for various purposes. Today these seem trivial advances but they were the springboard from which all our modern technologies have developed.

The transition from working stone to using metals was extremely slow. Bronze was probably not known widely until about 3000 BC. Iron, a metal that still dominates everyday life even today, came into use soon after. Working in metal together with the skills of pottery and carpentry, transformed man's potential as an engineer. He could make implements and utensils to improve the quality of his life, constructing the first complex tools and even the earliest machines – levers, pulleys, and simple pumps. As centuries passed man became increasingly skilled in adapting his physical environment. His houses could stand up to all weathers while even difficult soils could be cultivated with the right equipment such as the first plows.

Slowly civilization emerged and grew more complex and well-organized. But man's technological prowess was always limited by his skills at working with existing materials and his ability to find new and better

Above: the forging of bronze and iron during the Bronze Age. The use of metals like copper, and later bronze and iron represented the first real advance over stone as a working material. Also, the early forging of metal demanded real skill, whereas the fashioning of crude stone tools was within the capabilities of everyone.

Left: this ancient Egyptian pictograph shows the use of the earliest plows. They are scratch plows and are being drawn by oxen (horses were needed by the army). The plow first appeared in the fertile valleys of the Middle East. It probably developed from the simple plow that was dragged over the ground to break up the soil.

Above: polyvinyl baskets on sale in a Brazilian market. The plastic itself is harmless but the gas given off when it burns is so dangerous that exposure to even one part per million is risky.

ones. Such skills and discoveries were almost always based simply on experience and a trial and error approach to experiments. If a particular mixture of substances produced a more transparent kind of glass or a stronger metal, the ingredients were remembered and a tiny technological step forward was taken.

Even the rise of modern science in the 1600s did little to make people think about why materials behaved as they did. But as science revealed more about the structure of matter and concepts such as force and mass were examined more closely, some of the answers to the mystery of materials behavior began to emerge. The study of materials as a science in itself however was still delayed because it demanded much more than a narrow expertise. The first materials scientist had to be physicist, chemist, and engineer in one.

The rise of the modern theory of atomic structure in the early 1900s was the real impetus materials science needed. Once the basic structure of materials could be understood generalizations could at last be made about their properties – strength, hardness, brittleness, for example. The behavior of a particular material could be explained by the shape and structure of its atoms and molecules. Most exciting of all, this new insight made it possible for scientists to manipulate structures, to change the properties of materials, and to see new potential in them.

A dramatic example of man's new active role in the world of materials was the development of plastics.

By altering molecular compositions and even the physical shape of molecules, chemists could produce plastics with widely varying properties.

In the field of electronics, man's understanding of the behavior of electrons in solids revealed tremendous engineering potential in the materials known as semiconductors. Today the technology surrounding the use of tiny chips of silicon, etched with miniature electronic circuitry has revolutionized everyday life. Pocket-sized calculators and desk-top computors are only the beginning of a vast range of remarkable devices made possible by the latest semiconductor advances.

The pace of change in modern materials technology is so rapid that a lesson vital to our future is being learned only by painful experience: all materials exist in a "materials cycle." Man takes ores, hydrocarbons, oxygen from the Earth and its atmosphere. He then extracts, refines, purifies, and converts them into useable materials for various engineering needs. In one way or another, man uses the Earth's resources to make products on which today's world depends. But when the lifetime of his products is over, they return to the earth or atmosphere in an often unreuseable form. They are lost to the materials cycle and our resources of materials is steadily depleted. Even worse, product waste can also pollute the environment making it both unpleasant and dangerous to life. So as man becomes the master of materials, the pressure on the environment grows. For the first time man is confronted with the hazards of a high-technology society and more than ever before must question the long-term implications of his actions.

201

Ceramics in Industry

The technology of ceramics began thousands of years ago when man first began shaping clay into utensils. Today men still shape clay but the world of ceramics has become increasingly dominated by a complex new technology having little in common with the craft of the potter or the maker of fine china.

Ceramics are excellent insulators and for this reason they are widely used in the modern electrical industry. Porcelain – made from clay, feldspar, and flint – is used in light bulb sockets, switches, and in components for such devices as radios and televisions. It is also employed to support high-voltage power lines. The bigger the voltage it has to insulate against, the larger an insulator must be. In some overhead lines current is transmitted at up to 800,000 volts and single insulators large enough to cope with this are extremely difficult to manufacture. In most cases, therefore, composite units made of numerous porcelain sections are produced. Most common of these are the strings of disk insulators used on power lines over much of Europe and America.

In addition to their insulating properties, ceramics also remain strong even at very high temperatures.

Traditionally this has led to them being used as refractory bricks lining the interiors of industrial furnaces. But increasingly technologists are developing new kinds of ceramics with still higher resistance to thermal shock. For example, exotic blends of ceramic and metal called cermets have been produced for such specialized uses as cladding the inside of rocket engines.

One of the most promising of the new ceramics is silicon nitride, a combination of silicon and nitrogen, two of the most common elements to be found on Earth. It emerged as an engineering material in the 1950s and its remarkable properties have led to a wide range of uses. Not only does it stand up to very high temperatures but it hardly changes its dimensions even when rapidly cooled. This means it can be plunged red hot into cold water without shattering. Most important of all, it lends itself to very precise shaping in the course of its manufacture. It is made by a pro-

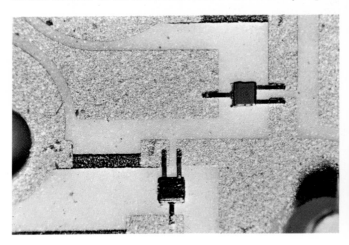

Above: ceramics are widely used in electronics. This picture shows beam-led transistors (dark brown) bonded to gold patterns laid down on a ceramic substrate. The blue strips are tantalum thin film resistors.

Left: ceramic disk insulators seen on power lines throughout Europe and the United States. Insulators like this must be rigorously for resistance before use. Voltages above 100 kilovolts were uncommon early in this century, but today insulators have to withstand voltage in excess of 1000 kilovolts.

Above: this electric arc steel furnace weighs nearly 100 tons. It is seen here being tapped. Ceramic refractories line the inside of the furnace to provide the necessary heat resistance. The walls of chrome-magnesite, and the ceilings are 15 per cent silica and 85 per cent alumina.

cess called reaction sintering, a two stage technique in which, in the first stage, the substance has the hardness and consistency of chalk. While in this state it can easily be fashioned into whatever elaborate shape is required. In the second stage, the substance is heated in an atmosphere of nitrogen and changes into a hard ceramic. This two-step process offers tremendous economic advantages over ordinary ceramics that assume their full hardness in a single process and then have to be machined by laborious diamond-grinding techniques.

The largest group of modern ceramics are those formed by combination with oxygen. These oxide ceramics include probably the most widely used ceramic of all: alumina, an oxide of aluminum. Far stronger than most other cermics, alumina has the usual properties of resistance to heat, chemical attack, and abrasion. It is therefore valuable for use as a tip for metal-cutting tools, for bearings and antiabrasion fittings on a wide range of machinery. The most advanced of the alumina ceramics is so pure that it is translucent. It is used as an envelope for high temperature lamps and specialized vacuum devices such as powerful electron tubes.

Another important group of new ceramics are the ferrites. Based on compounds of iron, ferrites differ from all other ceramics in one major respect: they are magnetic. In one form or another, they are used in a variety of electronic equipment but it is in computers that they have become most important. Most modern computers depend on thousands of tiny ferrite elements serving as the computer memory. Constantly magnetized and demagnetized by changing electric fields, their fluctuating magnetization is used as a means of storing basic data.

In nuclear reactors, special ceramics are being used as nuclear fuels. Conventional metallic uranium fuel has two main drawbacks: a low melting point and a tendency to swell when heated. Early reactor cores using metallic uranium often suffered melting or buckled fuel elements. Ceramic forms of uranium suffer from neither of these problems and still undergo nuclear fission producing heat. Nowadays, therefore, uranium dioxide or, more often, a uranium carbide is widely used as a reactor fuel. Alternatively, ceramic containers can be used to enclose vulnerable fuel elements protecting them from the high temperature of the reactor's core. These are often made of beryllia, an oxide of the metal beryllium. Its important characteristic is that it combines a resistance to heat with high thermal conductivity. This means it is ideal for conducting heat away from a fuel element while itself standing up to the stresses of extremely high temperatures.

Below: a photomicrograph ($\times 350$) of a ceramic can used in high temperature gas-cooled nuclear reactors. Thorium-uranium dicarbide fuel particles are within a layer of carbon (gray circle) and silicon carbide (white circle).

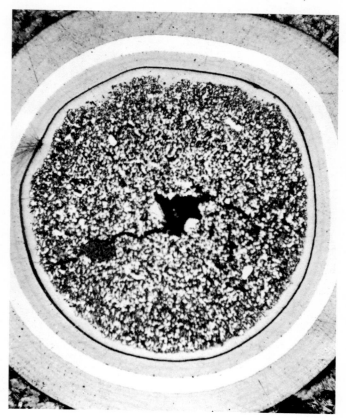

Glass

The basic raw materials of glassmaking have not changed over the ages. Soda, lime, and silica are still heated together to form the hard, transparent substance we call glass. But today various chemicals added to the mixture help to create a wide range of different types of glass and modern mass-production techniques have turned glassmaking into a major industry.

Glass has many advantages that make it useful for mass-produced goods. Cheap and easily made, it can be shaped into many forms by relatively inexpensive techniques. It has a hard, smooth surface that resists attack by most common chemicals and does not flavor

or contaminate food or other substances.

Modern mass-production processes rely on a number of ingenious machines, each designed to produce a particular kind of utensil or type of glass. The simplest, for making bottles or jars, consists of a mold into which molten glass is dropped. A suitably shaped plunger is then thrust into the mold and once cooled the now-solid glass takes on the required shape.

Flat glass is widely used for windows. Two important techniques for manufacturing it are the float glass process, where liquid glass forms a flat layer floating on a bath of molten tin, and the continuous casting process where molten glass is fed through rollers that press the glass into an unending flat sheet.

The uses of glass have widened enormously since the Middle Ages when the most sophisticated applications were in the first primitive telescopes and spectacles. The modern lens industry alone has made spectacular advances. Today computer-controlled grinding techniques produce precisely calculated curvatures in glass lenses for specialized use in cameras, telescopes, and binoculars. The optical glass itself can be made of unprecedented purity, virtually eliminating unwanted distortions.

One of the newest developments is the fast-growing use of optical fibers, flexible glass threads so fine that bundles of them can pass through the needle of a hy-

Below: a stage in the production of float glass. This picture shows the inside of the float bath. The hot glass is floating on molten tin. Temperature is controlled by heaters above.

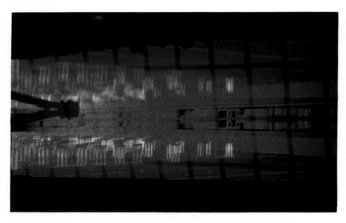

Below: specially toughened glass has had to be devised for the production of automobile windscreens. This windscreen is a glass/plastic/glass sandwich, and is assembled in a clinically clean air-conditioned room.

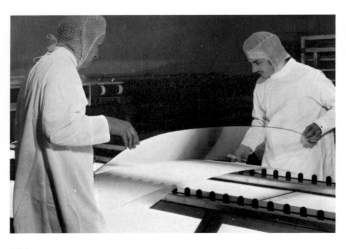

podermic syringe. Each thread forms a kind of pipe along which light is transmitted by repeated reflections from the pipe's walls. Although the light bounces from the walls the composition of the glass is such that no intensity is lost through them. Individual fibers used today range in diameter from four hundredths to four millionths of an inch. The flexible bundles, sometimes of hundreds of thousands of fibers, can even be arranged hornshaped so that an image transmitted through the small end is enlarged at the bigger end. The most common use for fiber optic systems is peering into inaccessible places. Engineers use them to examine the condition of the insides of machines and fuel tanks. Surgeons actually look deep inside their patient's body, for example by easing a fiberoptic tube down the patient's throat into their lungs or stomach. A future application currently being studied is to use optical fibers in computer circuits. Instead of electrical pulses passing through metal wires, light signals would carry information along glass fibers.

The high stresses that can be set up in glass are used for man's benefit in toughened safety glass for automobile windscreens. Curved glass sheets are made with their outer surface in compression and inner surface under tension. Since glass always fails in tension, the protective compression in the outer surface must first be overcome by an impact before sufficient tensile stress to cause a fracture can occur. Such toughened glass is therefore very strong. When it does break, the energy stored in the glass by the initial stresses, causes it to shatter into small rounded pieces that are not dangerous to the driver or passengers. Super-toughened glass has surface layers in very high compression. The forces are built up by a chemical process during which the glass is immersed in a molten compound of lithium. Chemically toughened glass can be extremely flexible and have a strength of as much as 100,000 pounds per square inch.

One of the best known of modern developments has taken place in the field of photochromic spectacles. Increasingly, sunglasses are being made that darken as the light intensity rises. Indoors the lenses are virtually untinted. But when exposed to sunlight they darken in moments. Photosensitive glass from which the special lenses are made, contain minute crystals of silver compounds that darken with increasing light intensity and, more important, recover their transparency when the intensity falls.

Once an ancient craft and now a fast-moving modern technology, glassmaking provides a continually expanding range of decorative and functional glasses for our modern world. Whatever future advances emerge, glass is certain to remain one of man's most versatile materials.

Above: glass flasks on the production line. The unwanted cone is removed from the neck by scoring the glass with a diamond, and passing the neck through a battery of gas jets. The heat cracks the cone from the flask's neck.

Left: members of the "Glasshouse" in London forming glass objects. They are producing a wide variety of ornaments and containers by blowing and molding. For small items a paste mold is used. Steel molds are employed for heavy pressed glass work.

Metals

Iron, one of the oldest of metals known to man, remains the most important metal in the modern world. For nearly 200 years it has been the backbone of industrial progress. Iron-working is itself an ancient craft and our most advanced foundries have done little more than mechanize, automate, and extend old practices to match output to the massive demands of today's highly industrialized society.

Research and development has not stood still in the constant search for improved engineering materials, however. Steel, basically, is an alloy of iron and carbon. In practice, because crude iron contains a high proportion of carbon, steel is made by removing most of the carbon from the iron. The precise amount of carbon in steel, rarely more than one percent, radically changes the type of steel produced. Steel varies from relatively soft, mild steel with little carbon in it to high-carbon steel, so hard it is made into cutting tools used for anything from shaving a man's whiskers to slicing through other metals.

Much development work is concentrated on the search for a supersteel, with a strength of around 500,000 pounds per square inch – twice the strength of the toughest steel in general use today. A major advance was the emergence in 1961 of the so-called maraging steels, now widely used for rocket casings and lightweight armor. These ultrastrong steels contain virtually no carbon at all. Instead, steel's traditional strengthening element is replaced by a range of metals in varying amounts. Top-strength maraging steel, for example, contains over 19 percent nickel, 9 percent cobalt, 5 percent molybdenum and a trace of titanium. Although very costly to make, the alloy is remarkably easy to machine and produces the valuable combination of extreme strength and light weight. Steels of this kind are still being developed and are in common use in aircraft undercarriages, special containers for high-pressure liquids and gases, and mechanical gears and drive shafts that need to be especially strong.

Before the new steels, aluminum was the only metal to offer man really lightweight strength. The large-scale process for extracting the metal from its ore, bauxite, was developed in 1886. Since then it has been playing an increasing role in engineering industries. In addition to being light, aluminum will not corrode, always protected in air by a tough surface coating of aluminum oxide. It is also a good conductor of electricity and sometimes replaces copper in electric wiring when lightness is important. Probably aluminum's most widespread use is in aircraft manufacture where although under some pressure from new alloys, its lightness, cheapness, and strength still make it a valuable building material.

Probably the most fast-moving area of metals research is in the field of heat-resistant metals. The most important of these in use today is titanium. Titanium, especially alloyed with aluminum, vanadium, and molybdenum, is widely used for building high-speed aircraft and the skins of artificial satellites.

Superalloys are now emerging as a new family of metals capable of standing up to working temperatures of up to 1000°C. More than 50 superalloy compositions have been developed but those based on nickel are the most widely used. Nickel-chrome, in particular, can easily withstand the disruptive and corrosive effects of ultrahigh temperatures. Called iconels, the first nickel-chrome alloys made gas tur-

Below: machine casting conveyors in a modern iron foundry. Iron smelting began in about 1000 BC, and it is still the world's most important metal.

bine engines possible, retaining their strength under the combined onslaught of high temperatures and burning fuel.

Superalloys such as nickel-chrome depend for their heat-resisting qualities on a tough layer of oxide forming on their surface. Although current maximum temperatures for nickel-chrome protected by a chrome oxide layer is slightly over 1000°C, research metallurgists are already working on a new generation of superalloys. Possibly using an extra-tough aluminum oxide coating for protection, the new superalloys may be able to withstand temperatures of over 2000°C, opening a whole new range of engineering applications.

Above: aluminum sheet rolling. The sheet is entering a 100-inch five stand hot mill. Aluminum or its alloys is widely used in the manufacturing and aerospace industries.

Above left: an armored vest of titanium plates covered with ballistic resistant nylon. Known as the "Composite Armored Vest", it weighs less than nine pounds.

Above: the development of alloys able to withstand high temperatures and severe stresses made possible the use of gas turbines on a large scale. Other materials could not stand up to such punishing treatment.

Left: this machine for welding titanium aircraft ribs follows a series of thin ridges and grooves. To prevent the titanium oxidizing the magnetically controlled welding tool is shielded by a hood containing inert gas.

Natural and Synthetic Polymers

The greatest revolution in man's use of materials has been the recent development of plastics. Although the scientific research that produced plastics is barely 50 years old, their manufacture is already one of the world's great industries. In fact it is hard to imagine life today without plastics. Now available in a seemingly limitless variety of properties, shapes, and colors, plastic is becoming one of the most widely used materials in man's history.

The essential difference between plastics and all other common materials such as salt, water, or sugar is in the structure of their molecules. Plastics molecules, known as polymers, are long chains usually consisting of carbon atoms attached to other atoms such as hydrogen, oxygen, and nitrogen. The individual molecular units of the chain are called monomers and the process of linking them into long-chain giant molecules is known as polymerization. The understanding of the chemical mechanism that makes it happen, has been one of the keys to the plastics revolution.

Polymers in nature have been known for centuries. Natural waxes and resins were used by the Ancient Egyptians as cosmetics, formal seals, and the first crude adhesives. Best known of the natural polymers today is rubber, made from a latex tapped from the bark of the rubber tree. The first important application of rubber was devised in 1823. Scotsman Charles Macintosh dissolved some raw rubber in naptha and sandwiched it between two layers of cloth, creating a raincoat fabric that still bears his name.

It was not until 1832 that the breakthrough was made that turned rubber into a really useful modern material. An American inventor, Charles Goodyear, accidentally spilled a mixture of rubber latex and sulfur on to a hot stove. He scraped it off and let it cool and then found that he was left with a flexible, springy material that retained its properties at all reasonable temperatures and resisted attack by

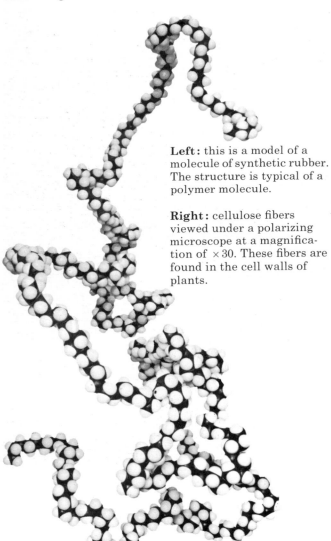

Left: this is a model of a molecule of synthetic rubber. The structure is typical of a polymer molecule.

Right: cellulose fibers viewed under a polarizing microscope at a magnification of ×30. These fibers are found in the cell walls of plants.

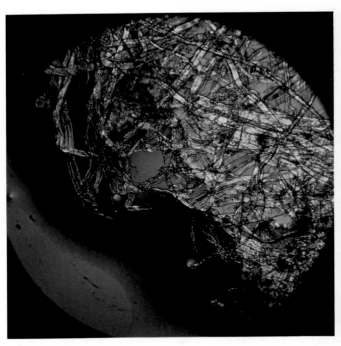

chemicals that would have destroyed raw rubber. Goodyear's process later became known as vulcanization and led directly to rubber's widespread use in automobile tires.

One of the most common natural polymers is cellulose, the substance that makes up the cell walls of all plants. A tough, fibrous, durable substance, it is valuable to man in a wide range of roles. Cellulose in wood – comprising about half its solid content – gives wood the rigidity and firmness that makes it such a useful building material. As a major ingredient of jute and hemp, cellulose contributes to their fibrous strength making them excellent materials for making rope, baskets, and mats. Cellulose in flax and cotton makes them ideal for turning into a woven cloth that is tough, washable, and easy to dye. Cellulose in paper provides the essential qualities of durability, firmness, and an absorbency that enables it to soak up and retain ink.

The first man-made polymer to be produced on a commercial scale was celluloid. It was actually a derivative of cellulose and was made in the late 1800s by American printer John Wesley Hyatt from cotton waste treated with camphor and nitric and sulfuric acids. It quickly became popular and was used for a wide range of articles including photographic film. In the years that followed other early plastics were made from plant fibers, including cellophane and the first man-made fiber, rayon. But none of these was completely synthetic. All were derived from natural substances. The first truly synthetic plastic was not made until 1908 when the Belgian-born chemist Leo Baekeland combined two simple chemicals, phenol and formaldehyde, to produce a tough plastic that he named Bakelite. Today most polymers are completely synthetic, chemists even being able to duplicate nature, producing an artificial form of rubber.

Plastics are classified according to the way they behave when heated. There are basically two kinds: thermoplastics and thermosetting plastics. Thermoplastics soften and melt when heated, like wax or butter. When cooled they regain their original hardness. Thermosetting plastics do not soften when heated but eventually blacken and char.

Thermoplastics are the most common kind of synthetic. One of the most famous goes by the trade name of Nylon. Strong and flexible, the first nylon fibers were manufactured in 1934 by Wallace Hume Carothers, a chemist working for the giant United States corporation, Du Pont. The first mass-produced nylon article was a new, cheap form of women's stockings. When they went on sale in May 1940, women stood in long queues to buy them. Over four million pairs were sold in the first four days and before long "nylons" had become synonomous with "stockings." Nylon is still the most widely used synthetic fiber. It is used in the sails of racing yachts, parachutes, hairbrushes, and certain kinds of machine bearings. The wide and almost instant success of nylon paved the way for a growing range of mass-produced plastics and synthetic fibers. One of the most important of these was Orlon, a material whose fibers have a permanent

Above: a rubber worker tapping a tree in Malaysia. Rubber was first developed as a useful material by Charles Goodyear in 1832. Rubber is now being increasingly replaced by synthetic substitutes.

Above: nylon stockings reappear in the United States after World War II. Each nylon molecule has a "backbone" of carbon and nitrogen atoms, with oxygen and hydrogen atoms attached to its sides.

crimp, giving it an airy lightness that makes blankets and sweaters warm and comfortable. In addition, Orlon, as a type of polymer known as an acrylic, is tough and resistant to wear, making it ideal for trousers and jackets when spun into a tightly woven fabric.

Other important thermoplastics include the ever-popular polyethylene. Flexible, lightweight, and transparent in thin sheets, it is ideal for making bags and tubes for packaging goods. Polyvinylchloride (PVC), strong, flexible, and easily colored, is widely used for electrical insulation, transparent raincoats, and acid-resisting plumbing. In one of its many forms it makes an excellent substitute for leather and is widely used in upholstery.

A more recent development has been the substance known as Teflon or PTFE. Although a thermoplastic, it is unusually resistant to heat. It also has very little friction. This combination of qualities make it a perfect coating for the insides of "non-stick" cooking utensils. Its relative lack of friction has also made it valuable in producing artificial joints to replace diseased joints in surgery.

Thermosetting plastics are in less common use. One group, the alkyds, are good electrical conductors and valuable additives for paints. Melamine and urea plastics are heat-resistant, odorless, tasteless, and easily colored. This makes them ideal for "unbreakable" cups, saucers, and dishes and as a durable sur-

Above: this picture shows the wide range of colors available in plastic sheeting. The colors are added during the production of the sheets by adding pigments. Literally hundreds of synthetic plastics have so far been developed. One advantage of using plastic is its flexibility in any production need.

Right: plastic pipes after manufacture. These pipes are virtually unbreakable, and are becoming almost universally used in modern plumbing. Produced by extrusion, the modern plastic pipe is made of polyethylene, which is corrosion resistant. Large pipes are used in agricultural drainage.

210

Above: plastic cups being compression molded. The cups are formed on the die of a six impression tool. A temperature of 150° centigrade and a pressure of 2000 pounds per square inch is necessary.

Right: plastic film is widely used in the food packaging process. It is made by blowing hot plastic into a thin-walled tube. When the tube has cooled the plastic is cut into sheets of polythene film.

face for kitchen worktops, dining tables, and trays.

The technology of forming plastic articles is now as complex as the chemistry of making the pastics themselves. The two most important processes are extrusion and molding. The principle of extrusion is similar to squeezing toothpaste from a tube. Solid thermoplastic is fed into an extruding machine, ground into small pieces and pushed into a heating chamber. The now molten plastic is then forced through an opening of a particular shape, if it is a circular hole, a rod of plastic emerges. If it is rectangular, a ribbon or sheet is formed. If a circular hole is used with its center blocked, a plastic tube is extruded. Extrusion techniques such as these are used to produce garden hoses, drinking straws, plastic bags and strips of film. Synthetic fibers are made by a similar process. A hot liquid plastic is extruded through a plate punctured by numerous tiny holes. As the thin pipes of plastic emerge they quickly solidify forming the synthetic fibers.

Molding processes are of three kinds: compression, injection, and blowing. The principle is similar to casting molten metal except that plastic, because of its consistency, has to be forced into the mold. Compression molding is used mainly to form thermosetting plastics. The mold is made in two parts. The plastic in the form of granules is fed into the lower part. The upper part moves down to compress the plastic as the mold is heated. Injection molding is used for thermoplastics. Melted plastic is injected into a mold and

placed under pressure so that it takes up the shape of the mold. Once cool the solid plastic is removed. Blow molding is a means of making hollow objects such as detergent bottles. The technique involves blowing compressed air into a plastic tube inside a mold. As the tube expands it takes on the shape of the mold.

The technology of plastics is still evolving and in recent years a number of space-age polymers have emerged with startling properties. One plastic called Merlon is as clear as glass and so strong that a thin sheet can stop a bullet fired at close range. Among its uses is vandal-proof window panes and visors for helmets worn by astronauts. Other remarkable silicon-based polymers have been produced. Some are transparent and yet transmit little or no heat even at temperatures four times the melting point of steel. Their most important use has been in the heat shields of spacecraft. Another polymer, silicon-rubber, is produced in membranes as thin as one thousandth of an inch. Although completely water-tight, the membranes allow gases to pass through them. This means they can act as a kind of artificial gill. A submarine with view ports of silicon-rubber theoretically could draw its oxygen from the surrounding water, expelling carbon dioxide the same way. At present, silicon-rubber membranes are used in artificial lung machines to oxygenate the blood of patients during major surgery. But a wide range of future applications is likely in the exploration of the sea and the development of the first undersea communities.

Composite Materials

There is nothing new in the idea of blending quite different materials to combine and even improve upon their individual qualities. Thousands of years ago, the Ancient Egyptians added straw to their clay bricks to improve their strength while the Incas achieved the same effect by adding plant fibers to their pottery. Today, eskimos modify the behavior of the ice used for building their temporary dwellings by freezing moss into it. This makes it far less brittle and more easy to shape.

A more sophisticated version of the eskimo technique was attempted in the 1940s when inventor Geoffrey Pyke suggested using icebergs as huge aircraft carriers. Although his novel idea was eventually abandoned it did lead to the discovery of pykrete, a mixture of ice and wood pulp many times stronger than pure ice and so ductile it could even be turned on a lathe.

One of the best known composites in modern use is ordinary plywood. The early versions were veneers of wood glued with vegetable or blood-based glues. The poor quality of the glue meant plywood had no resistance to moisture and it was not until the invention of

Above: this boat is being built from plywood – a tough wood composite. Plywood consists of three or more layers glued together with the grains at right angles.

high-polymer glues that it became the versatile material it is today. With the new adhesives it is so watertight that it is even used in boatbuilding.

Reinforcing concrete with steel is also a longstanding form of composite technology. Although an excellent building material in many respects, concrete has very little tensile strength. This means it

Below: the TWA flight center, Kennedy Airport, New York. Reinforced concrete has been used in the construction of the 6000-ton roof, which is supported by four buttresses. The thickness of the roof varies from eight inches at the edges to 44 inches at the buttresses.

will break rather than stretch and cannot support heavy loads. An unsupported beam more than a few yards long will crack under its own weight. To give it the tensile strength it lacks, builders reinforce it with rods of steel or a mat of steel mesh. Such a technique has enabled builders to produce a tremendous variety of new structural shapes and has made buildings and

the mid-1960s, Rolls-Royce research teams worked out how to make huge savings in weight by using a carbon-fiber reinforced material known as Hyfil for building aircraft engine components. One of the big advantages apart from weight saving is the ease of forming elaborate components that can often be molded in one piece.

Above: glass fiber (white) and carbon fiber (black) bundles. Glass fiber manufacture today concentrates on glass wool and continuous filament production.

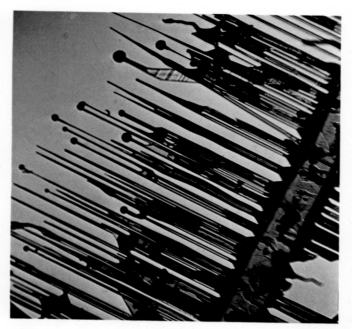

Above: aluminum oxide (sapphire) whiskers grown in a laboratory and photographed at ×60 magnification. They can support stresses of up to 1500 tons per square inch.

bridges far safer and more durable.

Modern composite technology is now concentrating on developing various kinds of special fibers to improve the tensile strength of construction materials. Everyone today is familiar with glass fiber – plastics reinforced with fibers of glass to stiffen and strengthen. The material was first made in the 1940s and is now widely used to manufacture automobile bodies, boat hulls, swimming pools, and bathtubs. The main attraction of glass fiber is the ease with which long, strong fibers can be produced. When freshly drawn, fibers no more than two ten-thousandths of an inch thick have tensile strengths comparable to the strongest steels available today.

More unusual fibers are now emerging, however, offering still greater reinforcement for a wide range of engineering materials. For example, ultrafine filaments have been drawn of such substances as tungsten, carbon, boron, and beryllium, the stiffest of all metals.

Boron fibers have been made by depositing a coating of the metal on a fine tungsten wire. Their earliest applications have been in reinforcing helicopter blades where the greater stiffness acquired can be used in making the blades long, so achieving greater lift. Carbon fibers are produced from continuous organic filaments such as rayon. When heated these can be carbonized into a continuous filament of graphite. In

The most promising innovation of recent years has been the use for reinforcement of minute fibers that form on the surface of crystals. These "whiskers" are incredibly strong and may lead to the development of new supermaterials that retain their strength even when white hot. The secret of the strength of these extremely fine whiskers lies in their near-perfect crystalline structure. They have virtually no weak points. In 1968, Soviet scientists produced a tiny defect-free whisker of tungsten that could sustain a load of 1635 tons per square inch. In America a whisker of sapphire (aluminum oxide) has been produced that is even stronger. The technology needed to combine these whiskers with the materials they are to reinforce is only now beginning to emerge. The most exciting possibility is to get whiskers to form, say in an alloy of two metals, as the alloy is solidified from its molten state.

The first applications of whisker-reinforced materials will be in the aerospace industry where extreme strength and heat resistance from lightweight materials will make possible radical new design developments. Everyday life may also be affected quite soon by the new technology. A new kind of dental filling is being investigated. Combining the strength of ceramic fillings with the softness of traditional metal fillings, whisker-reinforced gold may become a common material used by dentists in the near future.

Semiconductors

In the 1940s, scientists at the Bell Telephone Laboratories in New York began studying the properties of curious groups of substances known as semiconductors. They were given their name because they neither conducted electricity so badly as to be insulators nor so well as to be considered conductors. Working with substances such as silicon and germanium, the scientists found that their semiconducting properties could be enhanced by adding traces of impurities to them, such as boron or arsenic.

Transistors gradually took over most of the roles of the vacuum-tube valves, once they had been developed to withstand higher temperatures and a range of electrical frequencies, and reduced to almost microscopic size. Perhaps the most dramatic advance was in the building of bigger and more powerful computers. Extensive circuitry that might have occupied an entire building if constructed of old-fashioned valves, could now be contained in cupboard-sized cabinets. A new generation of computers emerged with hitherto unimagined calculating and data storage capacity.

A still greater revolution took place when, in the 1960s, transistors began to give way to electrical circuitry etched into the surface of small chips of silicon. Within a few years large numbers of individual circuits could be integrated onto a single tiny chip. Integrated circuits of this kind could pack the equivalent of over 30,000 transistors on to a single square inch of silicon.

The manufacture of silicon chips is now a massive,

Left: the forerunner of the modern transistor, the point-contact was invented in 1947. It has two pointed metal contacts touching the surface of a piece of germanium. A small positive charge applied to one contact increased the current flowing from the germanium to the other contact. This type was very quickly superseded by the junction transistor.

In 1948 the Bell researchers made a breakthrough that revolutionized electronic technology and, with it, everyday life. They discovered that semiconductors could be made into small, solid analogues of existing electronic components, such as valves. They called their invention the transistor. Within a few years, once bulky electrical appliances had become extremely compact. By 1953, tiny transistors were being used in hearing aids so small they could be fitted inside the ear. The new solid state devices, as they were called, were not only much smaller than their valve-built counterparts. They were also much tougher. Whereas dropping a valve radio would almost certainly smash the delicate vacuum tubes, a transistor radio would probably suffer no more than a broken case.

worldwide industry. Design teams are constantly working on new chips capable of different electronic functions. So far the biggest advances have been made in miniature electronic calculators. The most sophisticated of these fit into a brief case and perform all the basic functions of a small computer. Much of their size consists of their displays and the keyboard needed to punch in data and operating commands.

Silicon chips are also used widely in digital watches. Not only are modern electronic watches accurate timepieces, the chips they use also enable them to provide a range of alternative functions. Some actually double as tiny calculators. Another fast-growing industry made possible by silicon chips produces video games, played by linking computerlike electronic circuitry to a television. The most sophisticated

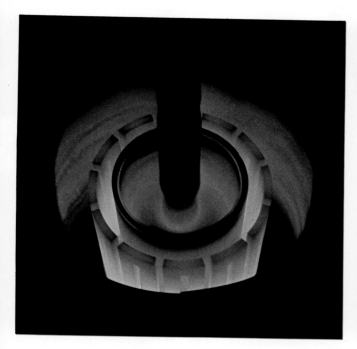

Above: a single crystal of silicon being grown. The crystal is slowly pulled out of a crucible of molten, highly purified silicon. Crystallization continuously takes place at its lower end. The use of silicon has made possible a revolution in electronics.

Above: an integrated circuit on a silicon chip. The circuit contains six logic gates and is formed on a piece of silicon less than a tenth of an inch square. Several hundred identical circuits were formed simultaneously on one piece of silicon.

of these enable a wide range of games to be played of varying sophistication, from a simple form of tennis to intergalactic wargames where players have to make complex decisions of strategy and tactics. Further developments include self-contained table-top units that enable games such as chess, backgammon, and bridge to be played against the logical powers of tiny silicon chips. One chess-playing unit currently available assesses several million possible positional situations on the chess board before deciding its next move.

Although silicon integrated circuits have already revolutionized both scientific, office, and leisure equipment, the pace of new development has barely slowed. In some areas, integrated circuits are a technology in search of an application. In wristwatches for example, an enormous variety of functions could be built-in by using advanced chips, but manufacturers are unsure just what the public actually wants and would consider useful in a wristwatch. Some experts estimate that over 25,000 applications for "microcomputers" are still awaiting discovery.

The growth of the chip industry has been phenomenal. The world market has grown from around 1,500,000 dollars in 1967 to an estimated 3,000 million dollars in 1980. New advances in chip design have been made so rapidly that some circuits are obsolete even before they have reached the market place. Latest techniques are actually enabling computers, themselves based on chip technology, to work out designs for new and more powerful chip circuitry.

Some of the most recent applications include their use by General Motors to regulate automobile

Above: a portable radio-telephone made by Motorola. It is chip-operated and equipped with a fold-down antenna. Called the DYNA T.A.C., it is similar in shape to an ordinary home or office telephone handset.

ignitions. Bell Telephones are working on a combined telephone, burglar and fire alarm, and intercom. Motorola have developed a personal communication system, involving a hand-held, chip-operated personal telephone weighing less than two pounds and having no cord. The future of the miracle silicon chip appears to be as limitless as the human imagination itself.

215

Superconductors

Superconductivity was discovered by Dutch physicist Heike Kamerlingh Onnes in 1911. A few years earlier, Onnes had amazed the scientific world by managing to liquify helium, achieving a temperature of about $-269°C$, just over four degrees above absolute zero ($-273.15°C$ or $-459.67°F$). Although a breakthrough in itself, the real significance of Onnes' success was the opportunity it provided for studying how materials behaved at ultra-low temperatures.

Onnes discovered superconductivity while studying the electrical resistance of mercury. Existing theory predicted that a metal's resistance would rise as it is cooled and Onnes' early results confirmed this. But as the mercury reached $-269°C$, its resistance suddenly vanished altogether. Onnes could barely believe his results. But he soon found other materials that behaved similarly. Lead, for example, became superconducting at about $-266°C$.

The implications of superconductivity were remarkable. An electric current, once started in a superconducting coil, would keep flowing round the coil indefinitely without any energy loss. In fact, in an early experiment a current of several hundred amperes was circulated in a superconducting lead ring for over two years without any detectable decrease of strength. At first some scientists thought they had discovered the secret of perpetual motion. But it was soon realized that a great deal of energy was necessary to keep the coil cold enough to remain superconducting. In fact, the cost of maintaining such a low temperature was a major obstacle to further studies. However, as refrigeration techniques improved, a wide range of new superconductors was discovered. Tin became superconductive at about $-269°C$, aluminum at $-271.9°C$, uranium at $-272.4°C$ and hafnium at $-272.8°C$. Today about 1400 different elements and alloys are known to be superconductors. The highest transition temperature found for a metal is that of technetium, which becomes superconducting at $-262°C$.

A major application of superconductors has been in producing extremely powerful electromagnets. A current of electricity passing through a coil of wire wound around a soft iron core produces a strong magnetic field – the greater the current, the stronger the field. But as more current is passed through the wire, heat is produced because of the wire's inherent resistance to the current. This limits the strength of field that can be produced. But in superconducting wires, electricity flows without producing heat and huge currents can be applied without danger of melting the coils. By this means the first ultra-strong magnets were developed. Unfortunately there is a limit on the field strength that can be achieved because superconductivity is eventually destroyed by powerful magnetic fields. The exact strength of field that a superconductor can withstand varies with the material. Much recent research has been concentrated on finding superconductors capable of tolerating very high fields. Some success has been achieved with alloys such as tinniobium and vanadium-gallium although problems still exist in producing wires from them for use in magnet windings.

Superconducting magnets are extremely costly to manufacture and run. So far their use has been confined to research into the behavior of matter in powerful magnetic fields and in manipulating the trajectories of subatomic particles in massive atom-smashing machines. The greatest single obstacle to the

Above: magnet rings in an assembly for plasma containment in a nuclear fusion experiment. Part of the superconductivity motor necessary for the experiment is shown. One of the most important possible applications of superconducting magnets would be in the operation of a new generation of highly efficient power stations.

wider use of superconductivity is the need for expensive refrigeration systems such as liquid helium. A few superconducting alloys have now been made with transition temperatures high enough for liquid hydrogen to be used. But although cheaper than liquid helium, hydrogen is still too expensive for large-scale commercial exploitation. This must await materials that become superconductors at about 77° above absolute zero, which can be reached by liquid nitrogen.

A massive effort is being made to produce such a new generation of superconductors. Some excitement

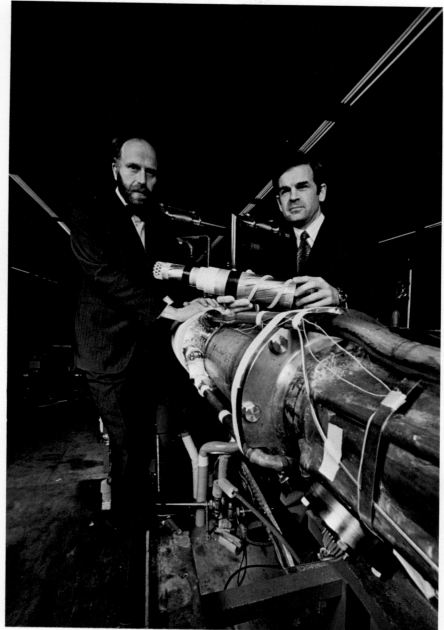

Above: a superconducting wire being made, a two inch diameter bronze ingot and its nobium rod inserts will become a superconducting wire only a tenth of an inch thick but nearly five miles long. When made, the wire will have more than 1300 continuous nobium filaments running through it. The technique was developed by Britain's Atomic Energy Authority.

Right: a superconductivity motor at the General Electrical Research Laboratory at Leatherhead, England. One of the problems of distributing electricity is that of resistance of cables carrying it. Over great distances a good deal of power is dissipated as heat. Superconducting cables and generating motors drastically reduce such losses.

was created in the mid-1960s by the American scientist W A Little who suggested that it might be possible to produce a plastic that remained superconducting at up to 1800°C! Unfortunately, few scientists now believe Little's suggestions will work in practice. So far the highest temperature superconductor synthesized is a compound of niobium and germanium, which becomes a superconductor at 23.2 degrees above absolute zero.

The possible applications of superconductors needing little or no refrigeration are numerous. Especially important is their potential use in the power generation industry. Superconducting magnets would make possible a new generation of highly efficient power stations, electricity from which could be distributed by superconductive power cables. In these there would be no need to use the high voltages of existing transmission systems. The power could be distributed without loss as very heavy current at low voltages. Less insulation would be needed and the entire grid could be extremely compact. It might even become economic to bury it underground, sparing the countryside the unsightliness of modern pylons.

Since all known superconductors are strongly repelled by magnetic fields, magnetic hovercraft may one day be developed, gliding like magic carpets over superconducting tracks Perhaps it may even become possible to develop superconducting pajamas enabling us to sleep comfortably suspended above a magnetic bed!

All such developments must await a breakthrough in either refrigeration techniques or superconductor chemistry, perhaps both. For the meantime, the science of superconductors remains one of the toughest but most promising challenges for modern technology.

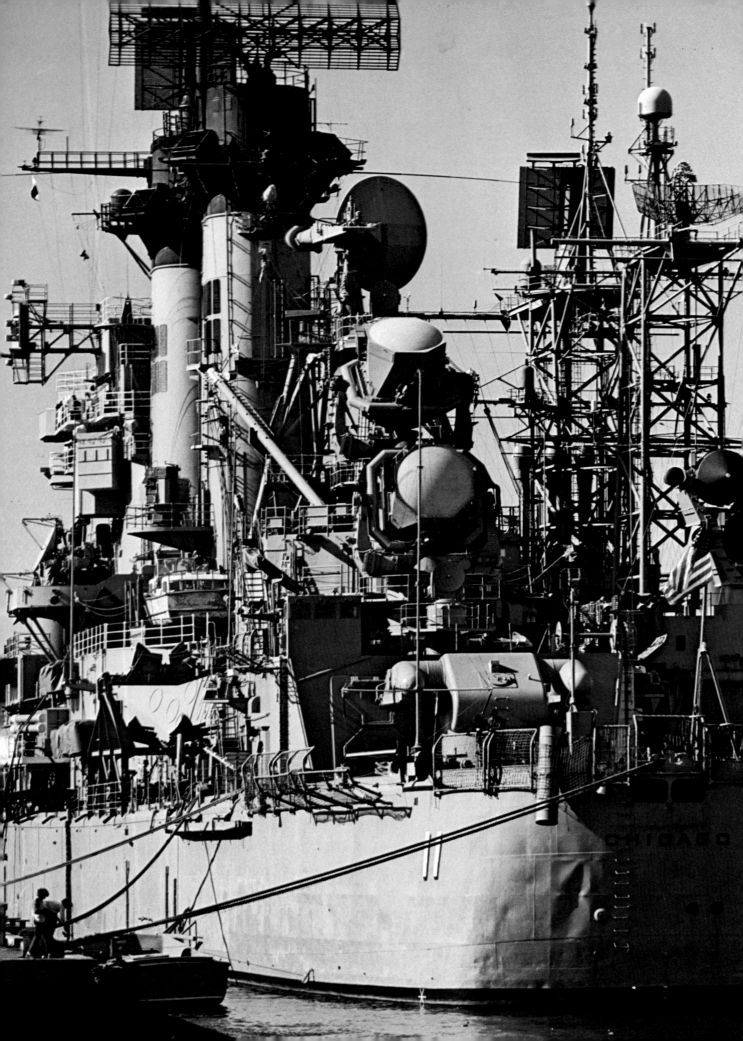

CHAPTER 9

MILITARY TECHNOLOGY

A high proportion of the research and development effort in our modern society is devoted to weapons. It is a sad reflection on human progress that this should be so. In primitive societies the proportion was probably close to 100 percent, but then at least the weapons had to serve for hunting as well as warfare. Today warfare has become so sophisticated that tools for the direct killing of individuals represent only a fraction of the total effort – and they are used in special cases such as operations against terrorists and primitive or civilian populations. In fact, many of the most advanced nations have been embarrassed by the unsuitability of their modern armed forces for use against rioting civilians, urban terrorists, or guerrilla warfare. There are still compelling reasons for the advanced countries to attempt to preserve the peace – or at least to prevent a full-scale war – by deterring aggression with the threat of retaliation by nuclear weapons. The argument has yet to be disproven.

Opposite: the cruiser USS *Chicago* moored at San Diego, California. Visible in this picture are the sophisticated radar systems and missile tracking devices. The *Chicago* is a guided missile cruiser. The world's navies began to employ cruisers after armor plate common in the 1880s.

A Brief History

Above: English Yeomen armed with longbows at the Battle of Poitiers in 1356. Disciplined forces of yeomen armed with longbows and employing new infantry tactics altered the balance of warfare in medieval Europe in favor of offence.

The earliest weapons were those that crushed, stabbed, or cut, and were designed either to be held in the hand or thrown. Later, devices were invented for increasing the velocity of thrown weapons, such as the sling, which increased the velocity of stones. Other ingenious devices for killing from a distance were the throwing-stick – the holder used to assist the throwing of a spear – and the South American bolas – two or three heavy balls attached to each other by cord. Perhaps the most remarkable of all primitive weapons was the aerodynamically designed boomerang, which when skilfully thrown either strikes its target or returns to the thrower.

About 20,000 years ago, during the Middle Stone Age, the bow and flint-headed arrow came into use. They multiplied the range at which a target could be hit. In Britain the six-foot longbow was perfected as a weapon and could be used with such consummate skill that arrows could be fired with deadly accuracy

Below: the Australian aborigine's boomerang is the earliest known aerodynamically designed weapon – used for both hunting and warfare. Varying in size between 12 and 30 inches, the boomerang returns to a skilful thrower when it fails to hit the target.

every five seconds over ranges of several hundred paces. After 1200 the crossbow also became important, firing a short bolt, quarrel, or stumpy arrow that often penetrated or crushed the armor that was increasingly being worn as protection against metal tipped weapons.

Development of fortifications resulted in the use of new defense measures such as boiling tar and inflammable oils poured on to besiegers from the walls above. The besiegers themselves used battering rams and large stone-throwing catapults known as ballis-

Above: the battering ram was one of the earliest methods of breaking through an enemy's defenses. Originally, rams were probably tree trunks, but they later became more sophisticated with a smooth shaft and heavy timber or metal knob at the front. This one is mounted on a chain to support the weight.

tae, mangonels, or onagers, which also became important in sea warfare.

By 1200 gunpowder was being exploited both as an explosive and as a propellant for rockets. The Chinese used rockets both as missiles and as carriers of fire. The use of these missiles, whose full potential was probably not fully grasped, was taken up throughout India and western Europe, where in the Napoleonic era Sir William Congreve brought them to the status of a steel-cased precision weapon. Until then, Europeans had concentrated entirely upon cannon. The earliest of these were crude artillery pieces filled with a charge of gunpowder and assorted projectiles, such as heavy arrows, stones, or balls of various materials. The cannon had a touch-hole in the barrel through which the charge could be ignited. Recoil was made sluggish by the sheer mass of the device. By 1300 breech-loading bombards or large cannon were in use, but muzzle-loading was much more common until the 19th century.

Early artillery pieces were cumbersome. Naturally enough attempts were made to produce guns for infantry, but at first these, too, were clumsy and extremely tiring to carry for a distance. They began as mere tubes, with a touch-hole for the powder at one end and a bell-mouth at the other into which the user could stuff lead-shot, nails, and stones. Often a forked rest was used to support the front of the barrel and improve the accuracy of aim. It was a major advance when around 1460 mechanisms were invented for igniting the charge. The mechanisms were called "locks" because they were first made by locksmiths. The earliest versions brought a piece of slow-burning match-cord into contact with powder but later this potentially hazardous arrangement was dispensed with and instead sparks were struck from a flint. At the same time as lock mechanisms were being improved, guns became lighter and more accurate, firing specially made bullets and, in the 19th century, increasingly having rifled barrels to spin the projectile and keep it pointing in the direction of travel.

By 1850 both artillery and hand-guns were not only all rifled but could also be breech-loaded with prepared ammunition. Percussion caps detonated by a sudden blow replaced the sparking devices, and in the first half of the 19th century it became increasingly common to use one-piece "fixed ammunition" in which the bullet was mounted in the end of a cartridge case housing a propellant charge and firing cap. It was then natural to progress to various self-loading schemes and ultimately to the automatic weapons of today.

It was the bullet that gradually made heavy armor for men and horses not worth the trouble, because it was impractical to wear armor thick enough to afford much protection. By 1720 armor was almost extinct, but in sea warfare it became even more important because of the protection offered by the great thickness of oak or iron, and ultimately special steels, that could be carried. Likewise in land warfare the stalemate of trench warfare in 1915, when unarmored men were pinned down in trenches by the murderous fire of machine guns, led to the tank, armored car, and other armored fighting vehicles of today.

Warfare today is dominated by electronics, in the form of sensors, communications, command and control devices, and guidance systems; and to traditional weapons have been added nuclear, biological, and chemical weapons able to erase human life from this planet.

Above: the flintlock was the successor to the matchlock pistol. The movement of the trigger produced two simultaneous effects: it lifted the pan cover and brought the flint against a piece of steel to produce the necessary spark.

Right: the squalor of trench warfare during World War I. Much of the war was static, with thousands of men dying in order to gain or defend a few yards of land. Those who were not killed lived for years in trenches like this.

Rifles and Handguns

The weight and numerous other drawbacks of early hand-held firearms prevented their widespread use until the 18th century, but it was the invention and gradual improvement of the percussion cap in the 19th century that revolutionized such weapons. There had already grown up two classes of gun, the long-barreled type for professional foot-soldiers that developed into the rifle, and the short carbine or pistol for the horseman. Many inventors tried to devise methods of making a repeating weapon, able to fire several bullets in quick succession. Some cumbersome guns had multiple barrels, but in 1835 Samuel Colt patented a "revolver" in which a cylinder, holding several (usually six) bullets, could be revolved to bring each bullet in turn opposite the breech end of the barrel. Even the first Colt revolver was a giant advance, though each chamber in the cylinder had to be separately loaded through the muzzle and barrel with the powder charge and bullet and then armed with a percussion cap at the rear; and the user had to cock the revolver afresh for each shot by pulling back the claw of the hammer with his thumb.

By the 1870s the almost universal use of one-piece "rounds" composed of percussion cap, case, and bullet had opened the way to much handier guns capable of several times the previous rate of fire. In some cases fresh rounds were fed in by a hand lever from a magazine under the barrel or breech, while the revolver was altered so that the chambers could be loaded from the rear and fired in rapid succession without the need to cock the hammer separately. After many false starts

Above: a 17th century English musketeer armed with a matchlock rifle. In order to fire his gun, the musketeer had to set a glowing slow match to the powder in the pan while taking aim. This was a tricky business, and it was almost impossible to fire such a gun from horseback.

the rifle mechanism was modified to a breech that could be opened and closed by a solid cylinder of steel called a bolt, worked by a hand lever, and locked in the forward position by rotating it into wedges to give a gastight seal. Driving the bolt forward automatically cocked a spring-loaded firing pin that was released by the trigger to strike the percussion cap on the base of the cartridge case. The bullet had left the barrel and

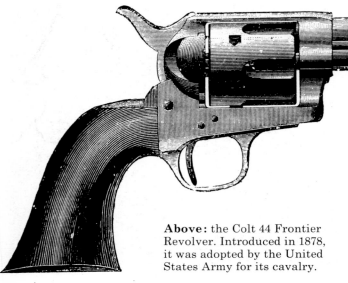

Above: the Colt 44 Frontier Revolver. Introduced in 1878, it was adopted by the United States Army for its cavalry.

the gas pressure dropped to a safe level before even the fastest rifleman could work the bolt to open the breech, expel the used cartridge case and, on the return forward stroke, feed in a fresh round. Sometimes rounds had to be loaded individually by hand, but it was simple to provide a magazine housing five, 10, or more rounds fed upward into the breech by a spring.

By the 20th century the rifle had become a reliable piece of precision machinery, firing standardized ammunition with a muzzle velocity of about 2450 feet per second, and capable of accurately aimed fire to distances as great as 3500 feet. A key factor in this extension of bullet velocity and range was the replacement around 1895 of traditional gunpowder ("black powder") by cordite propellant with its much greater gas pressure and absence of smoke. In almost every case the cartridge case was of solid-drawn brass, strong enough to withstand intense pressure without cracking, and the steel barrel was given much deeper rifling grooves to give a firm "bite" on the bullet and spin it at unprecedented rates. Britain especially concentrated on riflemen of extreme skill, who in conjunction with a carefully designed bolt mechanism could sustain 15 aimed shots per minute. This amazing performance gave deadly firepower to modest groups of men without the prodigious logistic demands for ammunition required by wildly fired machine guns; it also put off the introduction of self-loading rifles.

Toward the end of the 19th century German designers produced self-loading pistols in which the energy released by firing the cartridge was partly harnessed to work the mechanism of the bolt or breech-block and expel the empty case and feed the next round to be fired – as had already been done in the first automatic weapons. Throughout the 20th century the so-called "automatic" pistol – not automatic, but self-loading – has progressed in reliability and efficiency and like the rifle has seen progressive reductions in caliber and an increase in muzzle velocity. Virtually all modern pistols have box magazines for about eight to 13 rounds that slide up inside the handgrip. A few can be clipped to a rigid holster to increase the range at which aimed fire is possible; usually a pistol loses effectiveness at around 300 feet though the bullets are lethal to many times that distance.

Although the American John M. Browning produced a self-loading rifle in 1917 this could be used as a true automatic. The self-loading rifle was not common until World War II, and then only in the United States Army. Today almost all rifles are self-loaders, and most are automatics.

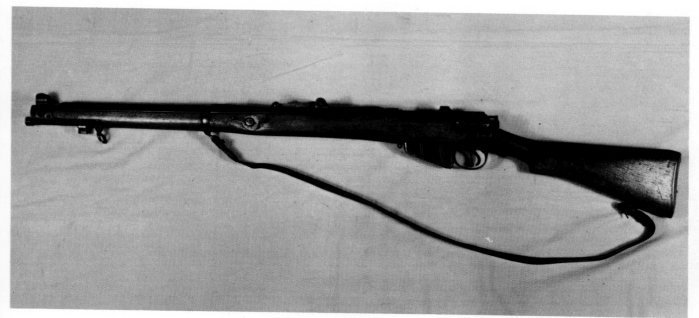

Above: the Short Lee-Enfield was standard issue in the British Army during World War I and for many years afterwards. This is the one used by Colonel T. E. Lawrence (of Arabia). Lawrence's initials and the date 4.12.16 are carved near the magazine. There are also five notches, the largest for a Turkish officer shot by Lawrence.

Right: a Browning Automatic Rifle being fired from a bipod by a United States soldier during World War II. It could also be fired from the shoulder or the hip. When set on automatic the Browning fired .30 calibre bullets at a rate of 550 rpm or 350 rpm. It could also fire single rounds. The gun weighed approximately 20 pounds.

Automatic Weapons

Throughout the 19th century inventors tried to harness the energy of the exploding cartridge to operate a gun's mechanism and load the next round. In 1862 the American inventor Richard Gatling bypassed the problem by building a gun with six complete barrels (10 in most later Gatling guns) in a frame rotated by a hand-crank. Several rapid-fire guns followed, but none a true automatic powered by the gun itself. It was no simple task to harness the forces. In a minute fraction of a second the propellant burns to form about 14,000 times its free volume of gas at temperatures well beyond "white heat." Pressure inside the cartridge case almost instantaneously rises to more than 22 tons per square inch. Unless the brass case fits accurately inside the strong gun breech it will rupture, endangering the user and jamming the gun.

The first really successful automatic was Sir Hiram Maxim's gun of 1884, which became the mass-produced Vickers machine gun of the British Army (and which remained in service into the 1970s). It was fed by rifle-type rounds inserted into a long fabric belt, from which they were withdrawn by the oscillating bolt. The front face of the bolt pulled out each round and slid it down to the barrel, rammed it into the breech and released the firing pin to hit the percussion cap. The force of the gas pressure on the inside of the base of the cartridge thrust it rearward, extracting the next round from the belt and ejecting the empty case underneath. For a brief period the bolt and barrel remained locked together; then with the bullet departed and the barrel pressure falling rapidly, it was safe to unlock the action and allow the bolt to open the breech. This so-called recoil method continues to be widely used, though the diversity of locking arrangements is astonishing.

A large proportion of automatic weapons using full-power rifle-type cartridges are driven by a method known as gas operation. A small hole is drilled through the wall of the barrel somewhere between the breech and muzzle, and connected to a cylinder able to withstand the intense pressure. As soon as the bullet has passed the hole the gas pressurizes the cylinder and hits a piston-rod connected to the mechanism of the gun. Despite these extra parts, and such problems as progressive clogging, or "fouling" by solid particles in the gas, and the difficulty of changing barrels, gas operation allows a gun to be light, have a fully locked breech and be reliable and adjustable in operation. A third method, and the simplest of all, is "direct blowback" action, in which the force on the fired case simply pushes it and the bolt to the rear. It is common on weapons using low-power pistol ammunition, but with rifle ammunition the blowback method would be dangerous unless the bolt was extremely heavy.

Below left: the Maxim gun was invented by Hiram Stevens Maxim, and the first time it was fired in anger was by the British during the Boer War. This Vickers version was the standard machine gun of both World Wars and fired at a rate of 650 rpm.

Below: munitions girls at the Inspection Buildings, Park Royal, London filling machine gun belts with bullets. Although much feared, machine guns in World War I were wasteful of ammunition.

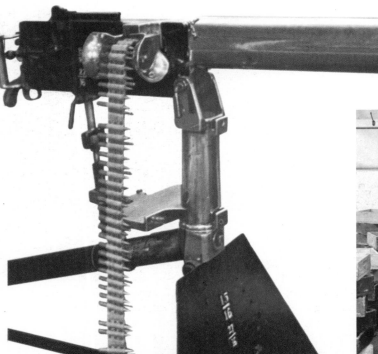

Above: a standard infantry Lewis gun, here in action during World War 1. This 47-round weapon was gas-operated. A portion of the propelling gas is directed via a port to a chamber with a moving piston. The pressure of the gas drives the piston and breach block to the rear.

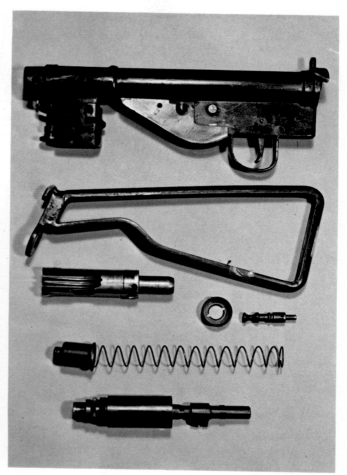

Above: this Sten gun was made by members of the Danish resistance in World War II. When dismantled as shown here the pieces could be easily hidden. The Sten was popular with the resistance throughout Europe, and most were dropped for their use by the RAF.

Automatics using pistol ammunition were originally all pistols, only a few of which could actually be fired truly automatically in a sustained burst. In World War I the same ammunition was used in a larger and more aggressive fully automatic weapon with a much larger magazine, and this soon became popular as the submachine gun or machine carbine. It is much used by urban troops, jungle fighters, crews of vehicles, and police where short-range firepower is more important than long range accuracy.

Full-power automatics were soon divided into medium or heavy machine guns, fed from a long belt and with a robust tripod or wheeled mount and sometimes a water-cooled barrel for firing long bursts without overheating, and light machine guns, fed from a small box magazine and with a small bipod near the muzzle for use on the run. In World War II there emerged the repeating carbine using lower-powered ammunition, small and light for use by paratroops or tank crews, and the automatic or "assault" rifle, using full-power ammunition. Today the heavy machine-gun has vanished, and nearly all modern automatics fire ammunition of smaller caliber than before (such as 4.85 mm, instead of .30 inch) but

Above: the Walther automatic pistol. This seven-shot weapon had a caliber of 9 mm. Introduced in 1930 it was widely used by the Germans in World War II.

with extremely high muzzle velocity. Most armies are hoping to use a standard gun mechanism that can be incorporated either into a pistol or, with a heavier barrel, light tripod and larger magazine, a machine gun. Wood, once common in all rifles and most automatic weapons, has given way to high-strength steel, light alloys, and plastics.

225

Artillery

By the end of the 19th century artillery universally had rifled barrels of high-strength steel, usually made in the form of tight-fitting overlapped tubes and in the largest sizes often being tightly wound with hundred of miles of high-tensile wire. Breech loading had become almost universal, the closure either being a vertically sliding block or a steel cylinder with interrupted screw-threads that could be hinged open for reloading, swung shut, and then locked tight by about a 60 degree turn. Both the gun and its ammunition vehicle, called a limber, were usually mounted on a pair of large wheels. During World War I the muddy conditions on the battlefields of France and Belgium soon demanded broad-tired wheels and a protective sheet of armor at the front. By this time the usual arrangement was to mount the barrel not only

on pivots, for elevation to the correct angle for the desired range, but also in a slide so that it could recoil relative to the chassis, the rearward motion being damped by air or oil hydraulic cylinders or strong springs. The rear part of the chassis, called the trail leg – or legs, because often there were two, hinged to open outward for firing – was fitted with a large spade that dug into the ground to hold the chassis against the recoil. A few guns were allowed to run back up a pair of sloping ramps behind the wheels, while others had their wheels raised or removed before firing so that they rested on a turntable.

Since World War I there have been several distinct families of artillery. The most common multi-purpose guns have calibers of around 3 to 4 inches, and fire fixed ammunition with a brass case into which is inserted any of a range of types of projectile. With a good crew of about six men they are capable of sustaining three to six rounds a minute of accurate fire. In recent years the weight of such guns has been dramatically reduced, from about two tons to half a ton, and most modern types can be transported by air and dropped by parachute. On the muzzle is a brake, usually of the multibaffle type, which blows gas sideways to reduce recoil. Unlike older guns it is possible to rotate the whole weapon through 360 degrees in a few seconds to fire in any direction to ranges of about 10 miles.

Below: "Anti-Aircraft Defenses" by Christopher Nevinson was painted during World War II. The impact of the anti-aircraft shell on its target is less important than hitting the target at all. In early World War II British anti-aircraft guns were hopelessly inadequate.

Below: a United States Army field artillery battalion in action in Korea, 1951. This gun crew is bombarding enemy positions, using a six inch caliber field howitzer. The mortar is a smaller, lighter, and more maneuverable version of a howitzer.

Right: the German World War II monster "Gustav" pictured in its firing position in a railway cutting. This massive gun was never found after World War II. "Gustav" was the modern successor to the German "Big Bertha" of World War I.

Howitzers are guns with large caliber but short barrels, firing extremely powerful shells at relatively low velocity on high trajectories for use against fortifications and other targets where it is necessary to have a large shell fall from above. A lighter version of a howitzer is the mortar, and though there have been some extremely large mortars the majority are small enough to be carried (sometimes dismantled into several parts) by infantry. The mortar tube is set up at an inclination of from 45 to 85 degrees. Finned bombs are dropped nose upward into it by hand, and fired by hitting the bottom and arching up to fall within a radius of about three miles. The complete opposite is the antiarmor gun, which fires armor-piercing shells at extremely high velocity over a flat trajectory. Another special purpose gun is the antiaircraft gun, which combines high velocity with high elevation; but this now survives only in the form of small rapid-fire cannon.

Most types of gun have been put on a powered chassis to become self-propelled – though hardly describable as a military land vehicle, the largest guns of all were mounted on railed tracks and used for heavy bombardment. The longest-ever ranges were almost 70 miles by the German "Paris gun" in 1918 and 94 miles with rocket-assisted arrowlike shells by the German "Anzio Annie" guns of 1944. The largest gun able to move on its own tracks was the "Karl" family of mortars of 1939, each weighing 125 tons and with a caliber of 23.6 inches. The biggest gun of all time was the "Gustav" used in 1942 to bombard Sevastopol; mounted on four pairs of rail tracks, it had a caliber of 31.5 inches, fired a shell weighing five tons a distance of 30 miles, and weighed over 1350 tons. Amazingly, this gigantic weapon simply disappeared in 1945 and has never been found.

Below: a howitzer belonging to the modern Swiss Army. Howitzers have a higher trajectory and usually a shorter barrel than other artillery of the same caliber. They will also have a flatter trajectory and a longer barrel than a mortar.

Bombs and Explosives

There have been major advances in what today are called "conventional" or non-nuclear explosives and in an increasing diversity of weapons.

For more than 2000 years the only known explosive was black powder or gunpowder, produced by mixing potassium nitrate, charcoal (carbon), and sulfur. In 1865 a safe way was found to make nitrocellulose or guncotton, and this became the basis for a wide range of modern explosives. Soon afterward the Swedish inventor Alfred Nobel discovered that the tricky liquid nitroglycerine could be absorbed into a porous material to make the safe solid known as dynamite. By gelatinizing nitrocellulose and nitroglycerine a British group made cordite, which since 1895 has been a standard propellant (a slow-burning explosive used in cartridges and shell cases to drive the projectile out of the gun). High explosives, which detonate with violence, could not be used for this task without shattering the gun. High explosive fillings are used in shells, bombs, and other devices where the maximum shattering or blasting power is needed.

Explosives are usually detonated either by an extremely sensitive high explosive, such as fulminate

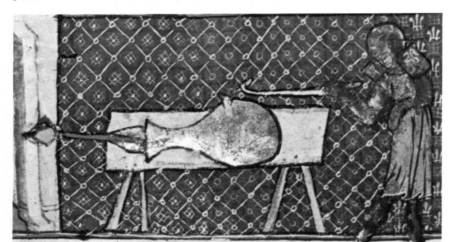

Left: the first known depiction of a gun in action is taken from a manuscript produced in 1326. The artist has attempted to show the bolt leaving the barrel at the very instant of firing. By the early 14th century three types of primitive gun had developed – the bombard, the mortar, and the culverin.

Below: the effects of high explosives can be clearly seen from what was left of Sackville Street, Dublin, after its buildings were blasted by British troops during the Easter Rising of 1916. The force of the explosions ripped out roofs, windows, and doors.

of mercury in a percussion cap struck by a pin in the mechanism of the gun or fuse, or by exploding bridge-wire, a fine wire suddenly made to pass a giant pulse of electric current. In either case the detonator sends a violent shockwave into the main charge which in turn detonates. In the open air some high explosives can burn without exploding. Obviously the tendency in the 20th century has been to use explosives that are safer to handle yet can be detonated with the greatest effect. An important basic composition is TNT (trinitrotoluene), which when mixed with ammonium nitrate yields Amatol; both are stable enough to be safely fired inside shells, detonating only when the fuse is triggered on hitting the target. In World War II several much more powerful explosives came into large-scale use, notably PETN (pentaerythritol tetra-nitrate) and another with a very long chemical name produced under such popular names as RDX, hexogen, trialen, and cyclonite. The latter family was standard filling for large blasting weapons by 1945.

Most explosive devices need some kind of fuse. The simplest hand grenade has a central fuse topped by a percussion cap that is struck by a sprung hammer released when the bomb leaves the hand of the thrower. The cap ignites a slow-burning tube or cord, which after 4 to 7 seconds explodes the detonator at the other end in the center of the main charge. Shells often have impact fuses with a safety device, which is sheared by the firing from the gun, thus making the projectile "live" and detonated by the free weight in the fuse on hitting the target. Antiaircraft and some other shells have heavy metal cases and scatter lethal fragments after the manner of shrapnel (shells that originally scattered lead shot invented during the Napoleonic Wars by a British artilleryman, Henry Shrapnel). The antiaircraft shell thus needed a time fuse, set to a given period of time-delay immediately before being loaded. In 1944 the United States' proximity fuse began to replace the time fuse in anti-aircraft and many other shells, with a remarkable miniature radio, able to withstand the shock of firing, which detonated the charge at the nearest point to the target. This device was later used in some modern missiles, which have sophisticated warheads with proximity fuses of several kinds.

Early aircraft bombs were merely fused canisters of high explosive but they soon diversified into: general purpose; heavy, steel-case armor-piercing; fragmentation for use against troops in the open; and blast or demolition with a light casing. In World War II the light-case bomb grew until in the Royal Air Force it came in multiples of 4000 pounds up to 12,000 pounds in a triple drumlike unit. The largest conventional bomb used in action was the Grand Slam, which weighed 10 tons and fell at supersonic speed, penetrating 100 feet into the ground and creating a local earthquake to shake its targets apart. Antisubmarine depth bombs and depth charges are large drums of high explosive triggered by a hydrostatic fuse actuated when the device has sunk to the correct depth. There

Above: minelaying off Iceland during World War II. The first known use of naval mines was at the Siege of Antwerp in the 16th century.

are numerous kinds of mine that can be placed on or buried in the ground, some of them arranged to fling a fragmentation charge to head height before detonating while others are booby-trapped to explode if they are tampered with.

In the late 1970s a large new family of explosive devices was coming into use for powerful blasting effect against large buildings and similar major targets. A large cloud of microscopic liquid fuel drop-lets is first scattered all around the target; this is then detonated. The effect is much greater than that of the same weight of solid high explosive.

Above: two South Vietnamese infantrymen of the 7th Division, Army of the Republic of Vietnam (ARVN). The soldier in front is armed with grenades. Grenades were used extensively in the 17th and 18th centuries, and are still employed today.

Chemical
and Biological
Warfare

Chemical and biological warfare have their origins far back in history. Many ancient civilizations resorted to the poisoning of enemy wells or rivers. And there is evidence of the deliberate spreading of such diseases as typhus, smallpox, and plague in order to weaken an enemy. The first use of irritants or toxins by a modern army was the tear-gas grenade, used by the French in August 1914. The Germans followed this by massive use of more severely irritant gases from the following October, and then the deadly gas chlorine on April 22, 1915. From then until the Armistice in November 1918 chemical warfare was commonplace and on several fronts. Various irritants, toxins, and blister agents were used, causing roughly 1,000,000 casualities of whom over 91,000 died (a smaller proportion than for casualties of conventional weapons).

Right: USAF aircraft spraying defoliants during the Vietnam war. Defoliants were ostensibly used to deprive the Vietcong and the North Vietnamese Army of cover. It also destroyed vast areas of food-producing land.

Below: World War I witnessed the first large-scale use of poison gases. These British troops are wearing the first primitive gas masks, which provided reasonable protection until mustard gas was introduced.

These early chemicals were used in crude ways, by spraying into a favorable wind or by being packed into artillery shells or mortar bombs to burst among the enemy. Similar chemicals were used by the Italians and Japanese in the 1930s. Gases were not used in World War II because each side was afraid of retaliation – possibly the first demonstration of the concept of deterrence. Research, mainly in Germany, had by the 1940s led to the manufacture of far more deadly compounds. Colorless and odorless, they were the first "nerve gases," so-called because their attack is on the chemical process by which the nerves command the body's muscles. They act very quickly, killing by both contact and inhalation and among other effects stopping the action of the lungs. Some gases irritate the mucus membranes in the windpipe.

In the immediate postwar years British research discovered the most lethal chemical agents so far publicly known. Related to the nerve gases, they were named "VE" and "VX," and are swiftly lethal even in microscopic quantities, whether inhaled or touched on the skin. Since then little information on this emotion-charged subject has emerged, though it is almost certainly true to suspect that research in the Western democracies has declined sharply for lack of inadequate funds. In addition to the diversity of lethal agents, there also exist many non-lethal chemicals whose effect is to irritate and disperse hostile crowds (tear gases), cause hallucinations (LSD and many other so-called "drugs") or merely make subjects

Below: South Korean police attempt to disperse a demonstration by heavy use of tear gas. Tear gas causes running eyes and coughing, which usually soon pass off, but can sometimes cause permanent damage. People suffering from chest complaints are particularly at risk.

docile and submissive (there has been controversy over alleged use of such preparations in prisons). Other chemicals have been used not against humans but to defoliate trees and other vegetation and so remove natural cover that previously screened hostile troops.

Unlike "chemicals," biological agents, such as disease-producing bacteria and viruses that invade the human body naturally, have never been used in warfare – so far as is known – since the deliberate spreading of smallpox among North American Indians in the 18th century. The number of possible organisms is almost limitless, and not only do they have the ability to reproduce themselves but their toxicity is appalling. One ounce of some viruses could infect up

Above: a worker at a biological warfare research station. He is thoroughly protected against the risk of contamination. For good reason – a quarter of an ounce of Botulinus Toxin could kill about 60 million people. Security is tight.

to 28,000,000,000 humans, many times the world's population. But the astronomic hazards and moral implications, and possibly the technical difficulty of accurately controlling biological warfare have, thankfully, so far prevented its use. Nevertheless research almost certainly continues on a large scale, the objective in most countries being how biological warfare might most effectively be countered.

The use of radiation weapons is usually associated with nuclear devices working by fission or fusion, but instead of building these in the physically destructive form describable as bombs they can be constructed primarily as emitters of lethal radiation. There has even been research to determine the extent to which lasers could be used as high-intensity radiation weapons; but lasers can also help to detect the presence of noxious chemicals, so some comfort can be drawn from the fact that even in these frightening new realms of war there are still counterweapons.

Nuclear Weapons

In July 1945 the first "atom bomb" was triggered on a high tower at Alamogordo, New Mexico; it worked, exploding in a vast fireball many times brighter than the Sun. It contained a large lump of U-235 into which one or more other lumps were fired at the highest possible speed. A large part of the U-235 was annihilated in fissions, each atom turning surrounding U-238 atoms into a new element, Pu-239 (plutonium) which in turn causes fresh chain reactions. On August 6, 1945 a B-29 Superfortress, *Enola Gay*, dropped a U-235 fission bomb, *Little Boy*, on the Japanese city of Hiroshima; three days later another B-29, *Bockscar*, dropped a bomb using a different fission method, *Fat Man*, on Nagasaki. Both bombs had the explosive energy of about 20,000 tons of conventional TNT.

compact and efficient, operating on the "implosion" principle in which shaped charges of conventional explosive, all triggered together in a sphere surrounding the fissile material, suddenly compress the latter into a tight ball that sustains a chain reaction made more complete by the intense surrounding pressure. The new H-bomb essentially added a thick sphere of such material as lithium deuteride around the fission bomb, with the implosion charges forming an even larger sphere on the outside. The fission bomb triggered fusion reactions in the deuteride, turning hydrogen atoms into helium. Compared with the fission reaction the amount of energy liberated was many times greater. Instead of thousands of tons, millions of tons, or megatons, were needed to compare the yield with the energy of conventional explosives, and in 1961 the Soviet Union detonated a thermonuclear weapon with a yield of no less than 60,000,000 tons of TNT. Today such mighty warheads are needed only for destroying nuclear-resistant targets such as underground missile sites. To attack cities it is much more effective to use many small warheads.

In 1960 a new kind of nuclear weapon became of interest because of its ability to kill troops inside armored vehicles. The enhanced radiation bomb, commonly called the neutron bomb, uses a low-yield fission reaction to trigger a fusion reaction, which

Left: The Manhattan Project set up in June 1942, led to the production of the atomic bombs "Little Boy" and "Fat man" which destroyed the Japanese cities Hiroshima and Nagasaki in August 1945. Uranium 235 was produced at Oak Ridge, Tennessee.

Right: a French nuclear bomb exploded over a Pacific atoll in Polynesia in 1972. Although exploded in a remote area, the bomb presented a fallout risk to Pitcairn Island. By 1972 most nuclear powers had agreed to ban testing in the Earth's atmosphere.

Above right: an Atlas intercontinental ballistic missile (ICBM) being launched from Cape Canaveral, Florida. The thermonuclear warhead in the now obsolete Atlas missile can destroy cities the size of New York.

In 1952 the United States tested the first bomb of a far more destructive character, the fusion or thermonuclear bomb, often called an H-bomb because deuterium, an isotope of hydrogen, is used to make it. By this time ordinary fission bombs were much more emits almost four-fifths of its energy in the form of penetrating neutrons that are deadly to humans over a wide radius. The NATO armies have for many years eyed the neutron bomb as a possible way of halting a hostile armored onslaught.

Missile Systems

Right: Nazi Germany was in the forefront of rocket development in a last-ditch attempt to win World War II in Europe. The V-2 was used extensively against targets in Britain and Belgium during 1944–45. Powered by a mixture of liquid oxygen and alcohol, the V-2 was the predecessor of the modern ballistic missile. Prominent in German rocket research was Wernher von Braun. After World War II von Braun went to work for the United States, where he played a major role in that country's space program, initially working for the Army.

Though they are not normally classed as missiles, the naval torpedo was the first of the self-propelled semi-guided weapons. The British inventor Robert Whitehead built his first modern torpedo in 1866, with a pendulum and gyro for stabilization. For 70 years torpedoes were stabilized to maintain their preset course and running depth, but in World War II new types appeared with guidance toward the target. Some headed for the target automatically – or "homed" – by acoustic means, some homed by magnetic methods, and others were steered by so-called "command guidance" by electrical signals transmitted along fine wires from the parent vessel. Today most torpedoes are acoustic homers. Some listen for sounds coming through the water from the hostile ship and steer toward their source. Others send out intense sound, or acoustic signals and head for the source of any reflections. Some larger torpedoes are wire guided, and a few modern types use wire and acoustic methods together.

In World War I many aerial guided or stabilized missiles were built, most in the form of miniature aeroplanes packed with explosives and steered by wire commands or merely stabilized by an autopilot.

Below: a V-1 being hauled to its launching catapult by German troops. During the summer and fall of 1944, the Germans unleashed an intense V-1 offensive against London and the southeast of England. Many V-1s were brought down by RAF fighters and the others were not very accurate. Nevertheless, they killed more than 2000 British civilians.

Above: recent wars have demonstrated the need for pinpoint accuracy in hitting ground targets from the air. Targets as small as a single truck could not be reliably destroyed by dropping bombs. The rocket-propelled air-to-surface missiles, like the one pictured in action here, offer pinpoint accuracy. The aircraft, or a companion, aims a laser beam at the target. The missile scores a direct hit by steering onto the laser energy.

None saw active service, and such work lapsed until World War II when there was renewed effort to improve on missile designs in Germany. In the autumn of 1943 two air-to-surface missiles went into active service with the Luftwaffe. A 3454-pound weapon, FX1400, was developed which consisted of a large armor-piercing bomb with radio command guidance acting on antennas on the tail. In its first action against the Italian navy – to prevent it falling into Allied hands after Italy had surrendered – it sank the battle ship *Roma* and severely damaged others. The Hs 293 was the first in a diverse family of miniature airplane missiles with similar radio command guidance and in most cases boosted by a rocket motor. Some of the Hs series were designed to hit ships below the waterline. Others were long-range gliding anti-ship missiles, whereas the unique Mistel (mistletoe) family were really rebuilt Junkers bombers fitted with a gigantic warhead in place of the crew compartment. Each was guided by a piloted fighter riding on top until near the target.

Two of the most notorious missiles of history were the so-called V (vengeance) weapons, V-1 and V-2. The former, whose real designation was Fi 103, was a Luftwaffe cruise missile (a cruise missile is a pilotless bomber with wings) which had no guidance beyond autopilot stabilization and thus was useless except against such large targets as London and Antwerp, both of which were hit by more than 2400 of these devastating weapons. V-1 was driven by a cheap pulsejet that was started by a large airhose; the missile was then shot off an inclined rail and flew at low level at almost 400 miles per hour for distances up to 150 miles, diving to the ground after a preset distance. The V-2 was the first large ballistic, or wingless, rocket, planned by the army as a mobile extension to artillery. This amazing rocket stood over 46 feet high when raised upright for firing and weighed over 28,000 pounds. It was propelled by a rocket chamber burning liquid oxygen and alcohol that gave it a thrust exceeding 25 tons, to fly at supersonic speed (mostly above the atmosphere) to targets up to 200 miles distant. About 6000 were built – compared with 30,000 V-1s – and again London and Antwerp were the chief targets.

In 1944 the V-1 assault was blunted by radar-predicted guns with proximity-fused shells, the fastest fighters, and barrage balloons; but there was no defense against the V-2. Meanwhile, in the Pacific Japanese Kamikaze suicide attacks against warships (a few by the MXY-7 Ohka, the only true piloted missile to go into action) spurred frantic development of shipboard surface-to-air missile (SAM) defenses. The chief United States Navy program was called Bumblebee, out of which grew such missiles as Terrier and Tartar (both rockets) and Talos (which had a much longer range, resulting from use of an integral kerosene-burning ramjet). These went into United States warships from 1956, and advanced forms are still

Above: strategic missiles, belonging to the Soviet Union, ready for launch. The Strategic Rocket Troops take precedence over all other forces in the Red Army. The Warsaw Pact has a greater firepower in missile-launched nuclear warheads than the West.

Right: the interior of a Minuteman missile silo viewed from above. Minuteman I was officially declared operational by the United States in December 1962, and was later replaced by Minutemen II and III.

in use. The lastest successors are two forms of Standard Missile that are integrated into complex ship systems involving magazines, handling systems, launchers, radars, and computers. The first SAM to go into use was the Nike Ajax, operational in 1953, with radar command guidance by a system in which radars track the target and the missile and a computer drives the two into the same place in the sky. From this descended the vastly more powerful Nike Hercules and, after the expenditure of billions of dollars, the Safeguard antiballistic missile system, which the United States Congress finally halted in 1976.

In the immediate postwar years improved guidance and the fission, and later fusion, nuclear weapons, transformed the potential of the ballistic missile. In 1957 the Soviet Union had built a gigantic intercontinental ballistic missile (ICBM) with 32 rocket engines, which in October of that year launched the first Earth satellite. The USA in 1947 underestimated the possibilities and worked only on cruise missiles, such as Mace (600 miles) and Snark (5000 miles), but in 1953 a scientific committee was hastily convened to reassess the problem. The result was a crash program that resulted in two 1700-mile intermediate-range ballistic missiles, Thor and Jupiter, and two ICBMs, Atlas and Titan. Atlas was first deployed in 1960 in exposed launchers, then in coffins recessed into the ground, and finally in silos sunk deep into the earth and hardened against nuclear attack. Both Titan and

the larger Titan II were silo-emplaced. Titan, with over 9000 miles' range, could be fired at short notice from the bottom of its silo.

In the 1960s the United States Air Force deployed 1000 smaller Minuteman ICBMs with solid fuel, and, after studying mobile use on railway trucks, perhaps mistakenly opted for hardened silos. Since then the Soviet Union has so persisted in ever greater forces of larger, more accurate, and more deadly ICBMs that the deterrent capability of the Minuteman force has become degraded. Today Soviet ICBMs can knock out the 1000 silos and still have enough missiles left to dominate the rest of the world. An additional factor is the submarine-launched ballistic missile. The first examples were the Soviet Union's N-4 and the United States Navy Polaris, both in use from 1960. Again, the initial United States lead was wiped out by vastly greater Soviet forces, which include not only more numerous and larger missiles but also outstanding accuracy and range, approximately double that of any

Western submarine-launched ballistic missile. Yet
another complicating factor is that modern strategic
missiles can carry penetration aids (known as
"penaids"), to confuse or dilute the defenses and
multiple independently targeted reentry vehicles
("Mirvs"), warheads that can deliver a shower of
nuclear devices to precise locations at the same time.
In the USA there has been no money for a new ICBM
since the Minuteman of the 1960s, but the old idea of
a cruise missile (a modernized V-1) has been the subject of much sudden effort as a possible way of prolonging the life of the B-52 bomber, first flown in 1952.

Today all advanced tactical air forces use precision-
guided air-to-surface missile systems (ASM), with
various forms of guidance including laser homing,
radar homing, infrared heat-homing, radio command,
and TV homing or command. Anti-ship ASMs either
dive on their targets or skim the waves, and defensive
shipboard missiles are fast-reacting multiple-launch
devices able to home in seconds on incoming missiles

even if the latter are very close to the water (a situation
which until the 1960s virtually precluded radar homing). In the 1980s close-range ship defense combines
missiles, unguided rockets fired in salvoes, and rapid-
fire cannon, all under integrated computer control.

The first air-to-air missile (AAM) was the German
X-4, with wire guidance, which just failed to see action
in World War II. Subsequent AAMs have had radio
command guidance, IR homing on the heat of aircraft
engines and, most commonly, semiactive radar homing, in which the missile homes on the fighter's own
radar signals reflected from the target. Today the most
powerful AAM outside the Soviet Union is the US
Navy's Phoenix, with a range exceeding 100 miles. In
contrast, close-range dogfight missiles can be launched
in traditional-style combat and outmaneuver all
manned aircraft. The most common warhead for
AAMs is the continuous-rod type in which the explosive is wrapped in a steel rod that breaks up into
lethal fragments. Similar warheads, but usually much
larger, are used in SAMs. The latest in this category
are carried on armored vehicles and accompany an
army at instant readiness. One, the British Rapier, is
much smaller than other SAMs because it is accurate
enough to hit and penetrate its target before the warhead detonates.

Other missile categories are the anti-submarine
warfare (ASW) and antitank types. The former include
submarine-launched rockets and ship-launched miniature aeroplanes carrying a homing torpedo. The antitank missile invariably had wire guidance until
recently, when infrared or laser methods became popular. These missiles are often small enough to be used
by infantry (as are a few SAM systems), and are lethal
to all armor to a range of several miles. Invariably the
warhead is the shaped-charge type, with a forward-
facing hollow cone to focus the blast into a jet able to
pierce all known armor.

Military Aircraft

Until the start of World War I the only aircraft used in war had been balloons and kites – which many armies had found valuable as elevated reconnaissance platforms – and a handful of early airplanes used in Tripolitania and the Balkans for reconnaissance and to drop primitive bombs. In the opening months of World War I air operations were sporadic. A lone German flew his 80 horsepower monoplane to Paris and dropped leaflets, while French and British biplanes dropped small bombs on Zeppelin sheds. The Zeppelin airships were regarded as dangerous foes, and in fact for the next three years the Imperial German Navy used ever-better airships with great courage on high-altitude bombing missions over England, enjoying a degree of success that seems surprising to later generations. Antiaircraft guns were inefficient and hard to aim, there was at first no organized defense, and airplanes found it almost impossible to keep pace with the airships while climbing up to their altitude. By 1918 the Zeppelins were flying at 20,000 feet, inaudible from the ground, but by this time superior aircraft could shoot them down.

In the first month of the 1914–1918 war three British airplanes harassed a German aircraft and forced it to alight in British Army territory. On October 5, 1914 a French Voisin, a crude-looking two-seat airplane used for bombing and reconnaissance, went on patrol with a Hotchkiss machine gun (then a rarity in the sky) and shot down a German aircraft. In April of the following year a famous prewar French aviator, Roland Garros, went into action with crude steel deflectors to protect his propeller blades from a forward-firing machine gun. At least eight inventors had promoted schemes for such guns before the war, without any official interest. Now, with this simpler scheme, Garros quickly ran up a succession of aerial victories. Within days he was forced down (not in aerial combat) and his secret was out. A week later the Dutchman Anthony Fokker, the Germans' leading aircraft designer, was flight-testing a nimble monoplane fitted not with the copy of Garros' device that had been requested but a proper synchronization system that let the gun fire only when no propeller blade was passing in front. For the next year the Fokker Eindecker (monoplane) was so successful in the hands of such pilots as Oswald Boelcke and

Above: the machine-gun arrangement in the Fokker E.III of German World War I fighter ace, Oswald Boelcke.

Right: a painting on a lid of a snuff box shows an observation balloon over the Battle of Fleurus in 1794.

Max Immelmann – the first "Aces" – that Allied pilots were called "Fokker fodder."

In 1916 most fighter aircraft were biplanes, with rotary engines of about 100 horsepower and one or two rifle-caliber guns. By 1918 the engines were 200–300 horsepower and after trying highly maneuverable triplanes designers went back to the biplane. Two-seat fighters were useful, the rear cockpit having one or two pivoted guns and affording defense to the rear. Large bombers with up to five engines were much used by Russia, Italy, Britain, and Germany, with bomb-loads of several tons. Equally impressive flying boats backed up the blimps (non-rigid airships) used for ocean patrol and antisubmarine missions.

Until the mid-1930s development was modest, but after that time engines of 1000 horsepower were developed in partnership with all-metal "stressed-skin" structures that made the unbraced monoplane not only practical but markedly superior to the wire-and-strut-braced biplane. Often against the beliefs of the diehard pilots, fighters acquired enclosed cockpits, retractable landing gear, variable-pitch propellers, flaps, and much heavier armament. Britain opted for eight rifle-caliber guns along the wings, while other countries preferred smaller numbers of cannon of 20 or even 37 mm caliber (0.79 or 1.46 inches). In the Spanish Civil War of 1936–1939 fighter opposition was

Right: P-51 Mustangs of the USAF. The Mustang was one of the most effective fighters of World War II and was used to escort both USAF and RAF bomber raids over German occupied Europe from the beginning of 1944. On March 4, the US 8th Air Force attacked Berlin, escorted by Mustangs.

Below: the most famous light bomber of World War II, the de Havilland DH.98 Mosquito. Of lightweight wooden construction, the Mosquito was powered by two Rolls-Royce Merlin engines, and could attain a speed of 415 mph. It has a crew of two and a range up to 3000 miles.

often poor, and the reborn Luftwaffe concluded that 200 miles-per-hour bombers defended by three hand-held machine guns (one in front, one at upper rear and the other at lower rear) were completely adequate. The dive-bomber, which aimed by diving almost vertically onto its target, also seemed outstanding.

In World War II both beliefs were shattered. Determined fighter opposition by the Royal Air Force knocked down the Luftwaffe bombers in droves, and before long the once-dreaded Junkers Ju 87 "Stuka" dive bomber was replaced by the Focke-Wulf Fw190, a superb all-round combat aircraft that was also one of the best fighter-bombers. The Royal Air Force's surprise weapon turned out to be the Mosquito, an all-wood bomber so fast (400 miles per hour) it needed no armament. Rejected by the officials in 1938-40, it proved to be one of the most versatile aircraft of the war, and also became one of the greatest radar-equipped night fighters. By 1944 the P-51 Mustang was escorting the B-17 and B-24 bombers on the longest

Above: Unites States Army troops practice disembarking from a Chinook CH-46 helicopter. Built by the Boeing Vertol Company, the Chinook can carry between 30 and 40 people. Powered by two 3750 hp Lycoming T55-L-11 turboshafts, the Chinook has a cruising speed of 161 mph.

missions, and the Germans had lost control of their own airspace despite introduction of several very advanced rocket and jet aircraft.

In the postwar era the jet swiftly replaced the piston engine in fighters and bombers, and in Korea the main battles were fought by aircraft such as the F-86 Sabre and MiG-15 with sweptback wings and tail for speeds close to that of sound. In 1955 the F100 Super Sabre and MiG-19 exceeded sonic speed on the level, and a year later the first F-104 and MiG-21 were approaching double this speed. Guided missiles were carried for use against aerial and surface targets, the largest such weapons being able to fly nearly 1000 miles after release from the parent bomber. The standard United States heavy bomber from 1955 was the giant 240-ton eight-jet B-52. All planned replacements of these aircraft have been canceled, which means that the surviving examples of these aircraft will have to fullfil their role to the end of the century. While the B52 was in production, in 1953–62, it was

Above: a Grumman F-14 Tomcat fighter firing a Phoenix air-to-air missile in 1977. The Grumman Tomcat of the US Navy, is one of the world's most advanced fighters. It is capable of attacking six targets simultaneously with Phoenix missiles.

improved by being given new engines of greater power and efficiency, much greater fuel capacity, and totally new electronics and armament schemes. Several further rebuilding programs have kept the structure sound and enabled these old and potentially vulnerable aircraft to penetrate defended airspace. In the 1980s the B-52 force will carry cruise missiles with a range of over 1500 miles after release.

In the vital maritime roles flying boats and seaplanes have virtually disappeared, and large jet or turboprop landplanes now fly for 12 hours or more with flight crew of four or five and a tactical crew of six to eight, clustered around a mass of sensing and display systems which, with computer management, tell them and any friendly ships or other aircraft with them every detail of the sea or submarine situation. Even more remarkable are the airborne early warning and airborne warning and control system aircraft. These are flying radar command posts, able not only to study and keep track of every air vehicle (including missiles) within a radius of up to 245 miles, but also to control all friendly air traffic to hit the enemy, avoid collisions or hostile missiles, and return safely to base even after suffering damage or in bad weather. The airborne warning and control system is said to about triple the effectiveness of an air force, and to have a major effect on surface forces also, justifying a price tag of about 85 million dollars each. Even more costly are the giant airborne command posts, which are the national seat of government in time of crisis, with a crew of about 60 and extremely comprehensive communications system.

In the 1950s there was a strong move toward the semi- or completely automatic all-weather interceptor, able to destroy hostile aircraft without the pilot seeing the target. Today there has been a somewhat questionable return to the traditional dogfight with the pilot able to see the target and the use of guns (usually a multibarrel cannon firing up to 6000 shells per minute) and close-range snapshoot missiles. The latest such fighters have so-called control-configured vehicle technology; they are designed to be unstable, for quick maneuvers, stability then being supplied by electronic systems in the flight control circuits. Requirements for these aircraft are tremendous engine power and a large wing, whereas the tactical attack aircraft needs small engines (for long range) and the smallest possible wing (for smooth flight at the ground-hugging height where it may be possible to evade hostile radars and missiles). Some combat aircraft have variable-sweep "swing wings," spread out for take-off or subsonic cruise and folded back for low-level attack at supersonic speed. Specialized tactical types include the Harrier and AV8B with vectored-thrust engines, whose nozzles can point down for vertical/short take-off and landing (V/STOL) and at other angles for forward flight, braking or special maneuvers, and the A10 Thunderbolt II close-support aircraft able to survive considerable hostile fire and hit tanks and other battlefield targets with a very powerful gun and missiles.

Helicopters, naturally valuable as tactical airlifters, are also efficient tank-killers when fitted with special sighting systems and missiles. When they are equipped with such antisubmarine sensors as dipping sonar (for listening to noises in the ocean) and magnetic-anomaly detection (MAD) for measuring the distortion of the Earth's magnetic field by a submarine, helicopters are the most potent antisubmarine warfare aircraft.

Below: the world famous BAe Sea Harrier jump-jet. This V/STOL jet is pictured here just after its launch from a ski ramp. The Harrier has caught the attention of air forces and navies around the world. They recognise the potential of a combat aircraft that can use small areas.

Above: two RAF Lightning interceptors demonstrate their astonishing rate of climb. The first Lightning flew in August 1954. They entered service with the RAF in the spring of 1960, and are still in use today as trainers.

Land Vehicles

Apart from large wheel-mounted catapults and siege cannon, the spectrum of army vehicles through the ages consisted of little more than supply wagons, field guns, and ammunition limbers until well into World War I. Then several major developments happened. The armored car, which had previously existed as a modification to high-quality motorcars, emerged as a vehicle designed for the job, with wheels and springs tailored to the severe task of carrying an armored body over rough ground. The heaviest artillery appeared on railroad tracks, at first for reasons of mobility and later, as size increased, because there was no other way the guns could be transported. Most important of all was the tank.

This was a totally new concept which, unlike the earliest armored cars, was designed for use on rough ground away from prepared roads. Its development was spurred by the terrible stalemate of trench warfare that had become apparent by 1915. Barbed wire and quite modest strength in machine guns could halt or slaughter any force of advancing infantry, and prolonged bombardment by artillery prior to an attack caused as many problems as it solved. An officer of the British Royal Engineers, Colonel Ernest Swinton, saw that by assembling several existing

Above: a light tank of a German *Panzer* division during the invasion of Soviet Union – Operation Barbarossa – 1941. What made the German tank forces so formidable was their deployment rather than their armored strength. They were concentrated in specialized *Panzer* divisons. The element of surprise was exploited to the full.

Below: the best known Soviet tank of World War II, the T-34. These 30 ton medium tanks provided daunting opposition to the Germans. Its main gun was of 76.2mm caliber, firing a 14 pound shell at a velocity of 2450 fps. A 7.62mm caliber machine-gun was also mounted.

Below: men of the 5th Canadian Mounted Rifles clear an obstacle in a Mark IV "Mother" tank on June 12, 1918. The word "tank" was a code word, chosen because the contraption looked like a water tank. Tanks were first used by the British at Cambrai in 1916.

components into a new vehicle the deadlock could be broken. The components were the internal combustion engine, caterpillar track, armor, turrets or barbettes, and various types of gun. The only new feature was bringing all these together.

With the support of Winston Churchill, First Lord of the Admiralty, the first armored fighting vehicle was built in 1915. As Navy money was involved it was called a Landship, and as a security cover it was known as a "tank" (because it could be made to look like a large water tank). The cumbersome 31-ton monster was slow and unreliable, but the Mark I tank was a usable vehicle by mid-1916 and it was decided to use a few in the Somme battles. The first went into action on September 15, 1916. Had thousands been used the war would have been won on the spot; as it was, the enemy was merely given an instructive demonstration of the new vehicle. Britain built "male" tanks with shell-firing guns (invariably a 6-pounder) in each side barbette and "female" tanks with machine guns only. These early tanks were slowly moving armored boxes whose interior was hot, deafeningly noisy, and occupied by a driver, six gunners, and a commander. They were proof against machine-gun and rifle fire, but could be knocked out by heavier weapons and in any case suffered frequent mechanical failure.

By the end of World War I hundreds of tanks were in use with nearly all participants. Most were rather smaller than the original designs, and until the mid-

1930s most tanks remained fairly simple rivetted constructions unable to go faster than infantry and armed with machine guns and, in some cases, a 2-pounder. But then came great changes. An American designer, J Walter Christie, who found little interest in tanks in his own country, went to the Soviet Union, which from 1930 built "cruiser" tanks incorporating two of his important innovations. One was large rubber-tired wheels filling the gap between the upper and lower parts of each track, and able to be used even after the track had been shot off. The other was sloping armor, which curiously had been ignored by earlier designers. Through the Christie BT series the Soviet designers produced the T-34, first delivered for service in 1940, which many consider the best tank of World War II.

Tough, and well-protected by sloping armor, the T-34 had ample engine power, gun power, and speed, in an era when most tanks (especially the British) lacked all three. Starting with a 3.0 inch gun, the T-34 was altered to take an 85 mm gun in 1943 and, unlike all other armored vehicles of its generation, remained in production until 1964, the total constructed being in the order of 100,000. It provided a perfect basis for all subsequent Soviet battle tanks, the T-54, T-62, T-72, and T-80, which are outstanding vehicles with powerful diesel engines, guns up to 125 mm in caliber, and excellent equipment and protection.

No other nation has achieved such a long series of good tanks. Some built slow but heavily armored "infantry tanks" in World War II, while their cruiser tanks were weakly gunned and armored. The Germans, whose Panzers – armored units – were justly famous, led in fitting large guns on tank chassis to produce the self-propelled gun, and special antitank guns on similar chasis to create the tank destroyer –

Below: the British Stalwart amphibious troop carrier. This vehicle has a top land speed of 45 mph, with a water speed of six knots. With an unladen weight of eight tons, it can manage a land acceleration of 0–45 mph in 46 seconds. It manages $4\frac{1}{2}$ miles per gallon.

virtually a tank with extra gun power but less armor and sometimes open at the top. Special armored fighting vehicles, or AFVs, were developed for beach assault; engineer duties such as bulldozing, rescuing stranded tanks, erecting bridges, and digging trenches; airborne assault by transport aircraft or glider; and amphibious travel either with large canvas sidescreens or made truly amphibious. Modern amphibious tanks float and are propelled by water jets, whereas heavy battle tanks can be sealed to travel under water (for example, in crossing rivers) while breathing through a schnorkel tube like the latest non-nuclear submarines.

In World War II there were thousands of armored cars and also the relatively small, light, and usually open at the top scout cars which were used for reconnaissance. Armored cars often had the same gun turrets as tanks, and by 1950 some outstanding chassis were being used to carry armored-car bodies, troop carrier bodies, special command posts with extensive radio communications, and several other kinds of superstructure. The armored personnel carrier emerged as equal in importance to the tank, with less armor but more room, able to carry a crew of two and up to a dozen infantry at speeds up to 40 miles per hour over rough ground or across water. The same chassis can carry a mortar crew, antitank missiles, anti-

Above: a British armored personnel carrier (APC) on maneuvers at Warminster, Wiltshire. The first such APC was developed in Britain as early as 1917, but the concept failed to gain serious attention until late in World War II. The advantages of providing troops with mechanized protective transport were only then appreciated.

Below: Israel's convincing military victory in the Yom Kippur War was achieved by superiority in the air and rapid armored advance. The Israelis advanced into Syria to bombard sensitive targets and enemy formations after the air force had "softened up" points of enemy resistance before the armored attack.

aircraft system (often with radar) and sometimes other weapons such as a flamethrower. A related armored fighting vehicle is the mechanized infantry combat vehicle which is generally more compact than the armored personnel carrier and carries fewer troops but has firing ports for them to go into action from within the vehicle.

Today large battle tanks weigh about 50 tons, have engines of at least 1000 horsepower (either variable-compression piston engines able to burn any available fuel, or a gas-turbine) and not only a gun of at least 120 mm caliber in a 360° turret but also such equipment as nuclear, biological, or chemical-warfare gear, infrared night-vision systems, advanced optics, and gyrostabilization for sighting the main gun, and a laser for accurate target-ranging. Main-gun ammunition can be of the armor-piercing discarding sabot type, which penetrates enemy armor by sheer velocity, or of the high explosive squash-head type, which sends shockwaves through the armor so that lethal pieces shatter off the inside. Antitank missiles have shaped-charge or hollow-charge heads that punch through the armor with a high-velocity jet.

A small proportion of modern armored fighting vehicles carry missile systems. Some carry tactical bombardment missiles, such as the American Lance or Soviet Scaleboard. Others carry batteries of anti-

Below: the building of enormous intercontinental ballistic missiles (ICBMs) necessitated the development of equally impressive wheeled transports to convey them to their launching sites. Picture shows a parade of rocket troops in Red Square, Moscow, with specially built carriers transporting ICBMs.

armor weapons, while one important class carry anti-aircraft missiles, radars, or guns. The Soviet forces have for two decades had two excellent armored fighting vehicle-mounted multibarrel gun systems for antiaircraft use, while armored fighting vehicle-mounted SAM (surface-to-air missile) systems include Mobile Rapier, Roland, and SA-4 Ganef. Of course, almost all land missiles are at some time transported by vehicles. Even the largest intercontinental ballistic missiles may be towed on large trailers, or delivered to their silos in special vehicle-mounted containers that can be up-ended vertically to lower the monster weapon by pulley-block systems. Some self-propelled

Above: an armored fighting vehicle acting as a launching pad for a Lance missile. The Lance is a tactical bombardment missile. These are intended to be used to bombard sensitive areas of enemy territory without the risk of spreading the destruction uncontrollably. Lance was developed by the United States. A similar weapon, but larger, is the Soviet Union's Scaleboard.

missiles ride on vehicles that also provide accommodation for the launch crew. Wheels and tracks are equally popular for such work, and the Soviet Union improves cross-country capability by providing for cab-controlled adjustment of tire pressures to suit the terrain, and by making every possible vehicle amphibious.

Today the spectrum of military land vehicles is larger than at any time in the past. Ordinary trucks, able to drive every wheel, and so cross rough terrain, carry such superstructures as command headquarters, multibarrel rocket launchers, heavy cranes, and large surveillance radars. Self-propelled artillery can fire every kind of shell including nuclear, canister (for use against exposed infantry), smoke, and illuminating. Armored fighting vehicle chassis are at least as versatile, some famous designs being available in more than a dozen variations for different kinds of superstructure or weapon.

Naval Warfare

The earliest warships were canoes and rowed galleys that fought by coming alongside each other so that their occupants could try to board the opposing vessel. Fierce hand-to-hand fighting would ensue, until one of the vessels was taken. The oars of galleys projected too far for a boarding party to jump the gap, and such ships hit the enemy head-on. Ramming was soon an important tactic in its own right, a projecting reinforced bow sometimes being able to inflict such damage to the enemy vessel below the waterline that it would sink. Ramming and boarding remained important to sea warfare even after the invention of heavy catapults and guns able to inflict severe damage at a distance. Battleships with ram bows continued to be built until 1911.

The defeat of the Spanish Armada by the English navy in 1588 was the first in which one side used long-range gunnery and superior maneuver to defeat a fleet much stronger at close quarters. From then on the naval gun developed swiftly, and in the mid-19th century it became a breech-loading rifled weapon firing various types of explosive or armor-piercing shell and mounted in pivoting barbettes or turrets. By 1880 guns were so large (calibers exceeding 12 inches) that they were aimed by steam or hydraulic power. Ships carried immense weights of steel and wood armor, often several feet in overall thickness. They usually mounted only two (at the most four) large guns in deck turrets; the rest were in the traditional place along the sides. The British battleship *Dreadnought* (1906) revolutionized the design of capital ships. It has ten 12-inch guns in five turrets, each with a very wide arc of fire. It also had steam-turbine propulsion, giving a speed of 21 knots.

By this time the torpedo had begun to revolutionize sea warfare, bringing with it two classes of small ship, the torpedo boat and submarine, either of which could sink a battleship. The destroyer emerged as a fast killer of torpedo boats, but the submarine was to remain a menace. Battle-cruisers, important in World

Below: the revolutionary British battleship *Dreadnought*. The Battle of Tsushima in 1905 convinced the navies of the world that their ships had to be more heavily armored and better armed. *Dreadnought*, launched in 1906, was armed with 10 12-inch guns in five turrets.

Above: anti-aircraft guns on a Royal Navy destroyer on escort duty in World War II. Germany tried its best to cut off vital supplies to Britain by attacking the convoys from the air and by U-boat. The anti-aircraft guns were not very lethal until after 1941.

Above: aircraft carriers have been used in the past to transport aircraft to positions well within flying range of enemy targets. There is intense interest in smaller carriers with V/STOL aircraft. Picture shows a Hawker Sea Harrier on the deck of HMS *Bulwark*.

War I, were at least as large as battleships but sacrificed armor for engine power and high speed. Cruisers were a numerous class like small battle-cruisers, with less-powerful guns but equal speed in the region of 32 knots. But by the 1930s the traditional battleship, like the lumbering heavy tank, was being replaced by the modern capital ship, with a displacement up to 50,000 tons (74,000 in the case of the two largest, which were Japanese) and lacking nothing in speed, firepower, or protection. Until World War II capital ships tended to have six to ten main guns of 11 to 18 inch caliber, a heavy battery of secondary

guns of about 6 inch caliber and a few antiaircraft weapons. Experience in World War II showed that, while the secondary guns were seldom used, anti-aircraft protection was vital, and the giant Japanese ships just mentioned had half their secondary guns removed by 1945 but the number of 25 mm antiaircraft guns was increased from 24 to 146.

In both world wars the German U-boat (submarine) proved nearly decisive in destroying opposing merchant shipping. Though in the second conflict the U-boats had to contend with capable long-range aircraft, their operation in packs in mid-ocean, with good radio communications and schnorkel breathing pipes (so that diesel engines could be run while at periscope depth, and batteries charged instead of discharged) made them deadly and hard to kill. In an urgent program in 1943–44 a new class was introduced with vastly greater electric battery capacity and underwater power, raising speed when submerged from the traditional 8 miles per hour to 18.5 miles per hour. This revolutionized post-war submarine design, but the United States Navy had an even greater development that matured with the maiden voyage of the USS *Nautilus* in January 1955. It was the first nuclear-powered submarine, and this made it the first true submarine whose natural habitat was the deep ocean (previous submarines had merely been able to submerge for limited periods). Moreover,

Above: the Soviet Union's floating satellite tracking station *Yuri Gagarin*. This vessel is equipped to pinpoint the position of the many satellites orbiting the Earth, and listen in on any radio communications between them and the ground. Both NATO and the Warsaw Pact employ such vessels.

nuclear power meant that the stored energy could be tens to hundreds of times greater than with chemical fuel, so underwater performance could be much greater. The introduction of a fat, circular-section "teardrop" hull shape, first tested on USS *Albacore* in 1953 made the submarine even more lethal. The new shape opened the way to much higher speed. The teardrop hull as well as nuclear power dramatically altered the survival capability of the submarine, so that it could be destroyed only by accurately placed homing torpedoes. In turn this led to the development of antisubmarine warfare missiles and helicopters.

Yet a further totally new development changed the entire role of the submarine. Instead of being a warship intended to sink ships it became a secure base from which long-range ballistic missiles could be launched against distant cities. The missiles were compact solid-propellant vehicles with thermonuclear warheads, carried in upright launch tubes and shot out by gas or steam pressure. Breaking the surface of the ocean, their first-stage motors would ignite. Today submarines can also carry other types of missile, fired from their regular torpedo tubes. One species is an antisubmarine missile, which after horizontal firing turns upright to shoot out of the ocean, arch up in a ballistic trajectory for scores of miles and then reenter the water at supersonic speed above the hostile submarine, which it destroys with a nuclear warhead. Other missiles are winged cruise types, which after making the transition from sea to sky sprout wings, tail, engine inlet (to an air-breathing turbojet), and controls, thereafter flying hundreds of miles to their target.

Other cruise missiles, pioneered by the Soviet Union, have transformed the capability of the fast patrol boat. In World War II the fast patrol boat carried either torpedoes, quick-firing guns, or antisubmarine weapons, and had no capability against any distant target. Today the small missile boat could sink a battleship – if any existed – beyond the range of the latter's guns, and some can hit targets beyond the visual horizon. One of the few weapons that can retaliate is the helicopter with overwater radar and

antiship sea-skimming missiles. Another is the high-speed attack aircraft.

Airpower has revolutionized sea warfare as much as the modern submarine. Many major actions in World War II were fought almost entirely by aircraft, most of them flying from aircraft carriers. Today only the United States Navy has a large fleet of carriers, and these are all larger than any other warships in history. One, CVN-65 *Enterprise*, is nuclear-powered. All have a complement of 5000 to over 6000 personnel, carry 70–95 aircraft, and have no armament except two or three small boxes of Sea Sparrow close-range SAMs. They are today's capital ships, replacing the traditional battleship which became extinct after the last United States example, BB-62 *New Jersey*, withdrew from Vietnam in 1969.

The main emphasis in most major navies is on multi-purpose carriers for vertical/short takeoff and landing aircraft (V/STOL), much smaller than those needed for large conventional aircraft, and various types of frigate. The carrier of the future is likely to be of about 15,000 tons (though the current Soviet series are more

Above: one of Britain's Polaris submarines. They form part of NATO's nuclear armory. The main hazard they face is the hunter-killer type of submarine. Polaris was operational in 1960, and long ago was phased out of the US Navy as obsolete.

than three times this displacement), with a flight deck curving up in a "ski jump" at the bow so that STOL aircraft can safely become airborne with an overload of fuel and weapons. No catapults or arrester gear are needed, and these ships will be more versatile and less vulnerable than the giant carriers. They will have a major role in submarine warfare and in amphibious operations against hostile coasts. Rather surprisingly, the big Soviet carriers, officially known as "antisubmarine cruisers," have powerful missiles for use against surface targets.

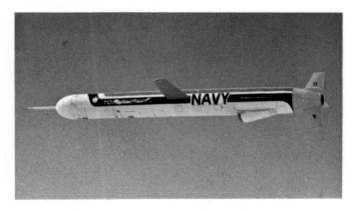

Above: a Tomahawk cruise missile in free flight just after its launch from an A-6 Intruder attack aircraft. Cruise missiles can fly low, parallel to the contours of the Earth. They try to avoid enemy radar.

The frigate or missile-armed destroyer is the chief warship in most modern navies. Usually displacing 2500 to 8000 tons, it has gas-turbine engines, several kinds of missiles and a high degree of automation for a small crew whose life-style is much more comfortable than in older vessels. Often there is only one gun, though modern naval guns up to 8-inch caliber can fire very rapidly – faster than one shell per second. In one respect modern warships differ greatly even from those of the 1950s: their electronic systems – made up of radars, counter-measures, other sensors, and communications and data-processing systems – are fantastically complex and powerful, putting completely new capability into ships that on paper may seem small and under-armed in relation to their cost and missions.

Above left: the early Polaris missiles had a preset guidance system, and could not quickly be re-targeted.

Above: the US Navy use the Boeing built Tucumcari hydrofoil for fast patrol-research. When foil-borne this 58-ton gunboat can reach a speed of over 50 knots.

Many navies also use vehicles other than displacement ships, notably hydrofoils and air-cushion vehicles, or hovercraft. Foil-riding ships are usually used in missile-firing roles and so far have generally been fairly small, around 200 tons. The naval air-cushion vehicles range from small inshore patrol craft, with automatic guns and wire-guided missiles, up to the large 3000-ton surface-effect ship used in antisubmarine warfare as a possibly unsinkable platform for helicopters and V/STOL fixed-wing aircraft.

Intelligence Systems

For centuries military intelligence had to grapple with the problem of acting on too little information. What information was obtained came from attachés resident in foreign countries, from spies, and from various other more or less reliable sources. Today the problem is the opposite one of handling a flood of detailed information obtained by networks that encompass the globe. It is a task demanding data-processing systems of immense capacity and speed, as well as large display systems, instant-retrieval systems, and many other devices – almost all electronic – that give information in the greatest detail not only to national leaders but right down to field and unit commanders. Obviously, the whole system has at the same time to be secure against hostile "listening-in."

The traditional methods of gathering intelligence are open publications of every kind (which can easily contain a key item of information), interrogation of prisoners, overflights by reconnaissance aircraft, and espionage agents. But while they continue to yield important information, the larger part of the major national intelligence machines is composed of electronic systems of amazing cunning and diversity that may orbit silently in satellites or listen from distant mountain peaks. An important facet of intelligence is the totally automated command, control, and communications network that in the major nations is not only nationwide but covers large areas of the globe and extends into space via satellite. Another vital part is the equally automated and fast-reacting system that provides warning of any kind of hostile attack, especially by strategic missiles.

The chief collection systems are generally Sigint (signals intelligence), Elint (electronic intelligence), Comint (communications intelligence), Rems (remote sensors), GSR (ground surveillance radar), weapon-locating radar, and a host of airborne sensors such as optical cameras, infrared, SLAR (side-looking airborne radar) and even human observation. Some sensors take traditional photographs, though using a variety of special films that defeat camouflage and reveal a wealth of information – and with such detail that the print on a newspaper can be read from a height of 10 miles. Many sensors give instant read-out at a distant base, typical information being the precise waveform, pulse-repetition frequency, and wavelengths of a hostile radar, together with its exact position. Infrared linescan not only provides a perfect picture of the temperature variation of hostile territory, hotter areas appearing lighter than cool ones, but can also show a spot where a vehicle was recently parked.

The Big Bird satellites put up by Titan IIID rockets for the United States Air Force are probably the most sophisticated reconnaissance satellites ever built, but it has been rare to have two launched in a year; in contrast the Soviet Cosmos series now exceed 1000 in number. Early "spy satellites" had to eject recoverable capsules, but today the information is sent back instantly by secure digital link for insertion into the

Above: A Russian Cosmos spy satellite. Both the United States and the Soviet Union make extensive use of satellites. During the Cuban missile crisis the Americans showed the Russians pictures showing how few missiles the Russians had on their soil. Russia backed down and withdrew missiles from Cuba.

Left: the USAF Lockheed EC-121 electronic warfare craft. It is slower than the famous U-2 or the SR-71, but has a greater range and maximum flight time than either. Flying at high altitude, the EC-121 can monitor communications for as much as 20 hours at a stretch. It is employed in many electronic roles.

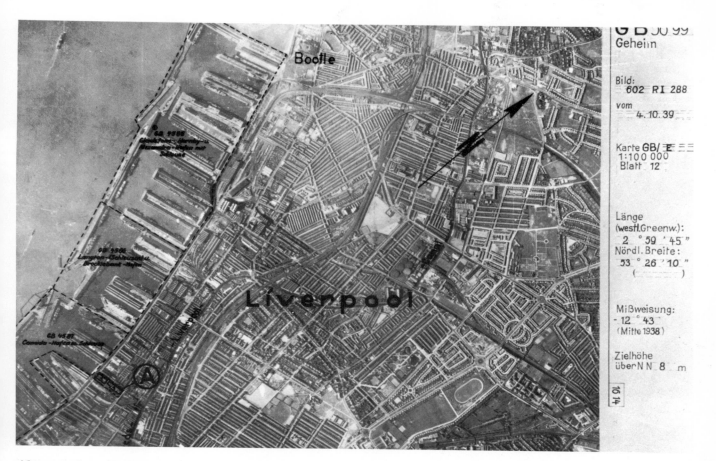

Above: a German World War II air reconnaissance photograph of Liverpool. It shows a section of the docks area, with two cold storage buildings east of Canada Dock. The area shown is more than 13,000 square yards.

national·intelligence data-handling system. When necessary, ground observers can "see" many of the things detected by the satellite with a delay of only a small fraction of a second, including not only current information but any difference compared with 12 hours previously.

Close-range surveillance can be provided by remotely piloted vehicles in the form of small aircraft so noiseless that their passage at low level on a still night is masked by the background of, for example, leaves rustling in distant trees. Such aircraft can also drop sensors that rest on the ground and transmit seismic data (for example, the identifiable vibrations of people or vehicles), acoustic data (noise of engines or voices) or even underwater or underground disturbance caused by various kinds of enemy activity. At the other end of the scale, large aircraft can be so packed with sensors as to detect, analyze, record, and transmit almost every detail of a hostile missile test seen from a distance of 100 miles.

The subject of intelligence transmission and military communications generally is a vast one in itself. Many kinds of signal can be sent by such clever electronics that there is no need for cryptography, but at the local and battlefield level the use of unbreakable ciphers and codes is as important as it ever was. Today the task of encoding, deciphering and, if possible, reading hostile codes is almost totally an electronic one. Small man-portable encryption boxes can do so good a job that even if the signal is intercepted it ought to be impossible to read its meaning – at least, until the message is so old as to be no longer important. But the main highways of intelligence are today so numerous, diverse, and carefully constructed that it would be extremely difficult to interfere seriously with a national network. Despite this, the leaders of great nations can still discount or disbelieve information, or incorrectly guess a nation's future intentions, so human frailty cannot be eliminated.

Right: Remotely Piloted Vehicles like this were used in the Vietnam war. They brought back film records, and sometimes relayed direct to base.

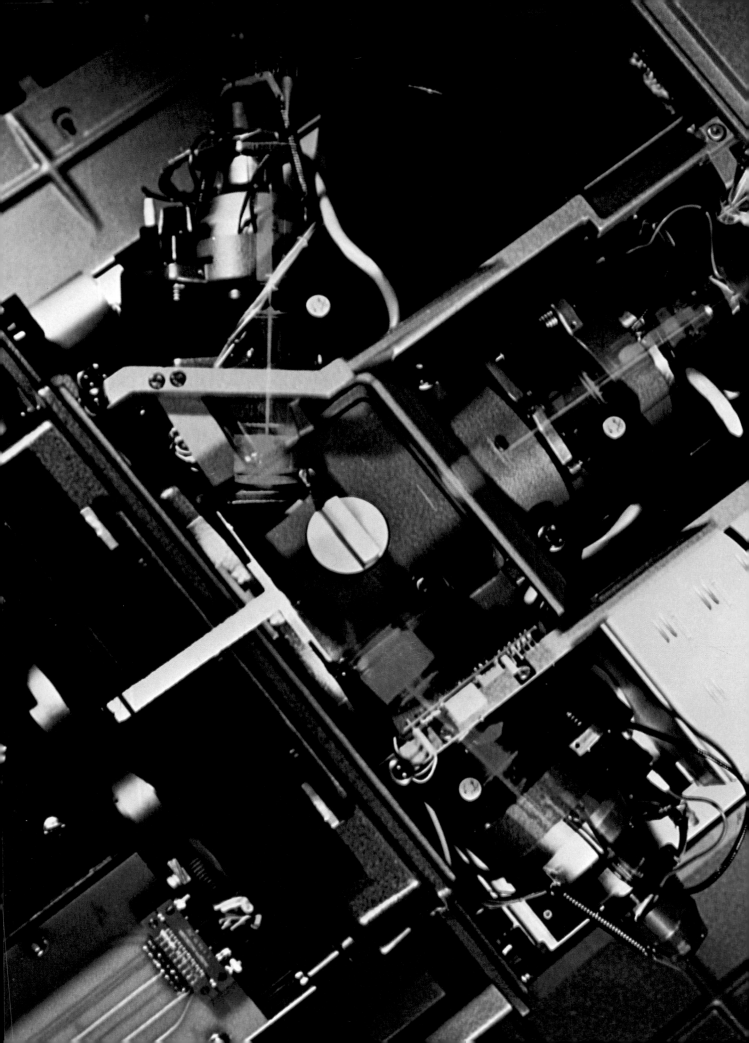

CHAPTER 10

COMMUNICATION AND CONTROL

Few of us today find much to wonder at in the machines that have taken over the heavy work formerly done by man or beast. Even the telecommunications that put us into immediate voice contact with someone on the other side of the globe rate no more than momentary admiration. With the widespread use of automation there has nevertheless come a feeling of unease over the possibility of machines usurping the vital human function of decision-taking. Some of the more fanciful writers have even voiced the fear that machines with intelligence parallel to that of man might some day emerge – perhaps even be able to outwit mere humans and take over their world. Scientists in the communications industry tell us that such fears are illfounded and that the benefits from the revolution of the electronic age will be even greater than those of the mechanical age, which was its necessary predecessor.

Opposite: a modern color television camera with the vidicon tubes visible. Color TV pictures are made up of red, green, and blue images. These are focused into three separate camera tubes. There are filters in the light beams, which correct their color and intensity.

A Brief History

Communication and control covers a vast range of topics central to human activity yet, because often intangible and hard to visualize, much more difficult to understand than technology based only on what is now universally called "hardware." Probably the difficulties began 150 years ago when the English chemist and physicist, Michael Faraday began to put the often naturally occurring phenomenon of electricity to practical use. Unlike steam pressing on a piston, electricity was hard to imagine. Even now radio, television, and computers are thought beyond the comprehension of most ordinary mortals. In fact, it is not difficult to explain in general terms how they work provided we do not delve too deeply. No scientist can fully explain electronics, nor even describe the electron on which most modern communications and control depends.

Far and away the most important single development in human communication was the invention of speech, with words having precise meanings. This entailed a new use of our voices, and – though we have no evidence of this – it was probably a gradual process over millions of years in the course of which grunts, squeaks, and clicks gradually developed into a complete spoken language. At some point, and again we have no information on when or where, there came a sudden giant mental leap as great as any single event in human history: it was discovered that some language could be written down.

Another invention equal in magnitude was another form of language: the ability to count, the invention of numbers, and, as a natural and marvelous sequel, the discovery of arithmetic. Although today man enjoys the benefits of a wealth of branches of mathematics, arithmetic remains the queen; and it is at the heart of the modern computer and of many other applications of technology.

Over a period of some 5000 years many languages evolved, and writing was transferred from clay to the more convenient papyrus and then to paper. In the 15th century came the invention of printing from movable type, and later it was discovered that pictures could be reproduced on printing machines.

The invention of the telegraph 150 years ago introduced the era of telecommunications but of course communication at a distance goes back thousands of years. Special forms of speech, such as Alpine yodeling and the whistling speech of La Gomera in the Canaries, were developed to be heard over long distances. Visual methods included hilltop beacons, American Indian smoke language, the Sun-reflecting helio-

Above: Begbie lamp and Heliograph Station at Bloemfontein during the Boer War. The heliograph had two adjustable mirrors. A beam of light from the Sun could be reflected in any direction.

Left: a Babylonian map of the world made in about 2400 BC. The map was made of clay, which could be transferred to papyrus. The text tells of the campaigns of Sargon I of Akkad.

precise position, and a vast number of other variables.

In all cases automation relieves humans of tedious repetitive toil. Almost always it also permits a process to be speeded up by tens or even millions of times. Indeed, many modern engineering design processes routinely involve calculations that would be beyond the capacity of the entire human race to perform unaided. The importance and influence of computers and telecommunications is impossible to over-estimate. Both make national frontiers largely non-existent. They improve our doctor's diagnosis, multiply the pace of new discovery and – contrary to a popular belief – virtually eliminate errors. Each day the volume of traffic along the automation highways increases, while costs sink to new low levels. It is one of humanity's few unalloyed success stories.

graph, and flag semaphore. After the invention of the Morse Code the heliograph was used to reflect sunlight as a series of dots and dashes. Even today Morse is sometimes sent by an electric lamp, especially when it is essential to transmit a message – between two ships, for example, in wartime – that cannot be intercepted by an enemy.

Automatic control has a shorter history, though simple centrifugal governors, incorporating spinning masses on pivoted arms which at excessive speeds applied a brake, were known at least 600 years ago to control the rotation of prayer-wheels in Tibet. By 1500 they were common on roasting spits, and in 1788 the modern form of self-stabilizing governor was invented by James Watt to control his steam engines. Since then almost every form of engine, and most kinds of machinery, have incorporated some form of automatic control with feedback to maintain a steady-state running condition. But automation goes far beyond mere control of speed. In almost every modern manufacturing process and countless other activities automatic control is exercised over weights, flows, thicknesses, operating sequences, shades of color,

The Written Word

Our knowledge of the earliest forms of written language is very scanty, and often mere guesswork. We believe that the first kinds of writing comprised mere sequences of pictures, each illustrating the subject under discussion. Obviously some subjects were difficult to portray, and stylized shapes had to be introduced. Before long, the theory runs, the pictures came to mean not only the original object but also something associated with it; thus, the Sun came to be the symbol for heat. The greatest advance came when written language ceased to be practical and instead became phonetic, based upon spoken sounds. Many important regions, such as China, never made this step but instead continued to write with an almost limitless number of pictorial characters.

There were many centers of early civilization, some of them doubtless unknown to us today, and each developed its own language in both spoken and (almost always, but with a few exceptions) written forms. One of the oldest known phonetic forms of writing is Sumerian cuneiform script, written with a triangular-section reed on clay tablets. The Egyptians combined phonetic and picture writing in the style known as hieroglyphics, and this became progressively refined and made more compact and versatile. Several races, such as the Indians, Hebrews, and Phoenicians, accomplished the major step of setting down symbols representing basic sounds, which form most of the vowels and consonants of modern language used by about half the world's population. This led to standardized alphabets used over large areas.

Today half the globe uses an alphabet based on that of the Romans, but numerals derived from Arabic. Many countries, such as the Soviet Union, Israel,

Above: in the middle ages, the monks were almost the only people who could write. This medieval French miniature depicts a monk laboriously copying out a book by hand. The printing press revolutionized communication.

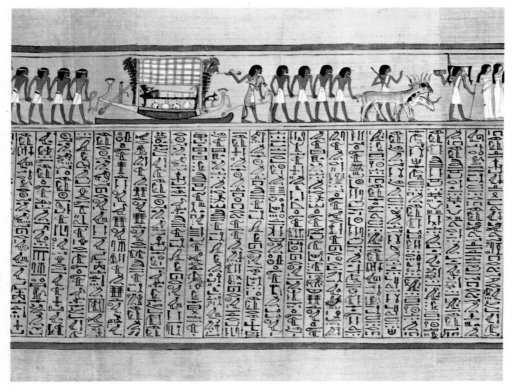

Left: hieroglyphics was the pictographic script of the ancient Egyptians. Inscriptions addressed to particular gods played an important part in the Egyptian cult of the dead. This plaque bears the name of King Anlami, son of Senkameniskon. It was taken from the foundation of a pyramid at Nur-Sudan.

256

Japan, and Arabian countries, use quite different alphabets, while in India there are several hundred spoken and written languages in current use. The number of languages is certain progressively to fall, because of the social and economic handicap of having a language not generally understood and not available on, for example, a typewriter or for communicating with modern computers. In the recent past computers have demanded special languages of their own, based upon English letters and numerals but generally unintelligible to the public; these special interface languages, such as Fortran, Cobol and Algol, are still important but are likely to be replaced by more easily understood languages as better computers are developed able to accept instructions in plain English or other modern everyday language.

As far as we know today, until the 15th century all written language had to be laboriously written by hand (though there were probably several accepted methods of making quick copies, especially in the days of clay tablets). Quill pens, fine brushes, charcoal sticks and other implements were used, but they were unable to make more than one copy at a time. But in 1409 the Emperor of Korea's type-foundry completed production of a large book printed entirely with movable metal type, and the technique was certainly in use in Korea earlier. Use of separate pieces of type was a tremendous improvement. A new artisan, the printer, could set up the pieces of type into a frame so that a whole page could be printed at once; and printed as many times as necessary, each impression

	Sun	Moon	Horse	Fish
1700 – 1400 B.C.				
776 – 250 B.C.				
250 B.C. – A.D. 25				
A.D. 25 – 220				
A.D. 380 – present day				

Above: this diagram shows the stages in the evolution of Chinese ideograms. The original simple picture is, in time, replaced with a shorthand version. It now takes years to memorize all of them.

being identical to the last. In 1456 there appeared one of the true masterpieces of printing, and a product able to stand with any other human achievement. A team led by Johannes Gutenberg printed 180 copies of a superbly produced Latin-language bible, with two columns of 42 lines printed in different colored inks on 643 leaves (vellum in 30 special copies, heavy hand-made paper on the others).

Below: Japanese children hold up New Year messages. The Japanese adopted the Chinese characters. Japanese has linked syllables, whereas Chinese has monosyllables ending in consonants.

Left: a type mould for a 300-franc note. Line engraving or intaglio printing was widely used for printing early bank notes and postage stamps. Intaglio printing works on the opposite principle to letterpress. The image is etched into the mould. Gravure developed from intaglio.

Below: an operator working a Monotype casting machine. With Monotype, a keyboard operator produces a punched paper tape containing all the letters and spaces to be set. A machine casts pieces of type and assembles them into lines. The Monotype caster "reads" the tape backwards.

Even today "hot metal" and traditional stored metal type are much-used in printing. Stored type requires thousands of standard-size characters to be stored, in different kinds of type-face, and in different sizes. The printer selects each piece of type with tweezers and makes up his text line-by-line, resting the type either in a small hand frame or on a bench or "composing stone", often tying the columns of text tightly with string. He checks the type (which is reversed left to right) by reading it upside-down, faster and more accurately than most people can read ordinary print. After approval of proofs the whole page is tightly locked up in a frame called a chase or forme, and in printing a book it is usual to print 8, 16, 32, or 64 pages at once, on a single sheet, which is afterwards cut up during the binding and manufacturing process. This procedure, called flat-bed letterpress, was almost universal until after World War II, except in the production of newspapers and mass-circulation magazines where, after approving each page, the whole mass of text and illustrations was cast in a curved slab of metal that was bolted to a high-speed rotary cylinder to print from a continuous strip of paper called a web.

Illustrations used to be carved by hand in blocks of wood, the principle being identical to letterpress type in that the parts to print were left at the original surface level while the white (unprinted) parts were cut away. Subsequently etched metal plates were used, where the picture was sketched with a fine metal point that cut through an acid-resisting wax coating. After immersion in an acid bath the wax was cleaned off, leaving etched areas that held ink. This was the opposite of letterpress. Often called line-engraving or intaglio printing, it was used for early banknotes and postage stamps and is still important in both these classes of printing. Another way of reproducing pictures used in letterpress printing is the half-tone process. The scene is "screened" and reproduced as a pattern of regular dots whose diameter varies with the dark or light tonal value of the original.

Larger dots run together forming dark areas, while white areas are characterized by dots that are barely visible. Coarse half-tone photographs are used on newsprint, and the dots can be seen clearly even with the unaided eye. Glossy magazines can use a much finer screen, giving a clear picture whose dots can be seen only with a magnifying glass.

Today most letterpress type is not picked by tweezers but cast as it is needed. There are two "hot metal" processes: Linotype, which casts a whole line at a time; and Monotype, which casts individual characters. Mechanization of typesetting has now led to computer-assisted photosetting in which the typesetter types each line, in many cases over a remote telephone link, and the computer stores the text, adjusts each line, puts hyphens in the correct places when a word has to be broken, and sets the whole text on photographic film. In major book production, as in the case of this book, each page is produced as a single piece of "final film" complete with text, pictures, and captions.

This book is not printed by letterpress. It is usual today for color printing to use either the photogravure method or lithography. In gravure printing a smooth metal surface is etched away on a microscopic scale, either over whole areas such as the inked part of a letter or numeral, or over a pattern of half-tone dots like inverted hollow pyramids. These recesses hold the ink. In litho printing the surface is flat, and the areas to print are differentiated by the mutual repulsion of greasy ink and water. Litho printing was discovered in 1798 by the Bavarian inventor Aloys Senefelder who was seeking a way to print music. He found he could print direct from flat stone, but modern litho uses aluminum, on which the text and illustrations are transferred photographically.

Color printing is sometimes done with inks of the desired color; this is the procedure used in many countries for multicolor postage stamps, which are magazine, but in multicolor stamp printing they can be seen on the sheet edges together with other information helpful to the printer such as check-dots for each color.

There are countless other aspects of communication by the written word. A few countries provide special attachments to telephones with which deaf and dumb subscribers can communicate with handwriting. Braille is a special form of writing for blind people, who "read" it with their fingertips. Deaf and dumb people have their own sign language using the hands, only unfortunately few non-handicapped people know it. Apart from printing methods there are many forms of copying and duplicating, some of which are unsuitable for illustrations while others, such as the inherently photographic "blueprint," are suited to large line drawings better than to text. The usual large office copier is today an electrostatic machine that relies on the phenomenon of photoconductivity;

Left: the Harris M-200 web offset press with the cathode ray tube control showing. The paper is run through the machine from a continuous roll. A web offset machine prints a book or magazine in sections of 8, 16, 32, or 64 pages. These sections can be printed and folded at the rate of 25,000 copies per hour or more.

Below: a telephone adapted for use by the blind. This subject is not only blind, but practically deaf. When the receiver is picked up the sender communicates by a sort of Morse Code, or by a pre-arranged signal. The subject rests his fingers on a pad, which vibrates in response to the signal.

usually produced millions at a time on a continuous web by litho or gravure printing in as many as ten different inks, including gold and silver. These two colors, and special metallic effects, are hard to reproduce in the normal method of color printing, in which the whole of each illustration is printed in sequence by four colors, yellow, magenta (a deep rose-pink), cyan (bright blue), and black. Each picture is photographed four times through filters that separate the color needed. These "separations" are then used to make four half-tone litho plates, gravure cylinders, or letterpress blocks, depending on the process to be used. The yellow printing is run off first, and then the other three, extreme care being taken to line up small marks on the paper to ensure that the images are superimposed correctly. The black printing adds the text and captions. The small marks for keeping accurate registration between the different printings are usually cut off during the binding of the book or

a charge is retained on the parts of a metal plate corresponding to the image to be copied, and special powder is transferred from these areas to each copysheet on resin-coated paper.

Computers

Below: one of the forerunners of the modern computer was this analytical engine designed by the British mathematician and inventor Charles Babbage (1792–1871).

The earliest computer was probably the abacus, with beads strung on cords or wires, which was in use throughout the Orient by 600 BC. This crude apparatus was a true ancestor of today's electronic data-processing installations because it works by a counting process. In 1642 the great French mathematician Blaise Pascal built a mechanical adding machine based on a series of gearwheels each of which had to make a 360 degree, or ten-notch revolution, to move the next wheel in the series on one notch or 36 degrees. The same idea is used today in automobile odometers and many other mechanical counters.

In 1812 an English mathematician and mechanical genius, Charles Babbage, set the world firmly heading to the "intelligent" computer. He saw that it should be possible to build an automatic computer, able to recognize numbers and simple calculating situations. Although he had no available technology other than clocklike machinery made of brass, he constructed a "Difference Engine" that worked out complex mathematical tables, and completed the design of the vital final step, the "Analytical Engine" able to accept input instructions and then proceed according to a series of logical instructions and sequences. Such a machine, able to work out a problem for itself, would have been identical in role to a modern computer but working by mechanical rather than electronic methods. Sadly, the government withdrew support and the machine was never completed.

Progress was then slow until in the Second World War the need for numerous complex calculations in such fields as artillery, rockets, naval warfare, aircraft design, aerial gunnery, and radar – to say nothing of the atomic bomb – put urgent pressure behind the creation of computers. At first these were enormously large, filling whole buildings. They contained millions of thermionic valves (vacuum tubes), consuming vast amounts of current and emitting considerable waste heat. In spite of their extremely poor reliability, and very high costs, these unwieldy machines were made to work. They were the first electronic computers, operating by the flow of electrons through conducting wires, empty spaces, or semiconductors.

Such devices exist in many forms. Those in use in modern business and industry are almost all of the digital type. A few, however, are analog machines, the simplest being the slide-rule: the user slides two scales alongside each other, with distances along the scales proportional to the quantities being manipulated, and

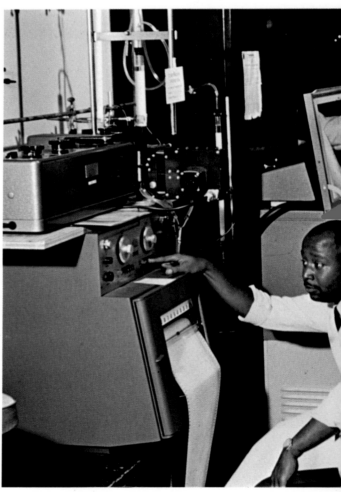

Above: the computer being put to the service of food hygiene. Here, officials of the United States Department of Agriculture are testing samples of meat. A computer cannot perform anything that large numbers of humans could not (given time).

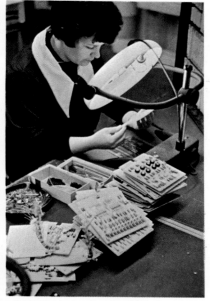

Below: electrical components being wired into a computer during manufacture. Computers are tailored to known functions and tasks.

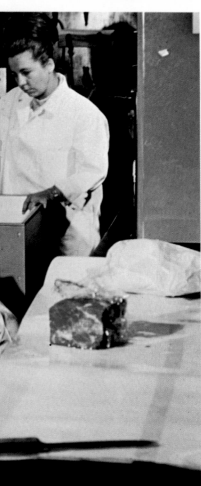

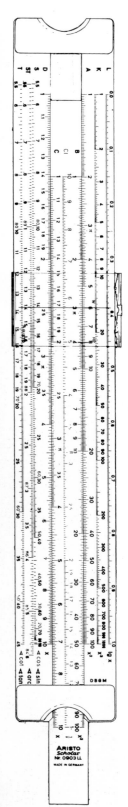

Above right: a slide rule like this is useful for making rapid calculations. It is made of two rulers marked with the same logarithmic scale. To multiply, lengths on the two rulers are added. In division the lengths are subtracted from one another.

reads off the answer at a particular point. Electronic analog computers usually work with varying electrical voltages. Hybrid computers are part-analog, part-digital.

Digital computers all work with individual numbers. They accept "input numbers" (such as $1+1$), process these according to a set of instructions called a program, and provide the operator with "output numbers," which are the answers or information sought ($1+1=2$). The computer may be of many kinds, made by many companies for many kinds of task, but invariably it is made up of the following parts: an input device; a central processor; a memory; and an output. There will probably also be at least one control unit, and possibly various kinds of interface devices where the computer is "accessed" by something else or used to control something else. All computers in common use have been given the acronym TOM, meaning totally obedient moron. They cannot actually think but merely follow instructions with absolute precision. The programmer, whose task it is to provide the machine with its set of instructions, has to have great skill and experience to instruct the machine completely and exactly, with every eventuality thought of.

After careful study and much advice nearly all large companies, laboratories, banks, airlines, armed forces, and similar organizations either buy a computer, lease one, or buy time on someone else's. If it is for a research laboratory the input devices may include a keyboard (like that of an electric typewriter), a punched-card reader, a punched-tape reader, a cathode-ray tube display, which is simultaneously input and output, and possibly special interfaces with instrumentation that measures something under test. Visual displays, such as a TV screen, can show either alphanumeric words and numbers generated electronically or even precise drawings of parts of a machine that the operator can alter with a light pen – a small photocell that detects light. When touched on the display at a particular spot it can have that region enlarged for a closer look, or allow the operator to reshape it in a way that inserts new numbers into the computer for processing (for example, to make the part stronger). But such "computer graphics" are a side-issue. How does the computer work?

Digital computers work with numbers, but they have to convert our familiar numbers into a different set called binary notation. Instead of counting in 10s binary numbers count in 2s. Thus, while 0 remains 0 and 1 remains 1, 2 becomes 10, meaning $1 \times 2 + 0 \times 1$. Three is 11 ($2+1$) and 4 is 100 ($1 \times 2^2 + 0 \times 2 + 0 \times 1$). The numbers 15, 16, and 17 are represented by 1111, 10000 and 10001. To bring all the numbers to the same length zeros are added to the front so that 15 might actually be rendered 0000001111. The computer reads from right to left, in effect saying "is there a 1? a 2? a 4? an 8? a 16? a 32?" and so on, each being another power of 2. The number 6553, for instance, becomes 1100110011001, which (read in ascending powers of 2, from the right) becomes "yes, no, no, yes, yes, no, no, yes, yes, no, no, yes, yes." We have changed our number 6553 into a sequence in which there are only two choices, and this is how the computer works.

Most modern computers have a memory of the magnetic-core type. The core comprises a large number of small flat "arrays" in a stack packaged into a box. Each array consists of hundreds of fine wires forming a mesh, and at every intersection a small ferrite (electrically magnetizable iron) ring is placed around both wires, and also around a third, diagonal wire. When the computer is programmed, each instruction is translated into small pulses of current that pass along the vertical and horizontal wires in the core. Every part of every instruction is "addressed' to one particular pair of wires, which cross at one particular ferrite ring. The two wires bring a pulse of current which, depending on the polarity of the two currents, either brings a 0 or a 1. Until otherwise instructed, that microscopic ring is thereafter a 0 or a 1. During the arithmetical processing the third wire "reads" the ring, sending back the 0 or 1 information. The process may appear clumsy, because 1100110011001 is longer than 6553 and appears more prone to error; but in fact the computer cannot make an error (unless it has suffered damage, which is unheard-of in normal civilian use) and processes so fast that the long numbers or "words" do not matter. The pace of the operations is kept absolutely exact by an electronic clock at possibly three million operations per second. In this way the long strings of 0s and 1s can be dealt with in the twinkling of an eye.

Special words have to be used in writing the program. Each computer understands only its own code or language. Although all are based on the 0 and 1, which are called binary digits, or "bits" for short, the program is written in "words" usually of 24, 36, or 48 bits. There are data words and instruction words, and sprinkled throughout are "check bits" for checking the code, and further self-checking bits and words to make sure the whole program is valid. Code numbers, incapable of being confused with any other part of the program, stand for such commands as START, CLEAR AND ADD, ADD, STORE, PRINT OUT.

Within the program are all the data and operating instructions, so that as each new item of information arrives at the input, by keyboard, punched tape (the holes being in transverse groups each meaning a letter or number or binary word), or other input, it is addressed to the right place and processed in the right way. The processing is done by microscopic logic and arithmetic circuits, a key element in which is a microscopic flip/flop device, or "gate," made in two halves, one the mirror image of the other; activation of either half deactivates the other. These route and process the data, multiplication being a sequence of additions and

Above: a computer printer in action. Machines like this print out information typed out on a keyboard, which may be many miles distant. They are widely used to provide things like sports results.

Right: the minute Random Access Memory (RAM) chip. This single chip can store up to 64,000 pieces of information. The development of the microchip is thought to pose acute problems of large-scale unemployment.

Above: a computer programmer at work. Modern industry has been increasingly dominated by the computer since World War II. The computer programmer is the vital link between humans and the machine.

Below: the IBM 2250 Graphics Display Unit. It has been developed for computer aided design. The consequences of a change in specification can be seen immediately by the designer, and the design altered.

division a sequence of subtractions and shifts. When necessary, large amounts of data are stored outside the computer in magnetic disks, drums, or tapes. In general, the larger the capacity of a store, the longer it takes to retrieve a particular item (though still measured in milliseconds).

Many computers spend their lives coupled into a large operative system, such as the radar and missile-control systems of a fighter aircraft, or a rolling mill in a steelworks. The two form a single entity, though the computer may need access to many different programs. Other computers form an integral part of a telecommunications system. When you book an airline ticket the counter-clerk probably has a remote terminal to a distant computer. Tapping out details of the required flight on the keyboard provides a correctly coded input instruction to the computer, which searches its memory and emits a stream of electrical signals that generate alphanumeric characters on the clerk's terminal display screen. The information will be immediately displayed as available flights, times, and seats available on each. If a customer at another counter somewhere else buys two seats, those will instantly be deducted from the print-out, avoiding double-booking. Many airlines are served by thousands of computer terminals in every city on their route network. Here the computer has little actual data-processing to do.

Automation and Control in Industry

Mass-production, and the replacing of human workers by machines is the immediate association when the word automation is used. The trend toward automation has been a real one throughout this century, and has important social consequences. At first, mechanization relieved man of much of his physical toil and then, with automation, he was relieved of even the necessary supervision of the machines. There has been much speculation on the amount of mechanization and automation in the future. A cartoon drawn in 1969 suggested that many people think complete idleness is the objective: told by a union leader that the new working week is Wednesdays only, the workers respond with "What, every Wednesday?" Another of

the same era was more perceptive: a worker, struggling to wrest a spanner from a colleague beside a vast computerized production line, shouts "You promised me last week it was my turn next." Social consequences in fact go much deeper than this, resulting in such anomalies as the fact that devices of amazing complexity can be produced very cheaply, whereas the most minor repair is exorbitantly costly.

Automation in fact extends to every aspect of life in the so-called developed countries. In general it is manifest in control systems of various kinds, but it is central to all modern planning, decision-taking, information storage and retrieval, flow-control and allocation of resources, simulation and analytical modeling, and almost any other activity capable of being expressed in numerical terms.

Simple automation consists of self-contained devices, often hand-held, that relieve humans of the need to keep adjusting a control. One simple automatic control is a thermometer, which measures the actual temperature and interrupts the supply when the desired level is reached. In an electric iron it is just a bimetallic strip made of two metals that expand and contract at different rates; when heated it arches into a curve, at the correct point breaking the electrical contact. This is called a closed-loop control because the system has a feedback that automatically informs the input of conditions at the output. Any outside disturbance is automatically counteracted. An open-loop system has no feedback. An iron made with such a control might automatically switch itself on and off to a regular time-controlled routine, but it would get too hot in summer and not hot enough

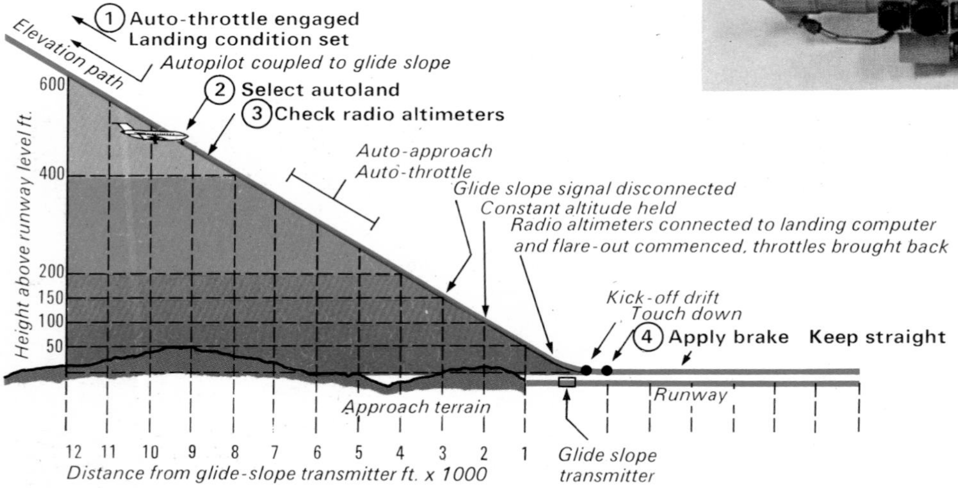

Above: the Simplex Activator Automatic Flight Controller. Flight controllers like this are almost universally used in civil aviation. There have to be strict limits on their control authority, allowing the pilot to take over when really necessary. **Left:** an automatic landing in progress. Operations **1** to **4** are performed by the crew. The others are dealt with by electronic systems aboard the aircraft.

in a cold room. It is the essential feedback that closes the loop that ensures that the output stays oscillating about the true desired value.

A slightly more complex system is seen in a jetliner approaching a landing strip. Inside it is an autopilot, the heart of which is a set of gyros whose axes attempt to remain fixed in space. If the aircraft were to change its attitude the gyro cases would rotate slightly around the spinning wheels to cause an "error signal" that would provide the feedback to close the loop and restore the original attitude. It would do so by sending a signal to one or more of the electrohydraulic power units connected to the various control surfaces, such as ailerons, spoilers, elevators, and rudder. The signal would change the position of the hydraulic valves, causing a ram to move the surface. This movement

Diagram labels

① Auto-throttle engaged
Landing condition set
Autopilot coupled to glide slope
② Select autoland
③ Check radio altimeters

Elevation path

Auto-approach
Auto-throttle

Glide slope signal disconnected
Constant altitude held
Radio altimeters connected to landing computer
and flare-out commenced, throttles brought back

Kick-off drift
Touch down
④ Apply brake   Keep straight

Height above runway level ft.

600
400
200
150
100
50

Approach terrain

Runway

12 11 10 9 8 7 6 5 4 3 2 1   Glide slope transmitter
Distance from glide-slope transmitter ft. x 1000

Glide slope transmitter

would itself progressively cancel the input signal, while the subsequent change in attitude of the aircraft would nullify the original demand signal from the autopilot. Finally, on the approach to the runway the aircraft descends down a series of radio beams. These are sensed on board and signals are fed to the autopilot so that any departure from the centerline of either of the multiple beams – a glide-slope for vertical guidance and a localizer for direction – would itself generate an error signal that would demand corrective action by the autopilot and powered flying control system. Thus the aircraft remains under positive control of a series of closed-loop systems each with feedback to ensure that the desired state is in practice maintained. Instruments display the situation to the flight crew, but they need not touch the controls. As in many vital systems some parts of this sequence are duplicated, or triplicated or quadruplicated, so that any dangerous fault, such as a sudden unwanted "hard over" signal to a control surface, is outvoted or overridden.

Modern manufacturing or packaging is almost entirely automated, and though impressive to watch, the technology is usually of a very simple nature. In the sequence of operations at a salmon cannery, for

Above: food production and packaging would be impossible without a very high level of automation. In this cannery, salmon are being cut into can sized pieces. A worker feeds the fish to the cutter. From this stage the finished product is produced automatically.

Below: press-forging by remote control. The hot ingot is held firmly on the anvil of a 1500-ton press by a railborne manipulator (*right*). This manipulator can move or rotate the ingot in any desired direction at the push button command of the operator.

Above: inside a modern ceramics factory. This automatic sequence control enables one man to control the processes of preparation, blending, forming, and firing on two lines with an output of 12,000 bricks per hour.

example, there is hardly ever a point at which a human need intervene. Preparation of fish, cutting into segments of desired weight, and placing one in each tin is simple to arrange. Fed with tinned steel sheet, itself produced totally automatically, the can-making line has no difficulty in cutting off rectangles, rolling it, soldering the seam, fitting the end and then, the fish inside, soldering or pinch-fitting the top to give an unfailingly airtight seal, which is automatically checked in a split-second. Printing the label is no problem, nor is cutting it off, wrapping it, and securing it with adhesive. The whole process is automation at its most primitive, with operations usually triggered by a piece of mechanism, or even the can itself, hitting a stop or lever. The next operation in turn makes everything move on one step.

This is mere process-mechanization, the crudest form of process control. There is little need to add a computer, because the system works and has few external disturbances. Humans still have to be present, to check that no fault exists and that the various lines are kept supplied with fish, juices, tinplate, solder, and wrapping. An empty can might get as far as being sealed, but automatic weighing would throw it off the line before it was given a wrapper.

More advanced forms of process-control insert at least one computer into the closed loop, usually in an "on-line" way in which the electronic data-processing (EDP) forms a permanent integral part of the system, operating "in real time," which means controlling events as they happen. Invariably there is at least one contact area with humans such as a control room

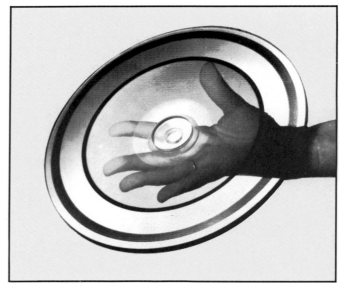

Above: this is a computerized language dictionary. 55,000 Russian words are stored in 700 hairline tracks on the dark outer ring of this glass disk. The computer can store the information of the largest libraries of the world in a space no bigger than a large cabinet.

with various displays and input devices like a keyboard, or a stand-by, or an emergency program that can be inserted to counter particular kinds of trouble. This would be the scheme at a large refinery, rolling mill, or chemical production plant. But automation goes far beyond this.

Consider the situation where a student wishes to find out as much as possible about a subject, such as mitochondria. An increasing number of libraries are automated: by accessing the sorted information through a computer keyboard it is possible to have a comprehensive list (possibly all the hard-back books and a high proportion of the reports in the desired language that exist on the subject) automatically displayed in alphanumeric characters. This can in turn be put on one side into a temporary store, for immediate recall, and a particular work accessed with a light pen. The actual reference may then be displayed, page by page, or if necessary (and for payment) parts of it can be immediately printed out in a take-away copy. The important fact is that in a matter of hours a researcher can have in his possession almost all the information available on the topic, whereas without automation he would spend weeks trying to track down what had been published.

What happens when a company wants to market a new product? It has to know the demand, how the demand is influenced by price or types of packaging, and whether the firm should go ahead with a full-scale plant to meet a world market, or with a smaller one for the national market only, or perhaps a trial marketing in a local area to test consumer response. The company immediately calls up its electronic data-processing manager, who inserts into his computer one of the special programs available for just this

Above: the computer is being increasingly used in medicine. They are employed in providing diagnosis. The patient answers a series of questions. Answers are fed into a computer, which then provides possible diagnoses.

Above: inside a modern glass factory. This computer controls the furnace which bends and stresses two glass components of car windscreens separately. This technician is monitoring operations through a closed circuit television and a visual display unit.

situation. The program tells the computer such things as average personal incomes, previous marketing experience, population preferences and habits where relevant to the product, likely costs of different sizes of plant (refined by the company to suit their own requirements) and such factors as government incentives in areas of high unemployment, return on capital invested for a wide range of other variables and the possible elements of risk. The results are presented within minutes of the refined program being inserted. The management can learn all they need to know about cash flows and likely profits, and can then

refine the answer by inserting different time-periods, different predictions for inflation, and different selling prices. The effect of moving the planned factory to different places can be studied, as can the effect of starting with a small pilot facility, or one designed to be scaled up, or building additional plants later or starting with a giant world-market plant right at the beginning. The promotional campaign for the product must also be studied, for example by seeing whether overall savings of a large plant might be nullified by difficulties in global advertising or in consumer acceptance in particular countries. Without automation the decisions would be based to a far greater degree on guesswork and hunches, with risks that today could topple a large corporation.

There are countless other roles for automation. One of the more interesting is the relationship between computers, doctors, and patients in clinical medicine. In the past doctors or specialists have used their knowledge and experience in knowing where to ignore a patient's description of symptoms and at which points actual tests are necessary. Today we are making the transition to a new era in which computers can study a patient's history, conduct a physical examination more quickly and more thoroughly than the traditional kind, where necessary order laboratory tests, then scrutinize the results, and, finally, handle the treatment. Already the vast majority of patients admitted to hospital can have their entire case managed by computer, the doctors and other staff keeping a watchful brief and looking after what might be called the human problems.

Prosthetics

The repair of the human body by fitting spare parts is one of the most exacting areas of modern technology. One of the sources of difficulty is structural strength, and resistance to fatigue. Another is the body's natural immunity mechanism, which makes it try to reject foreign material. Quite apart from these problems, the interaction between the body and a prosthesis or spare part raises questions of communication and control that are very far from being solved. As in most fields in which man attempts to emulate advanced living organisms, though the swift development of microelectronics has dramatically eased some of the difficulties, we are still a long way from being able to duplicate the capabilities of nature.

There are many kinds of prosthesis, ranging from a hip joint to a heart-lung machine. The hip joint is essentially an inert mechanical construction which provides physical strength to a part of the body where muscles and nerves are unimpaired. The heart-lung machine is the opposite – it has no structural function but performs the complex and vital task of supplying oxygen to the blood. A heart-lung prosthesis capable of being housed inside the body has not yet been demonstrated, though this is mainly not for reasons of communication and control. Heart-lung machines are at present large and cumbersome, as are artificial kidneys. The replacement liver will suffer from the same handicap when this has been developed. A machine outside the body can be connected to its own power supply drawing electric current from the main supply and controlled without much difficulty. It can often be made to work automatically with basic closed-loop control, in some machines not much more complex than that of a domestic refrigerator.

The situation is totally different in the case of an artificial hand or limb. Here the wearer wishes to perform an infinite variety of contrasting actions, such as left/right motion, up/down, to/fro and rotation. Also needed is the ability to exert infinitely varying force precisely matched to the job in hand. The human hand, but usually not the foot, also has a wealth of gripping and control methods involving the thumb and fingers in different arrangements and again with markedly differing forces. Not least, the human hand can sense what it is doing. Without being seen, it tells its owner if it is gripping correctly, or if the fingers or thumb are slipping, and it has all the body's normal feedback to warn of pain, electric shock, excessive heat, wet or corrosive environment, as well

as indicating its position and attitude. Even if it were possible to make a reasonable artificial hand, how many of these capabilities could be built into it?

Making a prosthetic hand has been done in many ways. Some look like hands, and are socially preferable to the wearer, but the simpler split-hook prosthesis is for most purposes stronger and more convenient, and for specific operations the best answer is a purpose-built prosthesis which, in its particular application, can even be quicker, stronger and in other ways superior to the supremely versatile natural hand. How are such hands controlled?

In the laboratory it is possible to connect wires or probes to individual human nerves along which pass the input demands to control muscles; but even in the late 1970s the ideal objective of coupling the original nerves into a prosthetic hand is still beyond man's capability, though recent developments have brought it nearer. Instead, the ordinary prostheses available in most of the advanced countries merely sense a

Below: a heart-lung machine. It takes over the functions of the heart by maintaining arterial pressure at a level sufficient for the patient's organs to function. The job of the lungs is done by the machine maintaining the right balance of oxygen and carbon dioxide in arteries and veins.

particular electric voltage, called an electromyograph (EMG) potential, inside a suitable muscle. Fortunately, this EMG voltage generated in the body varies approximately in proportion to the desired muscle output. A doubled voltage results in roughly twice the speed of movement or twice the gripping force. But there are still many problems. It is impossible by tapping a single input EMG signal to obtain a plurality of outputs, such as forearm motion with simultaneous rotation or gripping. Also, there is great difficulty in eliminating what the electronic engineer calls "noise", or unwanted signals, which may be caused by the electrical activity of quite different muscles, or motion and response of the correct muscle caused not by the brain issuing a command but an outside influence such as pressure by the brace attaching the false hand.

Recent developments have sought to drive the prosthesis by a greater number of EMG inputs, and to amplify these small voltages whilst reducing the "cross talk" or noise that can give unwanted responses. All the latest prosthetic hands retain electric drive but use electric power only for applying forces. Once the hand is in the desired attitude, or gripping with the desired pressure, the current is switched off. Straingauges or other force sensors must be built into the prosthesis to feed back the force being exerted, because not even an experienced user can judge this accurately enough (for example, to avoid crushing a delicate wineglass). Even so, an experienced user with a modern hand still finds difficulty in judging the correct force for an action that unhandicapped people take for granted. In picking up a sack, how much force will be needed? How tight should the grip be? A natural hand works by instant feel, automatically adjusting the grip to suit the load, but the prosthesis that can do this requires an external computer. Such a computer is easy, though costly, to build into a wheelchair. But a fully satisfactory prosthetic hand that operates like a natural one is some years away.

Below: a victim of the thalidomide tragedy being fitted with prosthetic arms. Prosthetic limbs are becoming more sophisticated all the time.

Bottom: electrodes attached to a subject and an oscilloscope testing a myoelectric hand.

Below: prostheses are artificial replacements for human organs, or human organs transplanted in the patient as replacements (homografts). Artificial spare parts are shown in the model on the left, and homografts are illustrated in the model on the right. Tissue rejection is still the major problem with many homografts.

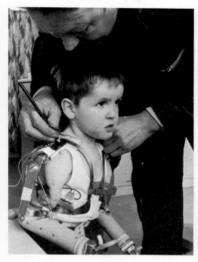

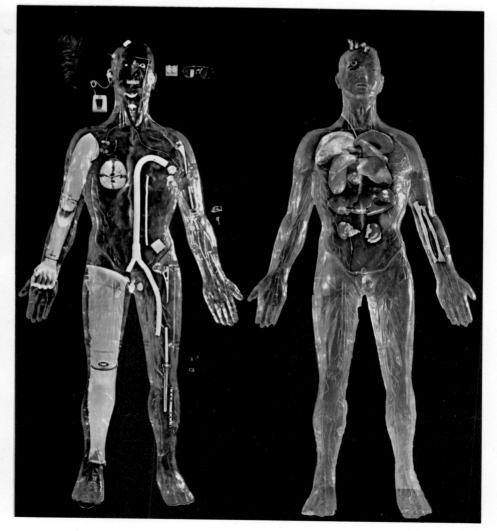

Cybernetics

The word "cybernetics" is derived from the Greek word meaning the science of the steering of ships. It has come to mean the broad interdisciplinary science concerned with control and communication in all life forms, machines, and organizations. Cybernetics involves such diverse topics as electronics, data processing, nervous systems, muscular systems, feedback, information theory, and deep mathematical concepts such as the spectrum that links perfect organization at one extreme to pure randomness at the other. Where hardware is concerned it involves every kind of manufactured structure other than wholly inanimate ones – such as statuary – and every kind of living organism.

In practice many cybernetic problems are relatively limited in scope, and many are concerned with the contact areas between humans and machines. For example, the first "swing wing" aircraft to go into service, the F-111, has a "trombone action" push/pull

Obviously, the objective was a control that was instinctive and foolproof. How difficult this was to arrive at is shown by the fact that an experienced test pilot later moved this lever in the wrong sense, causing a fatal accident.

More complex cybernetic problems are encountered in making any kind of intelligent machine. Whether the quality of intelligence is possible for a machine is itself arguable. The concept of the robot, the machine built like a human – usually of broadly the same size and shape but obviously a machine rather than a real person – has excited public interest throughout this century and been the subject of many books and plays, many of which involve either machines incorporating human or animal parts or humans with machine parts. With the important exception of human "spare parts" as replacements for diseased tissues no such development is known. But there are millions of more or less "robot" toys, and almost as many seemingly intelligent machines. Superficially there may not be much difference between, say, an intelligent toy car of 1939 and one of 1979, but in practice there is no comparison.

One of the popular toy cars of 1939 could be wound up to energize its clockwork drive and then placed on a table. Arriving at the edge, it turned left or right to avoid falling off. The secret was a small fifth wheel running laterally, which contacted the ground as soon as the front wheels ran over the edge. This is a simple mechanical system, with no intelligence. Today's counterpart has photocell "eyes" with which it can

Above: the F-111 "swing wing" attack aircraft. These five photographs show the wings at different sweep angles. Fully swept back wings are needed for high-speed flying, while the straight wing is preferable at low speeds.

handle in the cockpit with which the pilot can spread the wings out sideways or sweep them sharply backward. It was a basic cybernetic problem to decide which way the handle should work. Some pilots said it should work in the same sense as the wings: forward for minimum sweep of 16 degrees and back for maximum sweep of 72.5 degrees. Others vehemently disagreed and insisted that, as the throttles and other controls were traditionally pushed forward for maximum speed, so should this new control. Essentially simple, the argument nevertheless revolved around many aspects of psychology and human experience.

recognize its surroundings in a rudimentary way, steering itself along preprogrammed routes, performing simple preprogrammed tasks and, when its battery runs down to a preset voltage, just managing to drive to the charging point to plug itself in. A real-life tool in the same class is the so-called robot lawnmower or robot fork-lift truck, which has magnetic or photocell detectors that steer or halt the machine at preset locations identified by buried electric cables, lamps, or other signal sources. The machine does the work while the operator either watches or, more often, works a small control box to change the guidance system.

This is still primitive, and so are such common semiautomatic machines as the chair, box, or cart in or on which a crippled or paralyzed person can move about, step off the kerb or go up or down stairs. In most cases these have provision for some kind of human input, tailored to the user's very limited capability. Closely related machines are those used either to give normal muscle-power to a partly paralyzed person or amputee or, in a few cases, to give a normal person exceptional capabilities, such as jumping 33 feet from rest or picking up a mass of one ton. Most such machines have a large internal or external energy supply controlled by the movement of the wearer's limbs. Where a limb has been amputated it is a slightly greater cybernetic problem to couple the stump by mechanical or electrical (electronic) means to control the actuators that govern a mechanical wrist, hand, and fingers.

In the final quarter of the 20th century the research effort is devoted to giving machines not only some of the skills that humans learn in the first three years after birth but also the essential quality of adaptation and self-learning, so that from its original programming it can learn fresh skills. Many of these research efforts are devoted to perfecting the integrated cognitive system, which has the capability of identifying things. The task is considerable. Although the progress of microelectronics has increased processor-operating speed and, in some cases, storage capacity

Below: this robot "arm" has been designed to simulate as accurately as possible the action of a human arm and hand. It forms part of an industrial robot built to work side by side with human assembly line workers.

to well beyond human levels, the simple learning processes of humans are very hard to program into a computer.

Putting intelligence into the "eyes" of a computer has been attempted by such American giants as the Massachusetts Institute of Technology (MIT), Stanford, and General Electric, often with specific objectives in view. MIT solved such problems as how to identify basic Euclidean shapes such as a cube or cone. Many centers have since concentrated on the far more difficult problem of identifying irregular objects, such as an apple as distinct from an orange. Even inspecting a teacup and positively identifying it is a task involving millions of computer operations, with a sensor (such as a TV camera) inspecting the object from different aspects and the computer analyzing the output signal into contrasting light and dark regions and deciding where there must be discontinuities, edges, and holes. But once a working integrated cognitive system has been constructed it can often beat the human. For example, if one could be taught to recognize friendly or hostile aircraft it ought to be capable of being served by optical, infrared, and microwave inputs, with high power and magnification, that would enable it to do its job with absolute precision over distances out to the visual horizon by night or day, unaffected by rain, snow, or fog. At the moment the objectives are still some way from attainment – and friendly or malevolent robots are still confined to fiction.

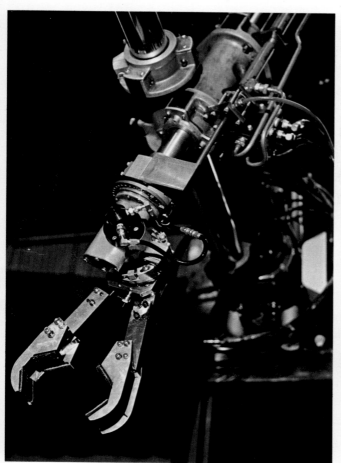

Above: industrial robots at work on a Fiat production line. Seen here on a spot-welding line is one of three Space Saver units. There are 22 other industrial robots on this line alone. Altogether, Fiat have more than 200 industrial robots at work in its plants around Turin.

Radio and Television

Above: radio was used as a propaganda weapon by all belligerents in World War II. This painting shows the people of Meuvaines, Normandy, listening to the BBC.

In 1865 the Scottish physicist James Clerk Maxwell mathematically calculated the behavior of various kinds of electromagnetic wave, even though such waves had not then been discovered. Today we know that electromagnetic waves cover a fantastic range of frequencies and wavelengths, ranging from a vast spread of radio waves up through microwaves to infrared (heat), visible light, ultraviolet, X-rays, and gamma rays. As Clerk Maxwell predicted, such waves are made up of oscillating electric and magnetic fields at right angles to each other. He predicted that waves of a length many times greater than that of light should result when electricity was made to oscillate in a conductor. A few years later, in 1879, the Anglo-American inventor David E. Hughes actually built a rudimentary radio transmitter and receiver and sent messages along Great Portland Street, London, but he did not immediately publish his findings and it was the German physicist Heinrich Hertz who in 1887 built a spark generator and receiver and showed how some radio waves are reflected from ionized layers in the atmosphere. Some travel in straight lines and go on into space, while others bend around the curvature of the Earth.

In the early 1890s the British physicist Oliver Lodge demonstrated practical radio communication. In 1901 the Italian electrical engineer Guglielmo Marconi detected the Morse letter S (three dots) in Newfoundland after transmission from Cornwall, and in 1906 the American physicist Reginald A. Fessenden transmitted a voice and music. Marconi's demonstration was radiotelegraphy, using a plain monotone signal of dots and dashes. Fessenden used radiotelephony or broadcasting, in which the desired sound, such as speech, is impressed on an electromagnetic carrier wave of a fixed frequency.

In commercial broadcasting the signal is generated in a microphone, where sound waves are converted

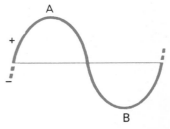

Above: a radio wave cycle. The distance between two crests, or troughs, is called the "wavelength." The number of cycles per second is the "frequency", measured in hertz.

Left: the radio transmitter and receiver built by David E. Hughes in 1879. He sent messages along Great Portland Street, London. Hughes is also known for his work on printing telegraphs. He also invented an early microphone, and made contributions to the theory of magnetism.

into a stream of small electrical signals. These audio signals are amplified (magnified) and impressed on a fixed carrier wave and carried along wires to a transmitting aerial (antenna). As the frequency of the waves is so high, from 500,000 to many millions of cycles per second, their oscillations in the transmitter aerial emit exactly corresponding waves through the atmosphere at the speed of light. Most radio waves are reflected from the ionosphere, as a perfect mirror-image, and then up again from the Earth's surface. Very short waves go straight through the ionized upper atmosphere and can be used for space communication, while the longest waves are refracted around the Earth.

Three instruments are needed to receive a radio broadcast: an aerial, to convert the passing waves back into electrical signals; a detector or demodulator, which eliminates the carrier wave and leaves only the original signal impressed upon it; and a speaker, which converts this signal back into the original sounds. Commercial receivers have to be capable of being tuned to pick up broadcasts on many frequencies. The whole radio spectrum, representing a major part of the total electromagnetic frequency spectrum, is assigned by international agreements to particular stations and particular duties. The longest radio wavelengths, in the very low frequency (VLF) band are used almost entirely for special military communications on a global scale between continents, aircraft, ships, and (rather remarkably) submerged

Below: a VHF and microwave beam antenna. One of the advantages of VHF transmission is that a great many radio and television stations can be accommodated without their signals interfering with each other. UHF waves lie between VHF and microwaves.

Above: a monochrome cathode-ray deciphers the image of the "electronic copy" of the scene produced by the television camera. Early TV was of 405-line transmission.

submarines. The slightly shorter wavelength medium frequency (MF) bands are used for regional or national broadcasts, while high-power short waves (HF or high frequency) are used for intercontinental commercial broadcasts not using satellites. Even shorter very high frequency (VHF) waves are used for the greatest number of commercial broadcasts and many other communications, such as in airtraffic control, to ships, by the police and many other local users. Even shorter ultra high frequency waves (UHF) are used for many special military communications, and also for television. The shortest waves of all are microwaves, which can be beamed like a searchlight from a special reflecting aerial; these are used for communications between towers, for space communications and for broadcasting with satellites, and for most kinds of radar.

Radar, a word derived from RAdio Detection And Ranging, describes use of radio waves, invariably very short microwaves, to detect the presence of solid objects and indicate their exact position. Waves are broadcast either generally or in a fine beam, and any that strike a solid object, such as an aircraft, are scattered, some being reflected back to a receiver near the transmitter. Though only a very small fraction of the original energy is received back again, precise measurements can indicate how long the waves took over their round-trip journey, which with other techniques can indicate the position of the reflecting object. Most radars generate a picture on a cathode-ray tube display which, depending on the design of the equipment, can be a maplike plan view (often with the radar at the center of concentric circles a fixed distance apart) or various kinds of end-on or side view. Modern radars can eliminate unwanted "clutter" from the ground or other interference, and many have special features such as the ability to measure the relative velocity of a target, or eliminate fixed echoes and indicate only those from moving targets.

Left: the original television apparatus built by John Logie Baird. This 30-line receiver was sending poor quality pictures over quite long distances by 1928. The scanning was mechanical by means of three disks rotating at different speeds. Definition was primitive, and the EMI-Marconi electronic scanning system took over.

Television is very similar to sound broadcasting, but as well as sound it converts a two-dimensional scene into electrical signals that are then transmitted as a video (TV picture) signal to be received by aerials on our homes. The first TV used mechanical scanning. In 1923 the Scottish inventor John L. Baird viewed a scene through holes cut in a spiral path on a rotating disk to give a succession of light or dark images that he converted into electrical waves. But in America in the same year the Russian-born Vladimir Zworykin applied for a patent for a far superior scheme using an Iconoscope, which scanned electrically. Gradually a TV broadcasting system was perfected, first used com-

dard is 625, which gives a clearer and better picture. At every point the brightness of the corresponding spot in the scene is converted by a photo-cathode into a stream of electrons, which fall on a target electrode to build a pattern of charges. This is in turn scanned by another electron beam, which is modulated by the pattern of charges to give the desired varying electrical signal for broadcasting.

The domestic TV receiver accepts both the sound and video signals. Because of the very high frequencies, compared with plain sound broadcasting, TV networks operate on much higher carrier frequencies, in the VHF and UHF bands, in order to accommodate enough different channels. One TV channel needs as much bandwidth as many radio or telephone channels, and color TV needs many times more than black and white. The received signal is processed and fed to scanning coils which deflect the beam in a picture tube, a special large-screen cathode-ray tube similar to that used in radar and many other devices. The electron beam is swept, or scanned, across the face of the tube in exact synchronization with the scanning

Above: TV broadcasting reopened in Britain after World War II on June 7, 1946. This picture shows girls from the Windmill theater, London, going through their dance routine in front of TV cameras at Alexandra Palace.

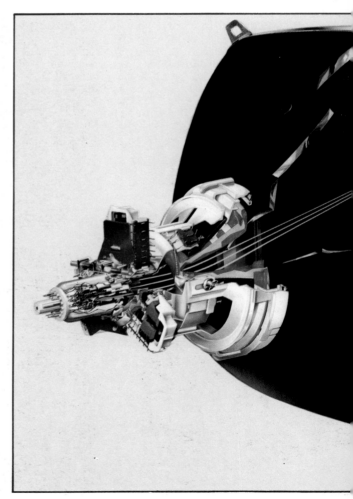

mercially in Britain in 1936.

Electronic scanning, first by the Iconoscope and then by the image orthicon, vidicon, or other TV camera tubes, surveys the scene to be transmitted in a series of zig-zag lines exactly like the human eye reading down a page of type. The original British TV had 405 parallel lines (joined by right-to-left diagonal back-strokes) while today's European stan-

of the camera tube in the studio. At every point on every line the received video signal modulates the intensity of the beam. As the electrons strike the phosphor coating on the face of the tube they cause light to be emitted. A brighter spot in the scene in the studio means a stronger video signal, more intense electron beam and brighter spot on the face of the tube. Thus, the receiver re-creates the original scene spot-by-spot on each line, and line-by-line from top to bottom. The scanning is so fast, all 625 lines being covered several times a second, that it looks like a faithful reproduction of the original picture.

Color TV is more complex, and needs far greater bandwidth. Basically, the scene is broken down into three colors, red, green, and blue, each of which is scanned and broadcast separately. The receiver has an extremely clever picture tube with three electron guns, one for each color, which are separately controlled by the three video signals. Each gun can "fire" at only a certain number of small spots covering the face of the tube (because of the presence of a very accurately made shadow mask with about 300,000 fine holes on which the three guns are focused). Thus the final picture is actually constructed from about 300,000 triplets each made up of a red spot, a blue spot and a green. The human eye cannot detect the individual spots and sees a re-creation of the original scene's hues.

Right: the world's first true pocket TV was developed by the British company Sinclair Radionics. Its two-inch screen is quite comfortable to watch as it is viewed from a shorter distance than would a larger screen. The picture is received in black and white.

Below: a color television cathode-ray tube. The picture is created on the screen by the fluorescence of a layer of material struck by cathode rays. These rays move quickly backwards and forwards across the screen. They produce a picture by traveling from the top to the bottom of the screen.

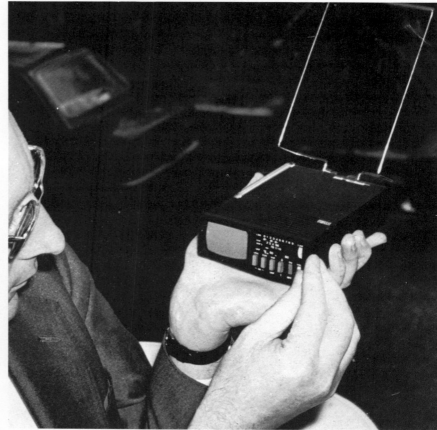

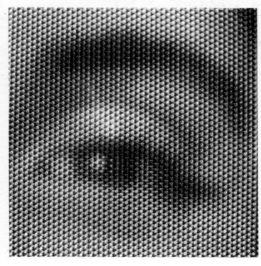

Left: a close-up of part of a color TV picture showing the human eye. The picture is formed by thousands of three-color dots – more than 250,000 on a whole screen.

The Telephone and the Telex

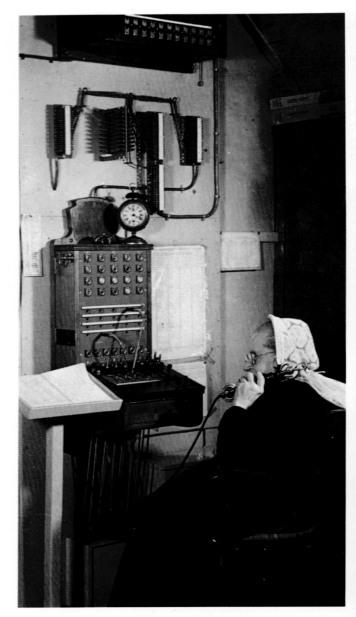

Today telecommunication is becoming so widespread and effective that many observers consider it is already cutting down the need for human travel, except to go on holiday. Already scientists, engineers, diplomats, and many other groups can have "tele-conferences" without actually coming together. Today's worldwide telephone network is one of man's largest interlinked operative systems, with direct electrical connections between almost every town on Earth, and in the advanced countries linking almost every dwelling. Now additional facilities, such as the ability to have a visual link rather like a personal TV, are adding to the possibilities. But the greatest revolution concerns totally new methods of operation. Already traffic exceeds that which could reasonably be handled by millions of separate insulated wires, and before long conductive wires for electricity are likely to give way to special beams of light. These will enable traffic to increase about one-millionfold.

There is little difficulty to modern technology in devising the elements of a simple telephone circuit. The caller speaks into a microphone built into a neat hand-set which also contains the receiver earpiece. This microphone is of the single-button carbon type. The incoming sound waves from the speaker vibrate a thin metal diaphragm that puts rapidly varying pressure on fine carbon granules contained in a cup with electrical connections. The changing pressure

Above: an early telephone switchboard in use in Holland. Telephones were considered a luxury for some time, and so switchboards did not have many lines to cope with. But as the number of telephones proliferated, manual switchboards grew to the limit of their operators' reach. Nearly all calls are now handled automatically.

Left: Alexander Graham Bell's telephone receiver. He exhibited this model and many others at the Philadelphia Exposition 1876.

Above: the steamship *Great Eastern* laid the first trans-atlantic cable to operate permanently in 1865. This ship was the only one of its time capable of carrying the 2300 nautical miles of cable necessary to complete the job. Designed by Brunel, it was launched in 1858.

Above: in 1900 a small army of operators was necessary to keep the Hamburg local exchange functioning, and this with a much smaller volume of calls than today. Each call was put through manually. Now, the vast majority are handled automatically.

alters the resistance of the granules, translating the speech into an electrical signal. Unlike radio this need not be impressed on to a high-frequency carrier wave but is simply transmitted along a wire. At the telephone of the called subscriber the signals are passed through coils which vary a magnetic field to vibrate the earpiece diaphragm to re-create the original speech. Although the sound volume is very small it is ample when the receiver is held against the ear.

The chief difficulty in constructing a telephone system was in devising a switching system. In the United States the Scottish-born Alexander Graham Bell secured his patent for a workable system in 1876, hours before a rival claim by another United States inventor Elisha Gray. Two years later an eight-line exchange opened in New Haven, Connecticut, followed by one in London. Girls in each exchange connected calls by inserting plugs into sockets. Some made mistakes, especially after the networks grew in complexity and came to link many exchanges. As early as 1889 another American inventor, A. B. Strowger, devised a semiautomatic method, called machine switching, in which each telephone incorporates a rotary switch controlled by a 10-position dial rotated by the finger. As the subscriber dials, electrical impulses in the form of a simple digital sequence reach the exchange and trigger electromechanical selector switches. These are rotary notching relays with contact wipers that rise on a vertical rod to the level corresponding to one of the dialed numbers and then rotate across a series of contacts to stop at the first that is not busy (the contacts by-passed are those already in use). If all in that row are occupied, the subscriber hears a "number engaged" signal. The final two digits are connected on a single

A model of A. B. Strowger's semiautomatic switching device. Built in 1889, it handled up to 100 lines.

selector, the first being the vertical slide and the second being the horizontal rotation. Strowger equipment became virtually the world standard from 1892 onward.

In 1926 a better method, the crossbar system, began to replace the old electromechanical Strowger, though the latter was so widely used in every country that it is taking decades to replace (by the late 1970s only Sweden had completed the task). Crossbar switching inserts intermediate controls between the pulse-train from the caller's dial and the exchange, storing the pulses in a register and simultaneously using a device called a marker to seek the best route for the call. Thus the actual switching time in the exchange is cut from as long as the caller takes to dial to a few milliseconds, and the equipment is then free for other calls. This greatly increases the capacity of each exchange, and also opens up valuable alternative routes which the mechanical switching could not do. In parallel with crossbar switching came improved telephones,

277

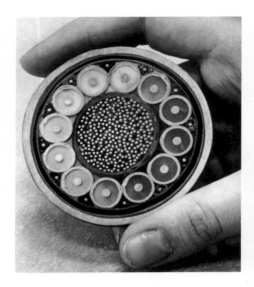

Above: a cross-section of a coaxial cable, made up of 12 smaller coaxial cables, surround conventional conductors.

Right: this Bell Telephone Company electronic switching system went into service in 1976. One of the great advantages over electromechanical systems is the avoidance of annoying hissing and crackling caused by dirty contacts.

such as the Bell System's neat Trimline with the dial in the handset and a rapid and more accurate push-button instrument called the Touch-Tone, which generates the same pulse-train but is more easily operated.

In 1960 modern electronics and computer technology, natural allies of any telephone system, came into the picture with the electronic-switching system. This at last escapes from the basically mechanical kind of switching system and instead uses a small number of extremely powerful central processors. In most electronic systems all subscriber lines are routinely scanned several times per second. The moment one line becomes live, by lifting the hand-set off the receiver base, the rate of scanning for that line jumps to about 100 per second. The incoming dialing pulse-train is stored for a brief fraction of a second while the computer's memory and processing circuits find the best route; then the switching is made in milliseconds, while the memory stores the information that the chosen route is occupied. The swift processing is important in enabling limited exchange equipment to handle far more calls. There are many other advantages; for example, if the called line is busy the processor notes the fact and puts the call through as soon as it is free. No human action is needed anywhere.

There are many other new developments coming into worldwide use to speed up and make possible increased telephone traffic without increasing the size of exchanges or numbers of linking cables. Early networks used ordinary insulated wires hung on

posts. These were progressively replaced by underground cables, which became increasingly complex as they were made to contain more and more conductors. Over long distances, especially with submarine cables, amplifiers or "repeaters" are inserted at intervals to boost the output (without ever feeding back to the input). In recent years the use of coaxial cable has allowed each line to carry up to 960 speech channels simultaneously.

Since the 1960s research has been in hand on transmission of voice messages along coded light beams. In this two important inventions, fiber optics and the laser, play a central role. With communications based on coherent light the tremendous reduction in operating wavelength and increase in frequency increases the traffic that can be handled by a single communications channel by thousands of times. Many light-based telecommunications systems are already being rapidly perfected. Until they take over there are several other significant techniques for relieving the strain on old-fashioned telephone wires.

Time-division multiple access (TDMA) or multiplexing, is one of these techniques. It relies on the inability of the human ear to detect silent gaps of less than 25 milliseconds (0.025 sec). With modern electronics it is possible to break each conversation into extremely brief 0.125 millisec fragments separated by relatively enormous gaps of 25 millisec. When all the capacity of a line is used no fewer than 2000 conversations pass along it simultaneously, none of the users being able to detect the very rapid sequence of breaks during which the other 1999 conversations are

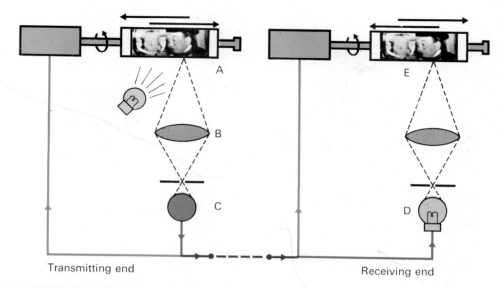

Right: to transmit a picture, the photograph is wrapped around drum **A** and illuminated. Light reflected from a small area is focused by the lens **B** onto the photo-tube **C**. This converts light intensity into electric current **red**. The picture signals combined with synchronizing signals **blue** are transmitted to the receiving end. Varying current controls the brightness of the recording light **D**. Image is reconstructed on photographic film wrapped around synchronized receiving drum **E**.

Transmitting end

Receiving end

Above: the Picturephone developed by the Bell Telephone Company. Variations on the visual telephones have been designed in many countries, but it is doubtful that they will be a commercial success. They can communicate complex information difficult to convey in words.

fitted in. Another technique is to dispense with a conductive wire and use microwaves, beamed by a paraboloidal dish aerial from high towers. Microwaves also enable all kinds of telecommunications, including radio and TV, to be beamed up to satellites for transmission around the world.

The first kind of telecommunications network was the telegraph, in which wires carried Morse code signals to a buzzer at the receiving end. Many inventors tried to find alternatives to the buzzer, the chief objective being an automatic printer of a readable message. Linking two typewriters over a telegraph or telephone line was not difficult provided the typist maintained an absolutely precise pace, without mistakes – which was not possible. Not until the 1920s did the "start/stop" coding method lead to the teleprinter. The typist does not inject electrical signals direct but instead types out a punched paper tape. This translates each character into a seven-unit code of electrical pulses or gaps, the first and last being START and STOP and with a five-digit code representing the character. (In this way the letter A becomes START-pulse-pulse-blank-blank-blank-STOP.) When the tape is fed into a high-speed reader it sends a stream of perfect digital signals, so fast that the line can handle 26 messages simultaneously. The key factor in making a workable teleprinter system was reducing each character to a standard length, with the seven-digit code, to permit sender and receiver to be keyed together. Teleprinters have the advantage of being "secure" in that government or military links cannot be interfered with or tapped by an enemy. Telex is a commercial teleprinter network in which each subscriber has a special number that others can dial. Each telex machine receives messages automatically, typing them out without need for an operator. The operator can send either straight from the keyboard or, at much higher speed, by inserting punched tape prepared and edited previously.

Another form of telecommunications is picture transmission by the facsimile or phototelegraphy method. In a technique rather like TV the picture to be transmitted is rapidly scanned and converted into a stream of signals. In exact synchronization, the receiving machine rebuilds the picture on a "dot for dot" basis.

Satellites and Communication

The Earth's spherical shape and global population make it difficult to provide adequate telecommunications to meet the enormously expanding demands of the growing population. Conductive links are expensive and are everywhere overloaded, HF radio becomes distorted and subject to atmospheric disturbance, and microwaves have to be beamed from tower to tower to gain the maximum distance within the available line-of-sight path. Clearly, the higher frequencies carry far more traffic, but there are limits to the practical height of towers. Then, with the ability to place artificial satellites in Earth orbit the problem was solved. Microwaves sent along the line of sight to a satellite can be beamed back to an area covering about half the globe. Today there are large numbers of comsats, or communications satellites, serving many military and civil purposes.

The first comsat was Project Score, placed in orbit on December 18, 1958 by the USA. This broadcast a recorded message and also relayed messages by voice and teletype. As it stored and re-broadcast messages it was an "active" comsat. The next, Echo, launched on August 12, 1960, was a "passive" satellite, a mere reflective balloon. This has the advantages of simplicity and the ability to reflect almost any available frequencies, but is useless for sustained intense traffic. Gradually over the next two decades comsats grew in size and power, until they are able to handle not only intercontinental telephone traffic and radio but even the enormous bandwidth demanded by color TV.

A unique comsat system tested in 1963 was Project West Ford, of the United States Air Force, in which messages were transmitted through a belt of 400 million very fine hairlike copper needles forming a belt around the Earth that passed roughly over the Poles. By this time the concept of the active repeater satellite was well established, with such programs as Courier, Telstar, and Relay, and today many hundreds of active comsats have been launched by most advanced nations. Some are low-altitude satellites, placed in many kinds of orbit, most of them eccentric elliptical orbits at a height of only a few hundred miles at the lowest point. More useful to most operators are the so-called geostationary satellites. These are placed accurately into an easterly orbit around, or near to, the Equator, at a perfectly circular height of about 22,245 miles. Such a satellite is "synchronous," in that it rotates in a period of exactly 24 hours, and thus always remains over the same place. Three synchronous satellites can cover almost the

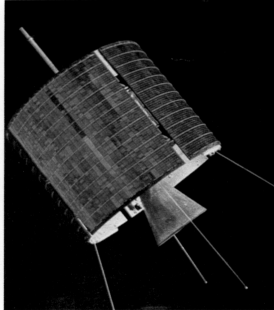

Above: the Early Bird communications satellite. It was launched on April 6, 1965, and began services in June of the same year.

Left: the fifth in a series of Intelsat IV communications satellites being launched at Cape Canaveral Florida by an Atlas Centaur rocket. Communications satellites are taken for granted today but were regarded with awe in the 1960s.

Above: this portable antenna relayed pictures from the Intelsat satellites. Color TV pictures of the 1968 Mexico City Olympic games were relayed by Intelsat.

Right: this satellite, popularly called the orbiting umbrella, provides India with its first educational television.

whole population of the Earth.

A comsat has to be a compact package to fit the available space in the launch vehicle, though the latest types are substantial units with considerable on-board electrical power. For example, the most important commercial comsats, the Intelsat IVA series, stand 23 feet tall, and weigh 3340 pounds each. These satellites, three of which are operating over the Atlantic and two over the Indian Ocean, are operated by an international multinational consortium. Each has 20 separate transponders that pick up signals, amplify them and re-beam them back to a particular receiving station on Earth. The satellites are accurately positioned to waste as little energy as possible, but nevertheless the outer surface of the main cylindrical body is covered in solar cells that convert sunlight into electric power. Each satellite in this family can handle up to 11,000 simultaneous telephone calls or 20 color TV channels, the usual traffic being a combination. Intelsat traffic links nearly 80 Earth stations around the globe.

Most comsats so far have been placed in orbit by the United States and Soviet Union, the latter having a large number of the military Cosmos series, the elliptical-orbit Molniya family, and the geostationary Raduga, Statsionar, and Ekran series. France and West Germany put up the first Symphonie satellite in January 1975, and a French Ariane launcher is to launch the first multinational European Communications Satellite in 1981. Other comsats have been produced in India and Japan. One of Japan's comsats is transmitting directly to special receiver aerials on individual dwellings, bringing economical TV to many remote islands where no reception was previously possible. Many other satellites are helping the undeveloped world by beaming radio and TV programs, chiefly of an educational nature, to places formerly without any proper schools or links with the outside world. Almost every developed country today has special comsat programs designed for broadcasting to underdeveloped nations, usually as part of an international team. The effect of bringing cheap and reliable communications links, including TV, to virtually every spot on the globe is already having a profound effect on people everywhere.

CHAPTER 11

SPACE TECHNOLOGY

Just like flight within the Earth's atmosphere, man's conquest of space was preceded by many myths, legends, and works of fiction. But the key to spaceflight has existed for at least 1000 years, because the gunpowder rocket was an accomplished fact long before any other mechanical propulsion system. Spaceflight, expressed in such terms as "flying to the Moon," became synonymous with foolish fantasy, so that an engineer or scientist who discussed it seriously risked jeopardizing his career. When the so-called Space Age suddenly began on October 4, 1957 it triggered intense worldwide interest, which was heightened by the launch of improved satellites (including one visible from Earth with the naked eye) and manned capsules, culminating in manned voyages to the Moon from 1969. Subsequently, the media have regarded space as no longer news. People hear little of the latest developments in space technology, which transcend in every way all that went before.

Opposite: the first Titan Centaur lifts off from Cape Canaveral, Florida. The Centaur stage failed to ignite, and the rocket was destroyed by the range safety officer. This type was designed for heavy unmanned payloads, and powered the twin Viking missions to Mars in 1975.

A Brief History

Though rocketry was a familiar means of propulsion centuries earlier, credit for the whole concept of using rocket power to escape the Earth's atmosphere is due to a humble Russian schoolteacher, Konstantin Tsiolkovsky, who in the final quarter of the 19th century not merely grasped the possibilities but worked many of them out with exact arithmetic. One of his mental leaps was to realize that a rocket would work in a vacuum – in fact it works better than in the atmosphere – and that there was a vacuum once one had climbed through the Earth's atmosphere. Another was that, instead of using gunpowder, rockets could be powered by mixed liquid fuels. He calculated many kinds of spaceflight trajectory, and showed how flight velocity would be increased by using a series of successively smaller rockets, each allowed to drop off as it was consumed. One of the most remarkable passages in his writing reads "When the velocity reaches [five miles] per second the centrifugal force cancels gravity. After a flight which lasts as long as the oxygen and food will allow, the rocket spirals back to Earth, braking itself against the air. . . ." Tsiolkovsky had described space flight more than 40 years before man escaped the Earth's atmosphere.

Though many of Tsiolkovsky's writings were published in a popular book in 1915 they had virtually no impact on the world at large, or even on the scientific community, where nobody bothered to take him seriously. But after World War I, rocket research began to make progress. Meanwhile, publications in Germany and the Soviet Union gradually began to appeal to a wider audience. In June 1927 the *Verein für Raumschiffahrt* (*VfR*) society for space travel was founded in Germany. It soon had over 500 members in many countries. In the Depression years after 1930 the *VfR* declined and was wound up in 1934. Many of its German members were by this time working on military rockets, and these led to the giant A4 (V2) which did more than any other project to advance space flight. Regular A4 firings exceeded heights of 60 miles, and when used for research on vertical trajectories after 1945 heights of 114 miles were attained. In 1949 a second, smaller rocket (WAC Corporal) was launched from the nose of a V2 to reach a highest point (apogee) of 244 miles.

In May 1946 the Rand study group of the US Army Air Force (US Air Force from 1947) completed a major report on an Earth satellite. But no funds were available to build it or its carefully planned launch vehicle. In 1948 the US Secretary of Defense said a few words about this in public, instantly drawing a

Below: a fictional view of Astronauts from a short story by an early master of science-fiction, Jules Verne. Verne's stories, although technically wrong, were surprisingly accurate predictions.

Below: the V-2 rocket was developed by the Germans to bombard London towards the end of World War II. V-2s were captured by the Allies, and their technology was exploited to build the world's first space rockets.

Above: the Soviet Union's *Vostok* space booster was developed from military missile technology. It was *Vostok* which launched the first manned spaceflight by Major Yuri Gagarin on April 12, 1961. Gagarin completed just on Earth orbit.

Above: NASA scientists anxiously await liftoff of a Saturn I rocket. On the extreme right is Wernher von Braun. He helped develop the V-2s for Nazi Germany during World War II in Europe.

violently hostile reaction in a Soviet newspaper which called an artificial satellite "an instrument of international blackmail." But many proposals were subsequently published, and on July 29, 1955 the US Government announced it would launch small satellites as part of the International Geophysical Year in 1957–58. Foolishly, the proven rockets were ignored and a completely new Vanguard vehicle was designed for the purpose. This had limited capability and poor reliability. Between 1955 and 1957 the Russians kept the world informed on progress of their own satellite. Little attention was paid to these announcements, so it came as a traumatic shock – especially to the Americans – when Sputnik I was placed in orbit on October

4, 1957. At 184 pounds, it was more than eight times heavier than the Vanguard satellite. A month later another Soviet rocket put Sputnik II into orbit. It weighed over 1000 pounds and among other things contained a live dog. In May 1958 Sputnik III went into orbit weighing 2926 pounds.

With much smaller launch vehicles the Americans fought back, and the media soon fed on the concept of a "space race" between two superpowers. On April 12, 1961 Major Yuri Gagarin of the Soviet Naval Air Force went into Earth orbit in the capsule Vostok I. A month later President John F. Kennedy announced the United States' intention of landing a man on the Moon and returning him to Earth before the decade was out. During the 1960s, while hundreds of satellites, probes, and other spacecraft performed increasingly useful functions, the gigantic Saturn V vehicle was created for the Apollo lunar mission, six of which were flown with complete success between 1969 and 1972. By this time scientists and engineers had turned their attention to the entire Solar System, while astronomers used spacecraft to study the Universe in a way impossible on the Earth's surface. Man is becoming increasingly experienced in living and working in space, and enormous space stations for many purposes including colonization can be planned on the basis of well tested knowledge. The cost of spaceflight is being dramatically reduced by reusable transport vehicles, such as the Space Shuttle. Spaceflight has become an everyday business for more than a dozen countries.

Below: the Vanguard TV-1 was launched on May 1, 1958 by the United States. The second pre-Vanguard test vehicle, it tested equipment for the Vanguard. The Soviet Union launched Sputnik I on October 4, 1957.

Propulsion Systems

Hundreds of thousands of military rockets were fired prior to 1850 powered by gunpowder. But the propulsion system for all early space missions was the liquid propellant rocket, in which two liquids, a fuel and an oxygen-rich oxidant, are carried in separate tanks, and fed by gas pressure or mechanical pumps to a thrust chamber. Some propellants are cryogenic (refrigerated), the most common being the oxidant liquid oxygen (lox) and the fuel liquid hydrogen. Others are hypergolic (igniting spontaneously upon contact), such as concentrated nitric acid, and either nitrogen tetroxide fuel or aniline fuel. There are many hundreds of liquid propellant systems, as well as an increasing number of solid propellant rocket arrangements which are the modern descendents of gunpowder.

As early as 1903, the year the Wright brothers first flew an airplane, the visionary Tsiolkovsky proposed the design of a spaceship with engines fed by liquid oxygen and liquid hydrogen. He even explained how the thrust chamber could be cooled by passing one of the extremely cold liquefied gases through a double-skinned wall. He also explained how thrust could be controlled by means of adjustable valves in the propellant feed pipes, and how a gyro could steer the vehicle through space by altering the angle of heat resistant vanes in the rocket exhaust. In every respect his suggestions described how space rockets would later be built, though engineers eventually learned how to avoid the need for graphite vanes in the hot jet and use different forms of thrust vector control (TVC) instead. The most common method was to mount the entire thrust chamber on gimbals (pivots).

The first large rocket, the German military A4 (V2), exerted a major influence. Its single fixed thrust chamber had double walls cooled by the alcohol fuel (a 75/25 mixture with water) about 8450 pounds of which was carried. The oxidant was liquid oxygen, of which about 11,000 pounds occupied a glass wool lagged tank between the fuel tank and the chamber. Both propellants were allowed to run down by gravity as the rocket stood vertical before launch. They eventually filled the large turbopumps that supplied the thrust chamber at high pressure. The pumps were driven by a turbine of about 680 horsepower running on high pressure steam generated by the violent decomposition of concentrated hydrogen peroxide mixed with permanganate solutions. At sea level this remarkable engine generated a thrust of

almost 57,300 pounds, the thrust increasing as the rocket climbed into thinner air to a figure of about 66,000 pounds at the limit of combustion. In extreme range missions this might occur at the burnout point, when the propellants were exhausted, but on operational A4 trajectories propulsion was cut off by the guidance system when the desired velocity had been reached.

Rocketry made rapid advances after World War II. Though lox/alcohol continued to be important, alcohol was replaced in most large vehicles by various refined blends of kerosene, such as the American RP-1. A basic measure of rocket performance and efficiency is specific impulse, abbreviated I_{sp}. This is the thrust obtained for unit rate of consumption of propellants, such as pounds thrust divided by pounds per second. Thus, the I_{sp} is expressed in seconds, though it is not a

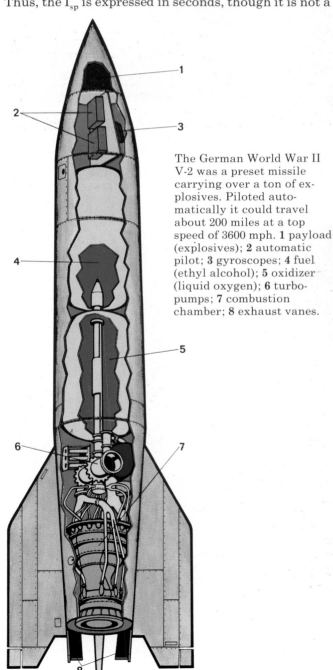

The German World War II V-2 was a preset missile carrying over a ton of explosives. Piloted automatically it could travel about 200 miles at a top speed of 3600 mph. 1 payload (explosives); 2 automatic pilot; 3 gyroscopes; 4 fuel (ethyl alcohol); 5 oxidizer (liquid oxygen); 6 turbopumps; 7 combustion chamber; 8 exhaust vanes.

Above: the enormous Saturn V lifts Apollo 15 from the launch pad of Cape Canaveral, Florida, on its journey to the Moon – July 26, 1971. 850,000 gallons of liquid hydrogen were stored in the foreground. Its second and third stages used liquid hydrogen and liquid oxygen.

Above: an interior view of the Atlas rocket. Originally designed as a ballistic missile carrying a nuclear warhead, Atlas was later to launch the Mercury one-man spacecraft into Earth orbit. Atlas is still being used as the first stage of space launchers.

Above: the huge liquid oxygen and liquid hydrogen tank of the Saturn V rocket undergoing static testing at Boeing's New Orleans plant. As a safety precaution these tests are viewed via remote control television. Stress testing was vital for the project.

measure of time. Specific impulse is also the total impulse (total burn time multiplied by average thrust) divided by the total mass of propellants consumed. With lox/alcohol I_{sp} is about 280, and changing to RP-1 fuel raised this to around 300. But other propellant combinations offer greater energy, and after solving many problems the lox/hydrogen engines came into general use with the Centaur launcher upper stage, raising I_{sp} to 390. Subsequently much larger lox/hydrogen "high energy" engines were used in the second stage of the Apollo Saturn 5 launcher. Higher energy still can be obtained with fluorine as the oxidant. Fluorine/hydrogen, for example, gives the excellent figure of 410. Some work has been done on intermediate oxidants, such as "flox" (70 percent lox and 30 percent fluorine).

An obvious disadvantage of all cryogenic propellants is that they are so intensely cold. Highly skilled engineering is necessary to make valves and pumps work without conventional lubricants. Materials that do not become brittle and crack in the intense cold have to be used. Once the liquefied gas has been loaded, it boils off continuously, and has to be kept topped up if the quantity and therefore range is not to diminish. Worse, the tanks cannot be kept at instant readiness over a period but must be filled before a mission – a procedure taking 30 minutes or more. This is a grave handicap for a military missile. For this reason non-cryogenic, "storable", liquids are used in most modern

Above: a solid fuel rocket engine with the launch gantry in the background. In solid fuel engines, the propellant burns at a known rate. Gases are generated under high pressure, and escape through the propelling nozzle. Enormous speeds can be attained.

liquid-rocket missiles, even though I_{sp} is seldom greater than 275 to 285. One such mixture is concentrated nitric acid and aniline or hydrazine. Another is nitrogen tetroxide and hydrazine.

Throughout the 1950s hundreds of laboratories also worked to improve solid propellant rockets, which in World War II had been small and usually loaded with diglycol or cordite. Gradually the smokeless powders, or double base (DB) propellants were joined by completely new composite propellants made up of an inorganic salt oxidant and organic fuel, intimately mixed and also mixed with finely powdered metal. The most common composite propellant is probably ammonium perchlorate and polybutadiene acrylic acid (PBAA), plus finely divided aluminum. This mixture made possible the submarine-launched Polaris and Poseidon missiles and the land-based Minuteman. Solid boost motors for spaceflight use the same propellant.

Design of motors varies greatly. In most missiles and space launchers the propulsion system also provides the means of controlling the flight trajectory.

TVC is still often effected by means of vanes or "jetevators" in the hot jet. But more common methods are either to gimbal the whole chamber in various ways or use a surrounding ring or liquid injection system (often injecting Freon, the common refrigerant), which deflects the jet at the required angle. With liquid propellants a gimbaled chamber implies flexible feed pipes, while with a solid propellant it involves either mounting the chamber on skewed rotary joints and steering by rotation, or else seating it in a spherical bearing able to remain gas-tight under high pressure when white hot. When liquid propellants are used, the tanks form the main part of the airframe. They are usually made of chemically milled aluminum but a few vehicles, notably Atlas, are of stainless steel so thin that the whole structure has to be inflated by gas pressure like a balloon. Solid cases are very strong, to resist the whole pressure of combustion. Some are welded of high strength steel while others are filament-wound in glass fiber, Kevlar, or graphite fiber.

Some of the largest solid motors are the United States Air Force 120 inch diameter segmented type, assembled from superimposed sections each weighing about 40 tons. From one to seven segments can be used to form a motor, thrust varying from 250,000 to 1,500,000 pounds. These giant solid motors have liquid injection TVC for vehicle control. At the other end of

Above: the US Air Force/NASA X-24B wingless lifting-body aircraft. It could reach a top speed of 1200 mph at an altitude of 90,000 feet.

the scale are many types of vernier motor. These are used for final trimming of velocity for the precise desired mission trajectory. The original monster Russian launch vehicle, used for the first intercontinental ballistic missile (ICBM) and to launch the first Sputniks and manned capsules, had 20 main engines and 12 small verniers, all firing together at lift-off (because in the mid-1950s it was thought to be unreliable to start rocket engines in space). Most chemical rockets can now be started with precision in the vacuum of space, and many can be controlled with a throttle lever. The main engine of the Space Shuttle is a high pressure lox/hydrogen chamber rated at 470,000 pounds in space. It is installed in a group of

Above: a technician provides lubrication during the machining of the main combustion chamber of the Space Shuttle's main engine. Weighing about 440 pounds, the chamber has a copper-alloy coolant liner and a nickel-alloy structural jacket.

three like the engines of a jet airliner. Controlled by a computer, it is designed to operate for a total of $7\frac{1}{2}$ hours accrued during 55 space missions. Between each mission the whole propulsion system is checked over by modified airline maintenance procedures. The power of this engine is immense. Each chamber's hydrogen pump, for instance, is rated at 75,840 hp.

There are many other systems apart from chemical rockets, some of very high I_{sp}, which can be used to give small thrust deep in space. In the nuclear rocket, liquid hydrogen is passed through a white-hot nuclear reactor and out through a nozzle. Radioactive heating has also been tried, and there are several kinds of electrothermal or electric arc thrusters. Among the most promising families are the charged particle motors, most of which accelerate a jet of ions (charged atoms missing one or more electrons) at great speed, the most common substances being cesium and mercury. The plasma rocket uses electromagnets to accelerate a stream of plasma or ionized gas. These low power thrusters are used for acceleration and control in regions far from Earth.

Above: an artist's impression of a nuclear powered spacecraft. The three large tanks visible contain liquid hydrogen propellant. Quarters for the crew are placed as far from the motors as possible, and in the shadow of the hydrogen tanks, which fulfil the role of a radiation shield.

Satellite Technology

Several early scientific forecasters had described how satellites and spacecraft might work long before such achievements became a practical proposition. They pointed out how a rocket, rising from the Earth, could be given greater and greater velocity until, at a speed that could be exactly calculated, the mass would never fall back but would escape the Earth's gravita-

ter is continually curving away at the same rate as the satellite accelerates towards it.

Satellites can be launched into any orbit round the Earth, at any height clear of the atmosphere. But they must obey the laws laid down by the astronomer Johannes Kepler in 1609 to explain the newly discovered motions of the planets. The two most important laws are: one, that the orbit is an ellipse, with one of the foci at (or, rather, extremely close to) the center of mass of the primary, in this case the Earth; and, two, the line joining the center of the primary to the satellite, called the radius vector, sweeps out equal areas in equal periods of time. Thus, in a highly elliptical orbit, the satellite has to move slowly at its further point from Earth (apogee) but extremely fast at its nearest, or lowest, point (perigee). In tables of satellite orbits the apogee and perigee are always given as heights above the surface of the Earth, but in calculating the velocities the true radius at any point

Right: retrograde orbits, as illustrated in the diagram on the left, circle the Earth counter to its revolution, whereas direct orbits (*right*) are in the direction of the Earth's rotation. The launch velocity of a satellite as well as the angle of its ascent determine the kind of orbit that will be achieved. Satellites are launched into a variety of orbits depending on the job they are intended to do. Polar orbits, for example, are inclined at an angle of approximately 90° to the Equator. They allow surveillance of practically the entire Earth. In synchronous orbit the satellite is about 22,300 miles above the Earth, and its velocity matches that of a point on the Earth's surface.

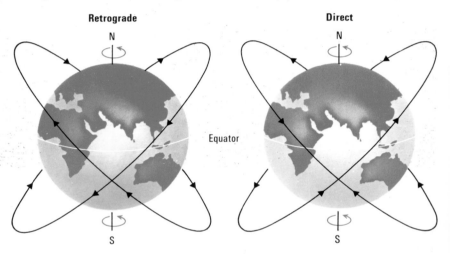

tional field entirely. One visionary, Tsiolkovsky, saw another possibility. If, after rising through the atmosphere, the rocket were to be turned through 90° to fly parallel to the planet's surface beneath, it would eventually reach a speed – calculated by Tsiolkovsky to be 17,850 mph – at which it would remain circling the Earth forever, with no need for further propulsion. It would behave exactly like a second Moon.

There are various ways of picturing the motion of a satellite whose mass is small compared with that of the primary body (in this case the Earth). Where the mass is large, such as the Moon, the position is slightly complicated by the fact that each mass influences the motion of the other. Thus, the Earth and Moon form a two-mass system with two bodies rotating around each other, the center of rotation being considerably displaced from the center of the Earth. But for an artificial satellite the difference in masses is so great that it can be assumed that the manmade device rotates exactly around the Earth's center. It can be regarded as a body continuously in freefall but unable to come closer to the planet's surface because the lat-

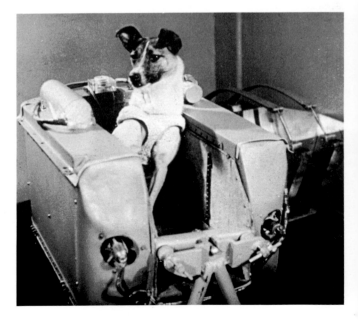

Above: both the United States and the Soviet Union put animals into space as a prelude to manned spaceflight. Laika, the first dog in space, was launched in Sputnik II in November 1957.

in an orbit must be measured from the center of the Earth.

Unlike the planets, which all move in ellipses (some of which are nearly circular), an artificial satellite can be launched into a perfectly circular orbit. In this special case the apogee and perigee are the same, but speed varies with orbital height. The lowest practical height is normally taken to be around 100 miles above the Earth's surface, because any lower level runs through atmosphere causing sufficient drag to slow

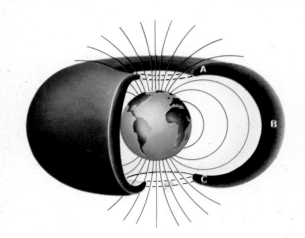

Above: diagram of the Van Allen Radiation Belt, discovered by James Alfred Van Allen. It is made up of subatomic particles trapped in the Earth's magnetic field about 250 to 6000 miles above the surface.

Above: the communications satellite Telstar was launched on July 10, 1962. It was developed by the American Telephone and Telegraph Co. at its own expense. They then paid NASA for the launch vehicle. Telstar was the first satellite to transmit color.

the satellite gradually, and as the speed falls the satellite inevitably drops closer to the Earth, finally spiraling in and burning up with friction in the denser atmosphere. This inability to recover objects from space without very special thermal protection was for a time a severe handicap. The second satellite ever launched, Sputnik 2, contained a dog which could not be recovered, and many subsequent satellites had to eject small instrument packages inside special "reentry vehicles" similar to those used with intercontinental missile warheads.

The orbit selected depends on the mission. Most early satellites were placed in roughly circular orbits. But some were very eccentric and the very first American satellite, Explorer I (put up by the previously ignored Army team on January 31, 1958), went out to apogee of 1155 miles and made the previously unsuspected discovery that Earth is surrounded by belts of electrically charged particles (named the Van Allen belts after their discoverer James Van Allen). Many satellites have gone into highly eccentric orbits reaching out hundreds of thousands of miles. Most of the military reconnaissance satellites, by far the most numerous single class with about 1500 launched

rises with orbital height, from a minimum of 4.97 miles per second at 100 miles to 6.68 miles per second at a height of 22,300 miles. This is a very special height, because when a satellite is launched into an easterly circular orbit roughly along the Equator at this height it remains over the same spot on the Earth below. The period at this height is exactly 24 hours, and the satellite is called "geosynchronous" or "geostationary". Communications satellites are usually placed in synchronous orbit so that they can continuously beam their traffic to their customers below. In any other orbit they would continually be going out of action, hidden behind the Earth.

Early satellites were simple. Sputnik 1 was merely a small radio transmitter powered by chemical batteries and sending out via four rod-like whip aerials. The surrounding envelope was a polished metal sphere. This was mounted atop the giant launch vehicle where it was protected by a payload shroud, or nosecone, during the accelerating ascent through the atmosphere. This nosecone was jettisoned once the satellite was free of the Earth's atmosphere. Gradually satellite engineering became more complex, though in a different way from aircraft. Few satellites have to operate inside the Earth's atmosphere, so aerodynamic forces such as lift and drag can be ignored. There is no need for any streamlining, though the whole satellite must fit into an often rather constricted space inside the payload compartment. This is normally on top of the launch vehicle, but large rockets sometimes carry additional payloads packaged along the sides. In these positions structural loads are sometimes increased, and heat-resistant shrouds must protect the satellite during launch.

Nearly all satellites are based on a light but rigid open-frame structure, often either a rectangular or hexagonal box, made of carbon or graphite fiber, light alloy, or other strong but low density material. In

Above: the 1978 World Cup soccer tournament held in Argentina was televised around the world in "real time." Real time is the actual time elapsed between an event occuring and the picture reaching the audience – a matter of a split second only.

in 1979, have been injected into polar orbits with an inclination (measured relative to the equator) of 70° to 100°. Orbits with inclination greater than 90° are called retrograde, because the satellite has an east to west motion – the opposite to the Earth's surface. Polar orbits, with inclination around 80°, are desirable for reconnaissance flights, for they enable almost all the Earth's surface to be covered. The orbital parameters are carefully chosen to gain the most frequent coverage of the interesting places, sometimes at the lowest (perigee) height to obtain greater detail.

The period (time per orbit) depends on apogee and perigee. In the lowest circular orbit at 100 miles the period is 87 minutes 42 seconds. Eccentricity makes the period rapidly longer, because of the lower speeds near apogee. But in circular orbit, the necessary speed

Above: this stainless-steel capsule contained nuclear fuel for the SNAP-19 generators, which were carried on NASA's Nimbus-B meteorological satellite, launched in May 1968. Heat from the nuclear fuel was converted directly to power to supplement that of solar cells.

the astronauts had to repair the damage and extend them manually.

Most satellites have to be not merely injected into the right orbit but stabilized in the correct attitude. Some are stabilized by being spun, at a rate varying from about 10 to 200 rpm. Spin motors, very small rockets or cold gas thrusters, are used to spin the whole satellite or a major portion of it. Sometimes repeated firings are necessary because the spin is slowed by Earth gravity and magnetism. Communications satellites (Comsats) must have their aerial (antenna) systems de-spun to remain pointed exactly at the correct receiver stations on Earth. Many satellites have attitude sensors, the most common being the infrared (IR) horizon sensor which keeps the craft exactly aligned in pitch and roll by sensing the sharp boundary between warm Earth and cold space. Yaw stabilization is then provided by spinning masses such as gyros, inertial wheels, or momentum wheels.

Above: this antenna was built to receive signals from the Soviet Union's two satellites in their Meteor space meteorological system. The satellite is equipped with an infrared sensor, making everything visible, even in pitch darkness.

Right: diagrams demonstrating the differences between the Tiros weather satellite and the more advanced Nimbus. Tiros was space-oriented. This meant that it was only possible for it to photograph the Earth at one point in its orbit. Earth-oriented satellites like Nimbus can take photographs of the Earth throughout their orbits. Weather satellites are now Earth-oriented.

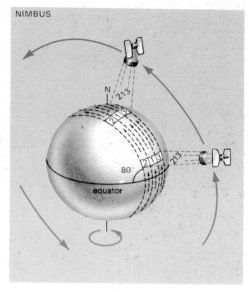

some cases special metals have been used, beryllium being especially suitable. On this basic space frame are hung all the operating devices, and in some satellites by far the largest items are electronic aerials and solar arrays. In almost every case there is a need for electrical power aboard, and only in a few missions of short duration will chemical batteries suffice. Some satellites have carried fuel cells, and others Systems for Nuclear Auxiliary Power (SNAP) reactors which generate electricity either directly, by thermoelectric or thermionic means, or by a mechanical steam-drive turbogenerator. For most missions the best answer is solar cells, which convert sunlight into electricity. Very large areas of cell are needed for substantial power output, and either the whole external surface of the satellite is covered with arrays of solar panels, or else they are made up into enormous "wings" or "paddles" so large that they have to be unfolded in space. Once, with the manned Skylab launched in 1973, a complete set of solar paddles was torn off by accident and the other failed to unfold and

A special self-stabilizing system is the gravity gradient "dumb-bell" which automatically aligns itself with the local vertical (the line from the satellite to the Earth's center).

The numerous kinds of satellite mission can be grouped under broad headings. Comsats have always been one of the most important classes, and the only one familiar to the general public. By 1979 more than 1000 million people were able to watch events anywhere in the world as they happened by means of satellite relay of TV signals.

The most diverse group of satellites are those for research. These can be subdivided into those used by engineers to improve satellite or space technology and those used by scientists to find out about the Universe. Both types are usually of moderate size, stabilized in attitude, and equipped with a computerized digital data handling system and radio telemetry. Often the radio is a two way link, the uplink sending human commands and the down-link sending output data and, in the form of millions of

293

Above: part of the state of Texas photographed from Apollo 6 138 miles above the Earth. Visible are roads linking Forth Worth (*left of center*) with Dallas (*right*). Northwest of Fort Worth is the Carswell US Air Force base and the General Dynamics plant. Satellites have an awesome potential for military surveillance.

digital "bits," pictures. Where the satellite is intermittently obscured by the Earth, data are grouped into relatively brief high-rate transmissions as the satellite passes closest to the ground station.

Thousands of experiments have been performed in space by satellites. Some have been concerned with investigations into cosmic radiation, micrometeorites, comets, planetary atmospheres, and more distant cosmological phenomena. Other experiments have been in an endeavor to perfect research and manufacturing processes in the weightless, perfect vacuum of space, where many things can be done easily that would be extremely difficult on Earth because of unwanted chemical reaction with gases in the atmosphere.

Of the purely civilian "useful" satellites the main categories, apart from comsats, are Earth surveillance and land use, disaster warning, meteorological and weather, air and sea navigation, and Earth mapping. Until the mid-1960s most satellites in these categories were to some degree still experimental, though the answers to some unresolved questions, such as the exact distances between the continents, needed for precise guidance targeting of intercontinental mis-

Above: satellites are being usefully applied in agricultural and geological research. Here, the Technical Records Officer of Botswana's Geological Survey and Mines Department interprets photographs taken by the ERTS satellite.

siles, had to be provided much earlier. To this end secret satellites were orbited by the Soviet Union and the United States before 1960.

Earliest of the meteorological satellites were the Television and Infrared Observation Satellite (Tiros) series of 1960–5. As well as providing detailed global photographic coverage, which gave completely new insight into clouds and especially tropical storms and hurricanes, the Tiros satellites sent back accurate measurements of Earth radiation broken down into

Above: NASA's Landsat-C Earth Resources Satellite. It was launched on March 5, 1978. Its job is to study the Earth's environment and resources from space. Areas as small as half an acre can be examined. Landsat-C can also detect temperature differences in vegetation, bodies of water, and urban areas.

Above: the 750-pound GEOS-C satellite was launched to extend and refine existing knowledge of the size, shape, and dynamics of the Earth and its oceans. GEOS-C was designed, built, and tested at the Johns Hopkins Applied Physics Laboratory for NASA. Satellites like this have been employed to study planets like Mars and Venus.

many frequency/wavelength bands. Next came the Nimbus series, which instead of being spun were statically stabilized and placed in polar orbit so that each time the satellite passed over an area of the Earth it would yield exactly matching mosaics of photographs and radiometer information. By 1979 there were many more advanced "metsats", such as the United States Air Force Block 5D series which continuously provides detailed global weather data not only to their USAF but also to civilian services.

Landsats, Seasats and Marisats, and Aerosats, are respectively helping users on land, sea, and in the air. Some are concerned mainly with precise navigation, but Seasat, for example, maintains a continuous readout of data on all the oceans, including such information as surface temperatures and winds, ocean currents, water temperatures, wave heights, ice conditions, ocean topography, and coastal phenomena such as erosion or deposition on beaches and reefs. This United States system is expected to have an economic value to that country alone of $2 billion per annum. The Soviet Union's Meteor series, of which 35 have been orbited since 1969, combine oceanography and meteorological data.

Possibly the most important single satellite application is that of managing the Earth's resources. The original Earth Resources Technology Satellite (ERTS) was launched in July 1972, and in 5$\frac{1}{2}$ years revolutionized the observation of Earth from space. Renamed Landsat, it led to later Landsat satellites which have proved increasingly useful. There have been countless valuable results that were not foreseen. Among the operational equipment the chief items are a Multi-Spectral System (MSS) and a Return Beam Vidicon (RBV). These conduct the most detailed investigations of crop growth, plant stress, land use, inland water areas, and urban areas. A diseased tree in a forest gradually sends back slightly different information (for example, its color changes when scanned at different wavelengths) and can be pinpointed and identified. A forest fire can be spotted instantly. Types of crop can be reported, as well as the health of the plants. Types of soil and moisture content are monitored. Full use of Landsat data by governments in the world's poorer countries could help multiply their food production. Related data, mainly not from Landsat but from other satellites, help in mineral prospecting, fisheries, and control of pollution.

Space Probes

Just as there is a minimum tangential velocity above which a body will go into closed orbit round the Earth and become a satellite, so is there a minimum (higher) velocity in a radial direction above which it will escape from Earth gravity entirely and become a space probe. At the Earth's surface this escape velocity is about 6.95 miles per second, or 25,000 mph. In practice the required velocity is slightly less, because it is attainable only after the spacecraft has traveled up through the atmosphere to a distance where the gravitational pull is much reduced.

There are many kinds of trajectory for a space probe. Some have gone to the Moon, and either impacted on the lunar surface or gone into orbit round it. Most have visited one or more of the planets of the Solar System. Planetary probes have a choice of three destinies: impact on the planet's surface, orbit round it, or continued travel in orbit round the Sun (thus becoming a miniature planet or solar satellite). A few probes have been launched straight into orbit round the Sun, while the longest voyages of all are "grand tours" in which a spacecraft travels to a distant planet and uses its gravitational field to swing round on a fresh trajectory to a second planet, and if possible a third and fourth.

In 1959 the Soviet Union opened this chapter of exploration with Lunik 1, 2 and 3 flown to the Moon, while the United States launched a series of Pioneers, one of which, Pioneer 5, entered a 311 day orbit round the Sun in March 1960. As early as October 1960 the Soviet Union tried to send probes to Mars, but it was 1962 before planetary exploration really started with the flyby of Venus from a distance of 21,643 miles by NASA's Mariner 2. This stabilized craft carried magnetometers, Geiger counters, cosmic dust detector, ion tubes, solar plasma detector, and microwave and infrared radiometers. The latter measured the surface temperature of the planet as 802.4°F, 212°F hotter than the melting point of lead. Like all space probes, Mariner 2 had directional aerials for telemetering data back to Earth in a stream of digital bits. Its electricity was generated by solar paddles.

The United States Mariner 4 of 1964 went to Mars and flew by at only 5400 miles to send back a mass of data and TV pictures. Then came the Soviet Venera craft which explored Venus, eventually soft-landing by parachute and reporting a temperature of 887°F and a carbon dioxide atmosphere with the crushing pressure of 90 Earth atmospheres. Later Venera probes showed contrasting TV pictures of either flat or violently mountainous landscapes, while in late 1978

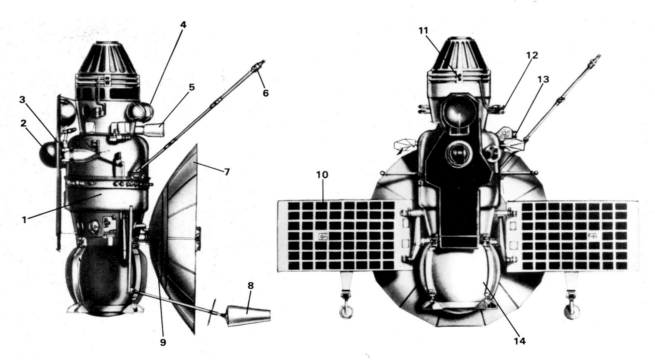

Above: the Soviet Union's Venera IV entered the atmosphere of Venus on October 18, 1967, and revealed that it consisted almost entirely of carbon dioxide. **1** Orbital section; **2** astro-orientation transducer; **3** constant solar orientation transducer; **4** gas balloons; **5** Sun-Earth orientation transducer; **6** magnetometer transducer and arm; **7** high-angle parabolic aerial; **8** low-angle aerial; **9** heat regulation system radiator; **10** panel of solar batteries; **11** engine adjusting device; **12** astro-orientation system micro-thrusters; **13** cosmic-particle counter; **14** detachable capsule. Venera had to withstand intense heat from Venus's surface of over 800°F.

Above: the surface of Mars photographed from Viking I in 1976. On the left of the picture is a dune of fine-grained material scarred by trenches dug by Viking I's surface sampler on an earlier visit.

several probes investigated the atmosphere and weather of Mars at four points of entry.

In 1973 Mariner 10 was launched on a trip that took close photos of Mercury, but the main effort was concentrated on Mars which received a stream of hard and soft landers, flybys and satellites. A major triumph was the arrival of the two United States Viking Landers in 1976, which not only sat on the surface transmitting pictures and data but also dug out Martian soil and subjected it to various tests which showed vigorous reactions, but no positive evidence of life. The thin atmosphere was found to be mainly carbon dioxide, solidified layers of which overlaid water ice in the great Martian icecaps. Detailed

pictures were obtained by Viking Orbiters of volcanos higher than Mount Everest and immense canyons surpassing any on Earth.

Mariner 10 visited Venus before reaching Mercury, but it is in the exploration of the outer planets that the "grand tour" becomes important, collecting a vast amount of information during a mission lasting several years. For these lengthy trips NASA has produced a new generation of craft called Voyagers, though these were preceded by two Jupiter-Pioneer probes launched in 1972 and 1973 which left the Earth at the record speed of 32,400 mph, and took detailed pictures and measurements of Jupiter 18 months later. The second of their Pioneer G, went on to reach Saturn in 1979. The Voyagers are to explore Jupiter, Saturn, and their satellites in greater detail. With electrical power provided by radio isotope thermoelectric generators (RTGs), these interesting craft have magnetometers mounted on long extensible booms, radio astronomy and plasmawave aerials. They are also equipped with a mass of other instrumentation including narrow and wide angle TV, cosmic ray detectors, various radiometers and spectrometers and a photo polarimeter. Voyager 1 reached Jupiter in March 1979 and Saturn in November 1980. Voyager 2 reached Jupiter in July 1979 and is to go on to study Saturn in August 1981, Uranus in January 1986 and Neptune in September 1989. As they will escape from the Solar System both carried messages, from the Presidents of the United States and the United Nations, in case they ever reach another civilization.

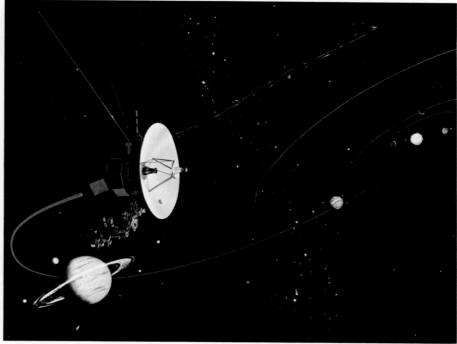

Above: Jupiter's Great Red Spot photographed from the space probe Pioneer 11. White cloud streams north of the Spot from right to left of the picture, and from left to right south of the Spot. The triangular regions at either end of the Spot seem to be where these two streams converge.

Above: an unmanned Voyager spacecraft flies past Saturn in 1977. The two Voyager craft are due to pass Jupiter in the summer of 1979, and fly by Saturn again in early 1981. They are then due to pass Uranus. The two craft are identical, both dominated by a large dish antenna needed for long-range communication with Earth.

297

Living in Space

Early space missions were dictated by available launch vehicles, and it was possession of the ICBM derived Vostok launcher that enabled the Soviet Union to orbit the first Earth satellite in 1957 and the first human in 1961. The Cosmonaut was Major Yuri Gagarin of the naval air force, who on 12 April, 1961 made a single orbit at 65° inclination with apogee of 196 miles in 108 minutes. His craft was the massive Vostok 1 weighing 10,418 pounds, most of which was the cylindrical service module supplying vital air, electric power, and other services to the small sphere in which the Cosmonaut worked. This sphere had a diameter of 90.5 inches and weighed 5290 pounds. While in orbit it was also attached to an equipment module. To reenter the Earth's atmosphere, Gagarin oriented the craft with small thrusters, fired a retro rocket to slow the speed and bring Vostok out of orbit, and then separated the sphere which was protected to survive the searing heat of reentry. It contained an ejection seat with its own parachute, but Gagarin said later he remained in the capsule as it descended by its own parachute on Soviet soil.

On May 5, 1961, the United States sent Alan B. Shepard on a short sub-orbital trajectory reaching apogee of 115 miles and impacting in the Atlantic 297 miles downrange from Cape Canaveral. He flew inside a Mercury capsule, carried on a Redstone Army bombardment rocket only a fraction of the size of the Vostok booster. Made by McDonnell Douglas, the Mercury was basically conical and weighed 2987 pounds. On top was a long escape tower carrying rockets to blow the package clear if the launcher should explode on the launch pad. This was jettisoned in space, and to recover from an Earth orbit the occupant rotated into the correct attitude and fired retro-rockets on the base. The retropack was then jettisoned, and the heat of reentry was resisted by the blunt base of the capsule, coated with an ablation shield made of polymer resin and glass fibers. On February 20, 1962 John H. Glenn Jr flew three orbits in a Mercury launched on a much more powerful rocket, an Atlas ICBM.

But before then the Soviet Union had stolen other "firsts." On August 6, 1961 Gherman Titov had become the first human to spend more than a day in space, completing 17 orbits in 25¼ hours. On return he decided to use "the second method of landing." He ejected from his Vostok 2 sphere in his seat, and landed by parachute near where his capsule also made a soft landing. Vostok 3 roared into space from Baikonur on August 11, 1962 with Andrian Nikolayev. It was followed a day later from the same Cosmodrome by Vostok 4 with Pavel Popovich, though there was no attempt to come closer to each other than 3 miles. In long flights, of 64 and 48 orbits, these Cosmonauts telemetred such information as cabin atmosphere, electrocardiograms, electroencephalograms, pneumograms (breathing), galvanic resistance of the skin and electro-oculograms to record movements of the eyeballs.

On June 16, 1963 the first spacewoman, Valentina Tereshkova, began a mission of 48 orbits completed in 71 hours. In October 1964 three men, a pilot, a doctor, and a scientist, squeezed into Voskhod 1, and in March 1965 Alexei Leonov carefully checked his spacesuit and stepped outside Voskhod 2 to perform the first extra-vehicular activity (EVA), popularly called a spacewalk. Leonov had previously had nitrogen "flushed out" of his body by a long period breathing pure oxygen. In space his suit was inflated to 0.4 atmosphere and he continued breathing pure oxygen.

Below: the first man in space, Major Yuri Gagarin of the Soviet Naval Air Force, pictured just before lift off. On April 12, 1961 Gagarin successfully completed one orbit of the Earth in Vostok 1.

Right: Astronaut Ed White became the first American to walk in space in 1965. He was attached to a long gold-foil-wrapped oxygen line. Many regarded this spacewalk as something of a stunt, but it has since proved necessary for men to leave their space vehicles to carry out repairs. An Astronaut Maneuvering Unit and a life-support system were tested during these Gemini flights.

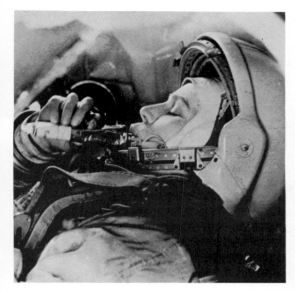

Above: Valentina Tereshkova became the first woman in space on June 16, 1963. Two days earlier, Valery F Bykovsky was launched into orbit aboard Vostok V. He achieved what was then a record manned spaceflight of 76 orbits in 119 hours 6 minutes. Tereshkova joined him in orbit aboard Vostok VI. She completed 48 orbits in 71 hours. Tereshkova had no pilot training prior to being accepted.

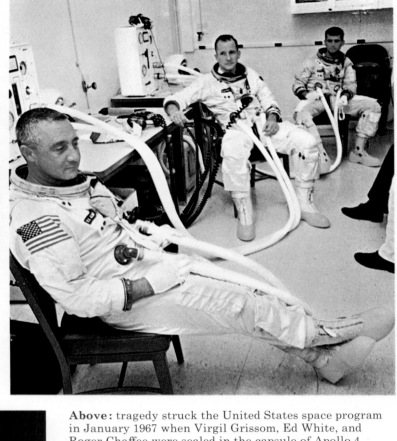

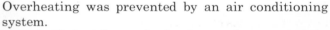

Above: tragedy struck the United States space program in January 1967 when Virgil Grissom, Ed White, and Roger Chaffee were sealed in the capsule of Apollo 4, which burst into flames. They were killed in seconds.

Overheating was prevented by an air conditioning system.

In the United States the flawless performance of the six one-man Mercuries was eclipsed by textbook missions by Gemini GT-3 to GT-12. This series of two-man missions accomplished prolonged EVA. The spacewalking Astronaut was attached to the end of a long goldfoil-wrapped oxygen line. A self-contained life support system and an Astronaut maneuvering unit (AMU) were tested during the Gemini series. And the first space docking between two craft in orbit was achieved. But a tragedy in January 1967 took the lives of three Astronauts training for the Apollo lunar mission: Virgil Grissom, Ed White, and Roger Chaffee were sealed in a capsule when it burst into flames on the pad. There was much argument over continuing to use pure oxygen atmospheres, but the changes were confined to other parts of the spacecraft.

Some Astronauts and Cosmonauts (the American and Russian names respectively) have spent many weeks at a time in the hard vacuum of space, where a single slip or malfunction can mean instant death. Nearly all the casualties have been due to accidents on Earth, or in final stages of Russian recovery systems. Weightlessness in space has caused no difficulty, and there has been no report of injury from micrometeoroid impact or any kind of abnormality in space fliers or their children. It was once feared that long-term weightlessness could be harmful.

Exploring the Moon

As the Earth's nearest neighbor the Moon was an obvious early subject for space exploration. The Russian Lunik 1 flew by the Moon in 1959, Lunik 2 hit the surface, and Lunik 3 took pictures of the hitherto unseen far side. Many Luniks failed to complete their missions, and the United States Pioneer series likewise had few successes. But the Ranger program did better. After two failures, Ranger 3 was launched in January 1962 hoping to take TV pictures and drop a small instrument capsule on the Moon's surface. But it was not until Ranger 7 was launched in the middle of 1964 that good pictures were obtained in the last 13 minutes before impact in the Sea of Clouds. Rangers 8 and 9 also sent back good pictures – boosting confidence for planning future missions.

On February 3, 1966 the Soviet Union landed an instrumented craft, Luna 9, on the Moon's surface.

Above: a replica of the Soviet Union's unmanned spacecraft Luna III. Launched exactly two years after Sputnik I, on October 4, 1959, Luna III obtained pictures of the reverse side of the Moon and transmitted them to Earth. This side of the Moon had not been seen before.

The 3490 pound spacecraft did not make a complete soft landing, but ejected a small 220 pound capsule in the Ocean of Storms where it bounced and rolled for a short distance. The capsule was self-righting. It opened four petals to stand upright, and then erected four aerials and began taking TV pictures and radiation measures. Luna 13 later studied lunar soil and found it of low density.

In 1961 the United States' intention to put a man on the Moon by the end of the decade greatly accelerated lunar research, and the next major project was Surveyor, built by Hughes Aircraft. These were robot stations designed to softland on the Moon. The basic design resembled a smaller version of the lander planned to take men to the Moon. It had three legs and a retrorocket braking system and high accuracy vernier jets controlled by a radar altimeter bouncing signals off the lunar surface. A really gentle touchdown was possible. Gross weight was about 2200 pounds. On the Moon the apparent weight would be one-sixth as much. Surveyor 1 landed in the Ocean of Storms on June 1, 1966 and sent 11,150 excellent TV pictures. Surveyor 3 landed in the same area and dug a trench for 18 hours while photographing the results.

Meanwhile, Boeing built five Lunar Orbiters which in 1966-67 achieved outstanding results in their mission of photographing the Moon from low orbit. Not only did the Orbiters send back impressively detailed pictures, they gave the NASA ground staff experience in controlling and communicating with vehicles orbiting the Moon, while the slight disturbances in the orbits provided the first detailed infor-

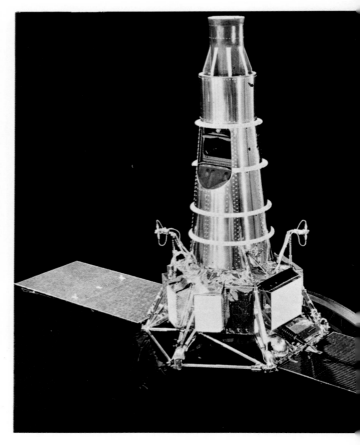

mation on the Moon's gravitational field. By this time the reliability of spacecraft had dramatically improved, and some of the Orbiters had to be deliberately crashed on the Moon to get them out of the way for planned later missions. During this program, in September 1967, Surveyor 5 brought hair-raising excitement and a new standard to mission teamwork when its helium system began to leak. Ceaseless toil with computers, continually updated, enabled changed flight plans to be substituted which eventually succeeded in bringing about a faultless soft landing – with helium pressure on the brink of failure. This craft analyzed soil samples. The following Surveyor 6 soft landed, then took off and landed again nearby to demonstrate the firmness of the Moon's surface.

During the 1960s the Soviet Union built a development of the Soyuz manned spacecraft called Zond, and it appeared certain that they intended to use it for manned lunar missions. Using a large new launch rocket called Proton a succession of unmanned Zonds were sent in orbit round the Moon. Some splashed down in the Indian Ocean, while others did an aerodynamic skip through the atmosphere to drop by parachute into the Soviet Union. But the Russians suffered several setbacks, including catastrophic accidents both on the Baikonur Cosmodrome and in flight. There was little real challenge to the American objective of a man on the Moon.

There are several ways of planning a manned lunar landing and return. Most of them are fraught with difficulties. Three were chosen by NASA for detailed study. One is the direct approach: a spacecraft is sent by multistage rocket to the Moon, soft lands and then returns to Earth. The craft would have comprised: an upper Command Module with the crew of three, navigation and communications; a large central Service Module containing nearly all the rocket engines for takeoff from the Moon and course correction on the voyage back to Earth; and a lower Lunar Module with a retrorocket for soft landing. This craft was calculated to weigh 150,000 pounds and need a launch vehicle with a thrust of 12,000,000 pounds. The second was Earth orbit rendezvous (EOR). An extra rocket, the escape stage, is added under the craft and it is launched with almost empty tanks into Earth orbit by the Saturn 5 rocket of 7,500,000 pounds thrust. A second Saturn 5 then takes off, orbits alongside the first, and transfers the missing propellants. The third method was lunar orbit rendezvous (LOR). A single Saturn 5 sends the spacecraft into orbit round the Moon. Then the lunar excursion module (LEM), later just called Lunar Module or "bug", separates and carries two crew down to a soft landing. The lower part of this module remains on the Moon's surface, while the upper part takes off, returns to lunar orbit and transfers the crew back to the Command Module for the return journey. LOR was found to be the best method. The entire mission was accomplished by a single Saturn 5 launch. The Saturn 5 is probably the most powerful rocket yet built.

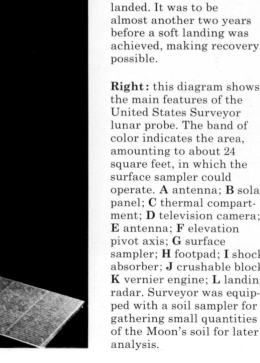

Left: the United States' Ranger 7 was launched in July 1964. It transmitted 4500 pictures of the Sea of Clouds on the Moon's surface before it crash landed. It was to be almost another two years before a soft landing was achieved, making recovery possible.

Right: this diagram shows the main features of the United States Surveyor lunar probe. The band of color indicates the area, amounting to about 24 square feet, in which the surface sampler could operate. **A** antenna; **B** solar panel; **C** thermal compartment; **D** television camera; **E** antenna; **F** elevation pivot axis; **G** surface sampler; **H** footpad; **I** shock absorber; **J** crushable block; **K** vernier engine; **L** landing radar. Surveyor was equipped with a soil sampler for gathering small quantities of the Moon's soil for later analysis.

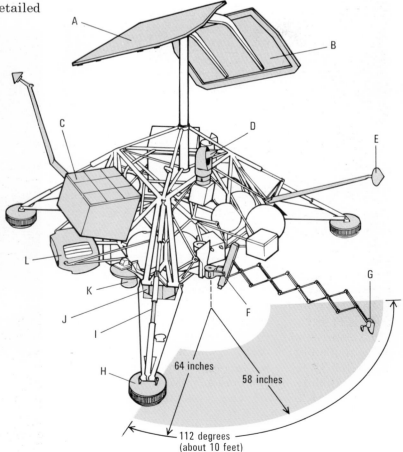

64 inches

58 inches

112 degrees
(about 10 feet)

Above: the command module of Apollo 10 in lunar orbit. Launched in May 1969, the lunar module of Apollo 10 made two low passes over the Moon. It descended to about 50,000 feet or 9½ miles above the surface.

By far the largest rocket of which details are known (though it is rumored that larger ones have been built in Soviet programs), Saturn 5 is a three stage vehicle designed to fly the NASA LOR mission, given the name Project Apollo. The first stage was S-1C, built by Boeing to lift 1,500,000 pounds to a height of 40 miles at a speed of 6000 mph. It had five Rocketdyne F-1 engines, each rated at 1,500,000 pounds thrust at sea level, burning liquid oxygen and RP-1 kerosene. Rockwell built the second stage, S-2, with five Rocketdyne J-2 engines, each rated at 200,000 pounds thrust burning high energy liquid oxygen and liquid hydrogen. McDonnell Douglas built the third stage, S-4, with a single J-2 engine. On top of this monster was placed the Apollo spacecraft, about 80 feet high and weighing about 95,000 pounds.

The Apollo 7 made the first manned mission on October 11, 1968. Walter Schirra, Walter Cunningham, and Don Eisele made an Earth orbital mission (163 revolutions in 260 hours) launched by a Saturn 1B – smaller than a Saturn 5. On December 21, 1968 Apollo 8 was launched by the giant Saturn 5. This had been planned as an Earth orbit mission, but Soviet progress caused a dramatic change of plan and Frank Borman, James Lovell, and William Anders were sent into orbit round the Moon. It was the first time that men had ever traveled far from Earth. Ten lunar orbits were made and many other objectives were met in a textbook mission. Apollo 9 was launched on March 3, 1969 in an Earth orbit mission to test the Lunar Module, not previously flown with a crew. James A. McDivitt, David R. Scott, and Russell L. Schweickart flew a 10 day mission during which the tricky maneuver was made of uncoupling the Lunar Module, maneuvering it round and turning 180°, and docking on the nose of the Command Module. Lunar orbits, at times 113 miles apart, were simulated on this mission.

Apollo 10, launched on May 18, 1969, took Thomas Stafford, Eugene Cernan, and John Young on the final "dress rehearsal." They flew into lunar orbit, released the Lunar Module, and docked. Commander Stafford and Lunar Module pilot Cernan descended in it to within 9 miles of the Moon (as there is no lunar atmosphere it is possible to orbit only just above the Moon). They then ascended back to the Command Module in its 69 mile orbit, having overcome a small fault which caused the Lunar Module ascent stage to gyrate wildly for more than a minute.

All was now set for a man on the Moon. On July 16, 1969 Apollo 11 lifted off from Cape Kennedy and went into lunar orbit three days later. On July 20 commander Neil Armstrong and LM pilot Edwin Aldrin entered the LM, leaving Michael Collins piloting the Command Module. The LM – *Eagle* – landed gently on the lunar surface in the Sea of Tranquility at 2117 British Standard Time (BST) on July 20. At 0356 BST the following day Armstrong stepped out on to the Moon, followed by Aldrin, and for 2 hours 15 minutes arranged experiments on the surface including a laser ranging retro-reflector for exact measurement of Earth/Moon distance, and a heated passive seismic installation for measuring Moonquakes and meteo-

Above: the first man on the Moon, Neil Armstrong, landed in the lunar module of Apollo 11 in July 1969. The phrase signaled back to Earth was "the *Eagle* has landed." In this picture, his partner Aldrin is operating a seismic detector. Manned missions have the advantage over more limited probes of greater flexibility.

roid impacts. A solar wind experiment was also performed. The Astronauts brought back soil and rock samples after a stay on the Moon lasting 21 hours 36 minutes. Leaving behind the LM descent stage, with a plaque recording the event, the ascent stage burned for seven minutes to put it close to the Command Module on its 25th lunar orbit. Before reentering the Command Module, Armstrong and Aldrin carefully vacuum-cleaned their spacesuits to prevent bringing possible contamination to Earth.

In November 1969 Apollo 12 brought Charles Conrad and Alan Bean to the Moon. Richard Gordon Jr remained in the Command Module. This landing was in the Ocean of Storms, close enough to Surveyor 3 for the Astronauts to inspect it and recover its TV camera. They set up the first operating Apollo lunar-surface experiment package (ALSEP), more advanced than the devices left by the previous mission. Powered by a plutonium-fuelled SNAP-27 generator, the package included a seismometer, solar wind spectrometer, magnetometer, lunar atmosphere detector and lunar ionosphere detector. Apollo 13, in April 1970, was commanded by James Lovell, who with Fred Haise (LM pilot) were to descend to the mountainous Fra Mauro region, leaving James Swigert in the CM. But a major explosion in the Service Module near the Moon caused a prolonged crisis, saved by using the unwanted LM as a "lifeboat" to replace the useless

SM to sustain the crew. Apollo 14 in January 1971 was better, despite computer failure and other problems. Alan B. Shepard and Edgar J. Mitchell spent 34 hours in the Fra Mauro area, leaving Stuart A. Roosa in the CM. A new feature was the use of a modular equipment transporter (MET), which looked like a golf kart.

In July/August 1971 Apollo 15 visited the Hadley Apenine mountains. David Scott and James Irwin spent 67 hours on the Moon (with Alfred Worden in the CM), during which the Boeing built Lunar Roving Vehicle (LRV) was used for transport. In April 1972 Apollo 16 took John W. Young and Charles M. Duke Jr to the Moon in the elevated Descartes region (T. K. Mattingley II being CM pilot), with long LRV excursions. A third LRV was busy in the Taurus Littrow region in the final mission, Apollo 18, in December 1972, flown by Eugene Cernan and Harrison Schmitt, with Ronald Evans as CM pilot. Geologist Schmitt selected a record 243 pounds of samples.

On November 17, 1970 the Soviet Lunokhod 1 eight-wheeled robot explorer landed in the Sea of Rains, opening a less spectacular program with Luna 17 to 24 which recovered various lunar cores and tested soil properties. Investigation of samples and analysis of lunar data is still continuing, and the ALSEP experiments stayed working on the Moon until they were switched off in 1977.

Above left: the Soviet Union's lunar station Luna-24 carried samples of lunar rock in an ampule in its reentry module back to Earth. Once recovered, the ampule was opened at the USSR Academy of Sciences. This picture shows the rock on display. Both the Soviet Union and the United States have removed small quantities of rock.

Above: Lunakhod 1 lunar vehicle, built by the Soviet Union. It was an electrically powered robot, equipped with television cameras to send pictures back to a control station on Earth. Lunakhod 1 was landed on the Sea of Rains by the Luna 17 carrier in November 1970. It travelled 34,588 feet on the Moon.

Space Shuttle

All spaceflight is expensive, particularly so until recently, when all launch vehicles were large ballistic rockets unable to be used more than once. Such extravagant use of resources would clearly make passenger space travel commercially unrealistic. The obvious answer was to devise a space transport system that could be used again and again, as in other forms of transport. The first such reusable space vehicle is the Space Shuttle, developed by NASA in the United States.

The chief element in the Shuttle – so called because it is designed to shuttle between the Earth and space – is the Orbiter. This is a spacecraft that is also an aircraft – provided with propulsion, wing, controls, and

Right: the Space Shuttle's Manned Maneuvering Unit is a propulsive backpack device which allows Astronauts to undertake extravehicular activity. Repairs and maintenance will be necessary during long periods in space.

Below: an artist's impression of an Orbiter Space Shuttle lift-off.

landing gear for operation both in space and on its return through the Earth's atmosphere to make a normal airfield landing. It takes off pointing vertically upwards, its main engines drawing propellants contained in a separate tank, larger than the Shuttle Orbiter itself. This is jettisoned when nearly empty. To lift the Orbiter and tank off the ground two large solid rocket boosters (SRBs) are attached on the sides of the tank. These boost motors are jettisoned after burnout and recovered for re-use. In normal operation the Shuttle will carry payloads of 65,000 pounds to low Earth orbit.

Produced by Rockwell, but incorporating a wing by Grumman, main fuselage section by General Dynamics, and tail by Fairchild Republic, the Orbiter has the appearance of an extremely short and fat version of the Concorde supersonic airliner. The low-mounted wing has a blunt rounded leading edge which has thermal protection. It is made mostly of aluminum alloy, like the capacious fuselage and swept vertical tail, with light but rigid honeycomb skins. The main central portion of fuselage is the payload bay, 60 feet long and 15 feet in diameter. On the insides of the giant roof doors covering this bay are radiator panels for shedding excess heat in space. The cargo is loaded and

securely attached, and the Orbiter positioned upright and mated with the tank and boosters.

At liftoff each booster gives a thrust of 2,900,000 pounds and the Space Shuttle Main Engines (SSMEs) comprise three Rocketdyne high pressure engines each rated at sea level at 375,000 pounds, burning liquid oxygen and liquid hydrogen fed from the tank. As the Shuttle climbs it gradually rotates away from

Above: the Space Shuttle Orbiter cut away to show its payload. Space Shuttle will allow the enormous effort and expense of spaceflight to be applied to many missions.

Right: Spacelab being manufactured by ERNO. The German company is leader of a European team.

the vertical, and as it passes through an altitude of 31 miles the SRBs are jettisoned and descend into the sea by parachute for subsequent recovery. Just before reaching orbit at a height of about 115 miles the propellant tank is released. At this height each SSME gives a thrust of 470,000 pounds.

Ahead of the payload (cargo) bay is the three deck crew module, with airlocks leading into the cargo bay and fuselage mid-section housing many of the Orbiter's complex subsystems. At the top level is the wide airline style flight deck, with dual controls. Behind are seats for mission specialists. On the middle deck are seats for three more mission specialists, bunks (or three additional seats), galley, "hygiene section", and many other items. The environmental control system and various stowage areas are housed on the bottom floor level. Airlocks allow crew members to walk through into the space-environment cargo bay or into a spacelab carried there, such as the four man laboratory being produced by a 10 nation consortium led by ERNO of West Germany which will be the first Shuttle-launched manned payload in 1980.

In space the Orbiter maneuvers with two engines, each rated at 6000 pounds, burning monomethyl-hydrazine and nitrogen tetroxide. Small corrections of orbit or attitude are made with a reaction control system (RCS) burning the same propellants as the

orbit maneuvering subsystem (OMS) feeding 38 small engines (14 round the nose, 24 round the rear, and 12 in each OMS/RCS pod) plus six even smaller vernier thrusters (two on the nose and two in each pod) for fine adjustment. In the cargo bay are lights, TV cameras, and a large retractable manipulator arm, controlled from the flight deck, for handling payloads. Some items will be left in space, some will remain on board, and others will have to be docked with existing stations or satellites, or even built on to them. One possibility is to use repeated Shuttle missions to build space stations much too large to be launched in one unit. Spacelab, the most important payload on early flights, will not leave the cargo bay but will remain in

Above: the space suit and rescue system developed for the Space Shuttle at the Johnson Space Center, Houston. **Left:** the Personal Rescue Enclosure is a 34-inch diameter ball with a life-support system.

space with the Orbiter for periods of seven to 30 days, thereafter being prepared for return to Earth with all crew back inside the Orbiter and the cargo bay doors closed.

To return to Earth the crew position the Orbiter in the desired tail-first attitude and fire the OMS engines in a deorbit maneuver which results in the craft plunging down back into the atmosphere. As it reenters it becomes heated by friction until major parts of the structure begin to glow. The Orbiter must be traveling nose first by this time, and increasingly the aerodynamic controls take effect to control the trajectory as well as the OMS. Unlike earlier reentry vehicles, whose ablative coating wore away and could be used only once, the Orbiter is protected by a unique thermal protection system designed to survive 100 missions. Most of the exterior surface, an area of 11,874 square feet, is covered in more than 34,000 tiles made of special quartz coated with reaction-curved glass, giving protection at temperatures up to 2300°F. Leading edges and the nose are covered in reinforced carbon/carbon, for higher temperatures still.

As it descends into the atmosphere and slows down through aerodynamic drag the Orbiter becomes an airplane. The pilots fly it with an advanced flight control system, with four-channel "fly by wire" (electric signaling) to four hydraulically powered spoilers on the wings and two rudders which split open to act as speedbrakes. After achieving reentry the Orbiter can glide a maximum distance of 1100 miles back to its airfield. It has a tremendous number of subsystems and equipment. As there are no main rotating engines, shaft power is provided by hydrazine fueled auxiliary power units, the three main APUs each being rated at 135 horsepower. Electricity is generated by fuel cells supplied with liquid oxygen

and hydrogen, and a storage subsystem provides for generating 1350 kilowatt-hours. There are four main subsystems making up the environmental control and life support system: an atmosphere revitalization subsystem, an active thermal control subsystem to maintain comfortable temperatures (emitting waste heat through the large radiators in and deployed from the cargo-bay doors), an airlock support subsystem and a food, water and waste subsystem to handle all body needs (including cooking) in the most efficient manner.

Nearing the airfield the flight Mach number is reduced from 10 (6600 mph) to 5 by use of the split rudders as a speed brake. This causes a nose-up trim change which is canceled out by depressing a large hinged flap under the main engines at the rear. Below Mach 5 the vertical-tail controls operate as speed brakes or rudders as necessary. The crew line up for landing with their numerous electronic systems including triple inertial navigators and a microwave scan landing system. The landing gears, with twin wheels and anti-skid brakes on each leg, are extended for a conventional touchdown which can be made at speeds up to 254 mph but is normally at 201 mph. The nose landing gear is steerable.

Among the unusual items of equipment developed for operations with the Space Shuttle is a new NASA Astronaut suit, and a novel Personal Rescue Enclosure (PRE), the first ever developed for recovery of individuals back through the Earth's atmosphere. PRE is a hollow sphere of 34 inches diameter, covered in three layers of material: urethane, Kevlar (a very strong filament), and a final thermal protective coating. There is a viewing port of Lexan transparent heat resistant plastic, and inside are simple life support and communications systems.

Left: the Space Shuttle Orbiter 101 Enterprise landing after concluding a 5 minute 28 seconds unpowered flight during the Shuttle Approach and Landing Tests carried out on September 13, 1977.

To take payloads from the Orbiter in low Earth orbits and put them in much higher orbits McDonnell Douglas is developing two types of Space Shuttle Upper Stage (SSUS). One is designed to handle payloads weighing up to 2450 pounds, which can be lifted into a transfer orbit and then, using small "kick motors", placed in geostationary orbit at 22,300 miles. The other is designed for payloads of up to 4410 pounds which can also be placed in geostationary orbit.

In early 1979 there were two Space Shuttle Orbiters, OV-101 and 102. The former, named *Enterprise*, was first completed with large aerodynamic fairings over the main engines and OMS pods. It was mounted, with landing gear retracted, on top of a much-modified Boeing 747, designated NASA 905. It completed taxi runs in February 1977. On February 18 it made a captive flight. Manned captive flights began on June 18, and the first free flight took place on August 12, 1977. On October 12, 1977 free flights began with the tailcone fairing removed. This program of general handling and landing tests at NASA's Dryden Flight Research Center in California was followed by comprehensive vibration and systems testing. The second Orbiter, OV-102, is to make a number of atmospheric test flights. Its first space mission is due to begin from Kennedy Space Center, Florida, in the second half of 1980.

Details of the Shuttle missions had not been published by early 1979, but a clue to their number is provided by the contract with United Space Boosters for refurbishment of the large SRBs. There are expected to be six development flights, extending to late 1980, and 21 operational Shuttle missions extending through 1982. Some of these are likely to carry military payloads, because Boeing has developed an Interim Upper Stage in four sizes for the US Air Force to take Shuttle cargos into high or geostationary orbit. A number of Shuttle payloads will be for scientific research. Commercial passenger payloads are likely before too long. Space Shuttles will probably be only for the privileged few, but wider use is possible.

Space Stations

Space stations are manned spacecraft intended to remain in operation over an extended period, in the order of weeks, months or even years. Such a station poses few problems not met in other manned space exploration. The hard (total) vacuum, intense solar radiation and other features of the space environment cause gradual changes in some materials. But structures mostly last better in space than on Earth if properly constructed of alloys and plastic tailored to this novel application. Of course, there must be a sustained and reliable supply of energy to keep the station going and provide a habitable environment, but much of this demand can be met by large solar arrays which turn sunlight into electrical power. Another problem is the human body's respiratory and nutritional cycles, and much care has gone into finding the best ways of recycling solids, liquids and gases for human comfort with minimal energy consumption.

Most of the effort on space stations has been made by the Soviet Union. Well over 30 large Soyuz space-craft and a few Progress unmanned supply ships have supported a series of space stations called Salyuts (salute). Each Soyuz normally had a crew of two, one a commander and the other an engineer or researcher. They occupied three modules or sections: propulsion and instrument (service module), laboratory/rest (orbital module), and descent (landing module). Weighing about 14,500 pounds these craft were placed in quite low Earth orbit at about 150 miles to dock with the much larger Salyut stations. These weighed about 41,700 pounds. Salyut 6, launched on September 29, 1977, was the first to have two docking ports so that it could be visited by two spacecraft simultaneously. One of the Soyuz craft which visited Salyut 6 was Soyuz 26, launched on December 10, 1977. Returning to Earth on March 16, 1978, it set a record for manned spaceflight of 96 days. While docked with Salyut 6 this crew was visited by Soyuz 27 and 28, as well as a Progress ferry which like the other Soyuz craft brought various essential supplies.

The first United States space stations were to have been military Manned Orbiting Laboratory (MOL) stations for the US Air Force, built by Douglas for complex surveillance missions (probably chiefly over the Soviet Union and Eastern Europe). Four years after the 1965 go ahead the MOL program was canceled to save money, and the team working on it switched to the Apollo Applications Program in which the S-4B upper stage of the Apollo manned lunar vehicle was converted into a space station at first called Orbital Workshop and later Skylab. Described as "rather like building a ship in a bottle", the work

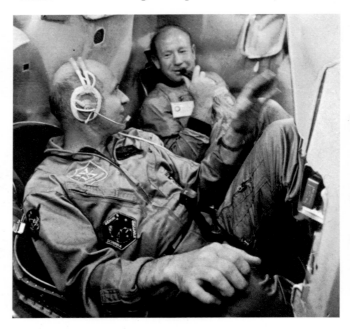

Above: there has been some cooperation between the United States and the Soviet Union over the development of space stations. Here, the Russian Leonov and the American Stafford train together in a Soyuz model.

Left: Soyuz 28 being installed in its launcher rocket. The crews of Soyuz 27 and 28 visited Soyuz 26 while it was docked with the space station Salyut 6.

involved cleaning out the spent rocket stage and completely converting it into a manned space station. The main workshop, with separate floors, was inside the former liquid-hydrogen tank. Skylab was to have been a five station program, but only one was launched, on May 14, 1973. It was badly damaged as it left the atmosphere during launch, and the first Astronaut team brought improvized tools and repair material with which they opened the one remaining solar "wing" (one had been torn off) and erected sunshades to cool the station and replace the missing micrometeorite shield. This meant prolonged extravehicular activity (EVA) and complicated structural work outside the station. Subsequent Astronaut teams spend 59 and 84 days at the station, which in 1979 was expected soon to reenter the atmosphere and burn up.

In 1979 the only important space station outside the Soviet Union was Spacelab. This was a series of well equipped and versatile stations being developed by a 10 nation consortium led by West Germany. Spacelab comprises a main laboratory, with "shirtsleeve" environment, to which are attached two external payload pallets for large experiments. Four payload specialists are to work in each Spacelab which is to go into space in 1980 in a Shuttle, remaining in the Shuttle Orbiter's cargo bay throughout the seven to 30 day mission.

There is a seemingly limitless list of useful tasks which can be done in space stations, many of them concerned with manufacture of superpure materials and pharmaceuticals. The expected next stage is to construct very large stations in space, possibly even using some lunar raw material such as titanium. One scheme which has been worked out in considerable detail is a gigantic station much more than a mile square covered in solar cells able to generate sufficient electricity to supply one of the world's larger cities. Though the station would have to be far from Earth, in geostationary orbit, very little power would be lost as the energy would be beamed to an Earth receiving station in the form of microwaves. There are even more ambitious studies for stations where as many as 10,000 people could live. Earth gravity could be reproduced by making the main inhabited portions spin, and paper calculations indicate that such a "space city" could be made almost self-supporting. But man is still a long way from finding the resources for such impressive projects.

Above: an artist's concept of what a 21st century space station might look like. It is envisaged that all the material necessary to construct such a colony would come from the Moon or the Astroid Belt. Manufacture and assembly would be carried out in space using solar power.

Left: the United States' Skylab was launched in 1973. It is equipped with eight telescopes as well as sensors of various kinds. With such equipment, and free of the obscuring effect of the Earth's atmosphere, Skylab proved an excellent platform for astronomical observation. The quality of observation was greatly superior to even the most powerful Earth-bound telescope.

The Universe from Space

Though man has probably no hope of actually voyaging beyond the Solar System, because of the huge distances involved, he has found that spacecraft are useful in observing the rest of the Universe in ways that are not possible from the Earth's surface. Astronomers and cosmologists have always been handicapped by poor "seeing", even with the aid of the best telescopes. This is partly because of particulate matter in the atmosphere. This diffuses the incoming light and also scatters light from sources on Earth. In an attempt to avoid this the main observatories are built on the peaks of mountains, far from any city, where there is clearer and thinner air, and less interference from artificial light at night. But there is still the problem of mass air movement in the atmosphere, and a phenomenon called airglow prevents the sky from ever being completely black. Not least, even the relatively small part of the atmosphere above tall mountain peaks filters out many electromagnetic wavelengths and allows an Earthbound observer to use only visible light and various radio wavelengths with which to study cosmological objects.

These difficulties are removed if the observatory is placed above the Earth's atmosphere. Then the "seeing" is perfect, and man can make use of the entire electromagnetic spectrum to study the radiation coming to him from all over the Universe. Studying the Universe from space has proved one of the most fruitful of all man's exploratory endeavors.

In the mid-1950s simple optical telescopes and spectrometers, the latter often adapted to infrared radiation, were launched by sounding rockets. But these up and down vehicles have the drawback of a very brief observing time. In 1957 a telescope was lifted to 16 miles by a balloon. Two years later plans were drawn up for the Orbiting Astronomical Observatory (OAO) – the first true space observatory. The second OAO, launched in December 1968, operated completely successfully, and sent back a vast amount of information from a reflecting telescope, and gamma-ray instruments, and experiments by the Universities of Princeton and Wisconsin which measured cosmic gas and dust and the stellar ultraviolet radiation, respectively. Star trackers pointed the OAO at the selected heavenly bodies with the extremely high accuracy of 0.1 second of arc (1/36,000°).

Hughes Aircraft has produced a series of Orbiting Solar Observatories (OSO) for detailed study of the Sun – each more advanced than its predecessor. The standard OSO form comprised a large "sail", continuously pointed at the Sun, containing solar arrays to generate power as well as a spectrometer and associated subsystems, and a spinning base called the

Right: a diagram of the Orbiting Astronomical Observatory 2 (OAO-2). Launched on December 7, 1968, OAO 2 was equipped with 11 telescopes. These were to study ultraviolet radiation which is blocked by the Earth's atmosphere.

A Sunshade
B Seven Wisconsin telescopes
C Balance boom
D Star trackers
E Four Smithsonian telescopes
F Solar panel

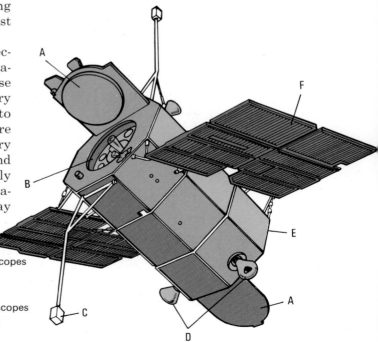

Left: the ISEE-B satellite being manufactured by the German company Dornier. Dornier led an international team to develope the ISEE Sun/Earth explorer satellite.

Above: an artist's concept of a space telescope. NASA has commissioned Lockheed to build an instrument capable of looking seven times deeper into space than anything built before.

"wheel" on which was mounted most other experiments. Typical instruments included a sensitive crystal spectrometer and polarimeter, mapping X-ray heliometer, system for mapping "soft" X-ray background radiation, cosmic X-ray spectroscope, high energy celestial X-ray system, an experiment for mapping the fine structure of the chromosphere that surrounds the Sun, and various ultraviolet spectrometers.

West Germany has also been very active in exploring the Sun. It has sent two Helios probes within 28,000,000 miles of the Sun. They were heated far beyond the melting point of lead on each close pass. They discovered the intense "rain" of micrometeoroids near the Sun, and the fact that their direction changes from time to time. The German aeronautics company, Dornier, led an international team in developing an International Sun/Earth Explorer (ISEE), the first of which was launched in October 1977.

Britain has put many useful cosmological experiments into space, using American launchers, in the Ariel series of satellites which investigated high energy astrophysics. Surprisingly little has been done in the Soviet Union to study the Universe from space, but the Prognoz (Forecast) series of satellites are studying the Sun and its influence on Earth and radio communications.

The United States continues to lead in space cosmology. High Energy Astronomy Observatory (HEAO) is a series of important spacecraft designed to study some of the most distant and mysterious fea-

tures in the Universe, such as pulsars, quasars, exploding galaxies and "black holes." All of these are particularly difficult to observe from Earth. The HEAO stations are being launched at about the rate of one a year from 1977. But perhaps the most ambitious space observatory yet designed is NASA's Space Telescope. It is being built by Lockheed for launch in 1983. The solar array and photon detector for this impressive observatory are being supplied by British Aerospace. When the Space Telescope is operating it will be able to see "objects 50 times more faint and seven times further away" than the limit for the best telescopes on Earth.

Right: The HEAO satellite was launched in 1977 to carry out a detailed X-ray survey of the celestial sphere finally to determine the existence of such mysteries as pulsars, quasars, exploding galaxies, and "black holes."

311

CHAPTER 12

TECHNOLOGY: A SUMMING UP

Technological advance has been much greater in the past 100 years than in the previous 2000. Despite this, most people – in all countries – remain technically illiterate. Their lack of understanding prevents them making informed decisions, and leaves the direction of technological advance largely in the hands of the technologists themselves. Most people are to some degree aware of the pace of technical development. They occasionally notice the introduction of new techniques, inventions, and materials. But the great mass of what goes on in the world's research laboratories remains a closed book.

Inevitably, many people view technology with suspicion and even fear. The media feast on technological crisis and controversy, real or imagined, concentrating upon such topics as atomic energy, solid wastes, radioactivity, tanker disasters, supersonic airliners, heart transplants, fluoridation, and the menace of the computer. But in fact every new discovery since time began has merely given man new options. How to exploit these new capabilities is a moral and ethical problem which man has found difficult to solve.

Opposite: Indians watch a troupe of travelling actors hard by the telegraph lines. The introduction of western technology into a completely different culture with different values has caused severe strains. More thought is being given to the appropriateness of types of technology to local conditions.

313

The Technology Debate

One of the fundamental problems with modern technology is that it has moved so far and so fast that communicating the results to the layman is extremely difficult. Telling someone they were flying in an airplane guided by a quad inertial navigation system (INS) with laser gyros (LGs) operating in the strapdown mode would convey as much information as a foreign language, perhaps less. Trying to explain how the LG worked would at once result in even greater confusion – and probably antipathy towards technology in general – unless the explaining was done with consummate skill. This communication barrier is bad for technology and for humanity. It means man is increasingly living in a world he does not understand.

In this volume care has been taken to communicate basic ideas in plain language, and wherever possible to impart a "feel" for some of today's technical problems, without delving so deep that the obscure jargon of the technologists becomes the only way to convey the information. Just occasionally the world population has become interested in some aspect of technology, such as manned flight to the Moon, and discovered that the subject is filled with fascination. Some people may have guessed that the same is true of all technology, could they only understand it.

To anyone fortunate enough to live and work in broad fields covering a lot of technology there is no doubt. Every technical development, every new way of harnessing matter and energy within the immutable laws of nature, is an additional chapter in the most exciting, most creative, and most beneficial story known to man.

Generally, the term "technologist" is – like "scientist" – used only by the layman. Workers would actually call themselves engineers, physicists, or geneticists, for example. The first group to undertake continuing research were probably those working in the field of medicine, where men have tried to find out about the body and its ailments for centuries. It was not until the 20th century that large numbers of researchers began to study such fields as dyestuffs, fertilizers, explosives, metal alloys, and radio. In World War II the growth of research and development (R and D) was meteoric. Not only was a fantastic amount accomplished but R and D was thrust into the public eye as a vital part of the national economy and recognized as something an advanced country needed in the fullest measure. Not until the 1960s was there increasing questioning of technology's directions and achievements.

In every area open to direct R and D the results have invariably been an endless succession of advances. Lighter and stronger or more heat resistant materials, cleverer semiconductor devices and microelectronics, more efficient conversion of heat into mechanical work, and thousands of other advances have been achieved. Even in the more difficult fields such as how living things really work, how light and other electromagnetic radiation actually behaves, and how the atom or the Universe are structured, as much is discovered each week as in the whole of the 19th century.

One of the chief results has been a vast increase in modern city dwellings and all that advanced people regard as a natural part of life: mechanical transport, telecommunications, hospitals with comprehensive equipment, and all the consumer products found in the home. This has not only made possible a frighteningly rapid growth in the Earth's population but it has greatly increased individual consumption of raw material and energy, and man's overall impact on the biosphere. Prehistoric man consumed only food and oxygen, but modern man has an insatiable appetite for almost all the elements of which our planet and its atmosphere are made. The rate at which mankind consumes this material is accelerating all the time. Many elements are being used up entirely.

It is gradually being learnt how to use material in ways that permit recycling, or at least repeated use of matter in different roles. This has a by-product in minimizing pollution, an objective in which progress

Left: the rise of the plastics industry had led to the problem of how to dispose of vast quantities of waste plastics that simply refuse to go away. Unlike paper and board, plastic will not rot. It is impractical and dangerous to burn it, as the amount is so great and the fumes are toxic. Much research effort is being put into developing biodegradable plastics.

is at last appearing to keep pace with growing problems. Better ways of using energy tend to cancel out the savings by increasing the demands. But the increasing use of atomic energy, and especially the fast reactor which "breeds" fresh fuel as it consumes the old, offer hope that man will avoid a global energy crisis as the workable reserves of oil finally run out early in the 21st century. Fast reactors could provide nearly all man's energy needs, but at present there are so many fears about both the safety of reactors and the problem of radioactive wastes that many people refuse to accept the assurances of the experts.

To a considerable degree it is technology's success that has caused problems. Enabling more babies to survive has multiplied the Earth's population. Increasing individual demands are consuming non-renewable resources too fast. In the face of such problems, perhaps technology's greatest achievement is to have unceasingly improved not only the length of human lives, but also their quality.

Above: Portugal's capital Lisbon suffers from all the problems of rapid urban growth to be seen in developing countries around the world. An expanding population and migration from countryside to cities had led to shanty towns cheek by jowl with high-rise blocks.

Right: an atomic reactor at Plumbrook, Ohio. The supporters of nuclear energy assert that it is a cheap alternative to fossil fuels as a source of power. But there are widespread, and partly justified, misgivings as to the safety of nuclear reactors.

The Challenge of Technology

Most non-technologists know more about technology's problems and controversies than about its successes. Invariably, an unvarnished success is judged to have little or no news value by the mass media. But almost all applied technology's end products are uncontroversial successes that never make the headlines. These successes tend to be forgotten as attention focuses on the controversies, which are a small though important minority.

As man's technical capability has expanded, the consequences of his developments have loomed ever larger. By midway through this century, he was able to exterminate most forms of life on Earth, and of rendering the planet uninhabitable. There are enough nuclear, chemical, and biological weapons to destroy the world many times over. The well founded warnings about pollution, genetic manipulation, and unemployment caused by automation are familiar to anybody with ears to listen. Man's difficulty lies in deciding priorities for technological development. For example, controlled thermonuclear power could be one of the greatest boons to mankind, but many laymen would probably jump to the conclusion that it ought never to be allowed to happen. This is because many people are not convinced that thermonuclear power can be safely controlled.

It has been repeated in this volume that each new technical discovery merely gives man fresh options he did not previously possess. Nuclear power can be one of man's greatest benefits, and as oil reserves dry up could enable him to maintain and even improve the prosperity of the Earth. It has an impressive safety record. As far as is known, the number of workers killed by atomic energy since World War II can be counted on the fingers of two hands. In the same period more than 5000 coalminers have died underground and countless others suffered from bronchial diseases. Yet there are those who would actually prefer coal to replace nuclear power! An even greater killer than coalmining is the automobile, and certainly the present road system is an imperfect one. But the benefits of mechanical road transport far outweigh any drawbacks.

When the first television program was broadcast in Britain in 1936 it was naturally hailed as a great new development in communications and entertainment. Television has since become one of the world's most powerful media for education and opening up the lives of less affluent people to the whole of human life and knowledge. In many countries TV is satiated with violence, triviality, and wholly misconceived advertising. But this misuse of the medium is no argument against its original development.

Today there are many new developments that might be misused. Automation is not one of these. Its proper use and control poses few serious problems and with a positive outlook and common sense there should be only very occasional local difficulties with unemployment. But what about genetic manipulation and several other dramatic developments in molecular biology? It is probably right to reduce the incidence of malformed or handicapped children and to maximize the production of healthy humans, but the possibilities for misuse are terrifying. Even such seemingly harmless developments as major control of the weather, or dramatic prolongation of human life by chemical (or other) control of the aging process,

Below: these Chicano migrants in Colorado live eight to a room. By United States and western standards, they are desperately poor. But they still own a TV set, which suggests that it has ceased to be a luxury in the west.

Below: the British Post Office's Prestel. Information can be received on an ordinary domestic TV set after minor adaptation. Information on a wide variety of topics is stored in a central computer.

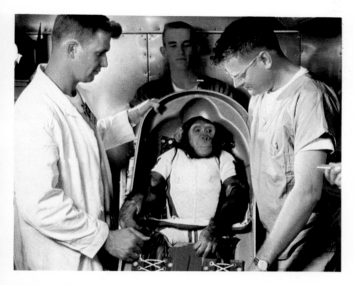

Above: Ham the chimpanzee was launched into space aboard a Mercury spacecraft in 1961. There are serious ethical doubts about the use of animals in research.

may bring serious unexpected problems. Again, what about breeding intelligent animals for use as low-grade labor? Man suffers few moral qualms over the traditional use of draft animals, but what about animals able to receive and act on orders? Molecular biology makes such a development likely in this century, and it could raise serious difficulties.

It is the apparent interference with brains and the innate nature of animals that is disturbing, just as in the reported use of tranquilizing drugs to pacify and control violent or uncooperative mental patients and prisoners. Some would say the argument hinged on whether the treatment was for the recipient's own good, while others would insist that the good of the community came first. Before long mankind can expect to be able to achieve direct electromechanical interaction between the human brain and computers, with dramatic speeding up in processing and probably a reduction in programing errors. By not very different techniques, it will be possible to impart information, in schools or anywhere else, by direct information recording on the brain. But, though in principle this is probably no different from traditional teaching, it raises fears about the manipulation of human beings, and possibly damage to their own character. In the ultimate extremity millions could come under the secret control of one megalomaniac individual.

It is clear that the future is going to demand human beings of great integrity, strong character, and firm decision-taking. Mankind should have sufficient faith in the human race to believe that the discoveries of the future will be used for good rather than for evil.

One of the skilled members of today's technology team is the technological forecaster. It is his or her job to predict what is likely to be accomplished by particular dates, in order to help in decision taking about investment and planning. Often it is a remarkably exact science, and without exception today's

forecasters expect the accelerating R and D trends to continue.

But the picture is always changing. Relatively less progress will be made in future by the "advanced" countries, for the simple reason that far more will be made by the others. There is scarcely a country that does not have sophisticated research laboratories in universities, industry, hospitals, and defense establishments. One has only to study the advanced technology in the exporting industries of such countries as South Korea, Hong Kong, Taiwan, and Singapore to appreciate that the original technological leaders may even be overtaken, especially where their economy is weak. To an increasing extent, man's discoveries and accomplishments will be limited by economic factors, which in turn are dictated largely by inflation.

Undoubtedly the first casualties will be the "big" projects whose direct benefit can be questioned. A second generation supersonic airliner to follow Concorde would be extremely difficult to launch, simply because the R and D costs would be so large. Likewise, more powerful machines for research into subatomic particles and other branches of nuclear physics are extremely costly and difficult to justify in terms of immediate benefit. (It may be that their benefit would be very great, but this cannot be predicted). Another questionable asset would be a giant astronomical observatory built in space. It could be started tomorrow but who would vote the funds?

The derogatory word "technomania" has been coined to describe the urge to build what are judged to be costly and glamorous creations of seemingly little immediate benefit to humanity. It is simple to adopt this posture, but deciding upon the question of true value to mankind is often extremely difficult. There are countless examples of past research that, contrary to what might have been predicted, have proved to be of very great and immediate benefit to all

Below: research laboratories are as much owned and operated by private industry as by governments. Laboratories are increasingly improving and inventing products, especially in food technology.

Above: research into a slew-winged aircraft. Developed at the NASA-Ames Research Center. The concept is to provide quiet supersonic flight by altering the geometry of the aircraft.

mankind. In future man is increasingly going to fail to know what he will be missing, because he cannot now afford to find out.

One of the few areas where this problem should be minimal is sociology, with related studies in psychology, crime, and other manifestations of human stress. This is because R and D in these fields is relatively cheap, and the problems are growing and becoming better identified. Much greater effort will be applied to solving the problems of tomorrow's cities, which are not only the obvious ones of structures, transport, energy use, and waste disposal.

Equally important to consider will be the influence of the urban environment on the inhabitants; the effects on culture and human behavior, and the startling variation between cities in such matters as crime, suicide rate, mental health, vandalism, and public attitudes to various kinds of change. The rejection of the modern urban world and the search for a simple lifestyle making minimal demands on the Earth's resources is an extreme view, but much more research is needed to find an optimum lifestyle for the future. At present, virtually no coherent attempt is being made anywhere to minimize consumption of resources or creation of pollution. In turn this is at least partly because technical problems usually rate poorly in a list of political priorities.

Ten years ago there were numerous exciting forecasts of the future of technology, nearly all of which took it for granted that what was technically possible would come about. By 1990 the more advanced countries were expected to have an automated environment that would guide road traffic (including blind pedestrians), link every shop and bank (to verify customers' credit-rating), translate from one language to another and printout newspapers or other information in the home. The votes of all adults would be instantly recorded, and sufferers would be enabled to

diagnose themselves (a very tricky process because the doctor knows what symptoms to ignore). Limitless other useful tasks would be taken care of. Today we are less starry-eyed, and though automation, along with other branches of technology, is ceaselessly moving forward, many major developments are moving at a snail's pace, or even in reverse, because of soaring costs.

Among the few major technologies where relentless pressure demands sustained progress, energy sources and conversion processes rank very high. Man's energy consumption is increasing so rapidly, in a world where fossil fuels are subject to absolute limits, that future technology must try even harder than in the past to reduce the rate of consumption growth, find alternative sources and, to the greatest degree possible, use available sources such as sunlight much more fully and efficiently.

Probably the most helpful spur to future technology is the ever-increasing understanding of matter, at the molecular level and below. This assists mankind's knowledge of both inanimate materials and living cells. At this level there is close kinship between the worker in microelectronics, engineering structures, and human tissues or nervous communications. With greater understanding comes greater capability, greater power, and greater responsibility in decision taking.

Above: molecular models of complex crystals. The increasing understanding of matter at the molecular level and below has been of great assistance in providing knowledge of inanimate materials and living cells.

Left: fossil fuels are being consumed at an increasing rate, and will one day be exhausted. There is only about 30 years' worth of known oil reserves left. These houses use the free energy around them. All their needs are provided for by solar and wind power.

Below: thirty years ago the Negev Desert was a wilderness. But Israel had a desperate need for agricultural land. Modern technology was put to work and a vast irrigation scheme started. Today, the Negev Desert is one of the most fertile regions of Israel.

INDEX

Opposite: the ISEE-B Sun/
Earth explorer satellite was
built by the German company
Dornier to explore the
magnetosphere and adjacent
interplanetary space. Satellite
technology has enabled
humanity to discover more
about the Earth and the
surrounding Universe.

References in *italics* refer to captions

A

A10 Thunderbolt II aircraft, 241
abacus, 260
Abbe, Ernst, 122, *122*
abrasives, 14
accumulators, 182
acoustics, *147*, 150–1, *150*
    missile homing systems, 234
acrylics, 210
    see also plastics
Advanced Passenger Train (APT),
    47–8, *47*
advertising, *255*
aero engines, 88
    mountings, 92
    on supersonic aircraft, 96
    use of carbon fibers, 213
    used on racing cars, 28
Aérotrain system, 49, *49*
afterburners, 90
    on supersonic aircraft, 96
agriculture
    aviation applications, 102, *103*
    benefits of research, *14*, 15
    early, 10
    early mechanization, 11
    early plows, 200, *200*
    Landsat information, 295
air-cushion vehicles, 80–1, *70, 71*
    see also hovercraft; tracked air
    cushion vehicles
air traffic control, *94*, 109–11, *110*
Airbus A300, 95
aircraft
    cargo, 98–9
    cybernetics in designing of
        controls, 270
    fuselage design, 93–4
    general aviation, 100–3
    increase in size, 15
    inertial navigation systems, 72
    intelligence-gathering, *250*, 251
    maintenance and inspection, 95
    military, 238–41
    noise levels, 169–71, *171*
    structural materials, 206, *207*
    structures design, 95
    subsonic, 92–5
    supersonic, 96–7
    swing-wing, *318*
    VTOL and STOL, 106–7
    wing design, 93
aircraft carriers, *247*, 248
airfields, construction techniques
    applied to roads, 32
    airfoils
    design, 93
    on racing cars, 29, *29*
airline booking systems, 263
airlines
    aircraft usage rate, 95

development, 88, *88*
airports
    ground control, 110
    layout, 108, *109*
    noise problems, 170–1
airspace, utilization at three levels, 110
alkyds, 210
alloys, space applications, 13
Alouette helicopters, 105
alphabets, 256
alumina, 203
Aluminaut submersible, 79
aluminum
    properties, 206
    sheet-rolling, *207*
Alvin submersible, 79
Ambassador bridge (Detroit), 51
American Indians, use of the travois, 22
amphibious aircraft, 102
amphibious vehicles, *36*, 37, *37*
    see also air-cushion vehicles
analog computers, 260–1
Analog Threshold Logic (ATL), 152
animals
    transporting, *30*, 31
    see also draft animals
answering machines, 153
antennae, 273
antiaircraft guns, 227
    on ships, 247
antiaircraft weapons, *226*
    guided missile systems, 237
antiarmor gun, 227
antitank missiles, 237, *237*
Antonov An-10, -12 and -22 aircraft,
    99
apartment buildings, 19
apogee, 290
Apollo program, 285, *299*, 302, *302*
    Applications Program, 308
    use of hydrox cells, 183
    use of Lunar Rover vehicles, 37, *37*
aqualungs, 80, 81
"aquaplaning" on wet roads, 33
Arabs, early inventions, 11
arc lamps, 129
arch bridges, 50
archer, demonstrating energy
    conversion, *174*
argon lamps, 129, *129*
Argosy aircraft, 99
Ariel satellites, 311
Aristotle, on nature of light, 116
arithmetic, 10
armor
    antiarmor gun, 227
    early countermeasures, 220
    for naval warfare, 246
    in sea warfare,   21
    on vehicles, 221
    personal, reasons for obsolescence,
        221
armored cars, 242, 244
armored fighting vehicles, 244
armored personnel carriers, 244, *244*
armored vests, *207*
Arnold, Dr. Thomas, 23

art, use of spectroscopy, *130*, 131
articulated trucks, 31
artillery, 226–7
    early pieces, 221
    see also firearms
aspect ratio, of aircraft wings, 93
astigmatism, 119
astronavigation, 72
astronomy, 11
    Copernicus' theories, 11
    optical, 120–1
    use of photography, 127
    use of spacecraft, 310, *311*
    use of spectroscopy, 131, *131*
Atlas rocket, *287*
Atomic Energy Authority (UK), *217*
atomic physics
    influence on materials science, 201
    use of photomultipliers, 132
    use of spectroscopy, 131
atomic radiation, detecting, 132
atomic structure, 12
Audi cars, 27
Australia
    "beef trains," 31
    differing railroad gauges, 39, *39*
Auto Union cars, 29
autogiro, 104
automatic braking, on trains, 40
automatic control, 255, *255*
automatic weapons, 224–5
automation, 11
    of trucks and tools, 270
    positive benefits, 316
    social implications, *262*, 264
automobiles, 24–7
    batteries, 182
    body design development, 26
    city cars, 54, 56
    early
        based on carriages, 24
        design standardized, 25
    electronic ignition, 215
    engine position, 26
    first car built, 24
    front-wheel-drive, *25*. 26
    limited success, 15, *16*
    mass-production, 11, 25
    misuse of resources, *16*, 17, *19*
    racing, 28–9, *28, 29*
    safety glass for windshields, *204*, 205
    safety in design, *26*
    small car design, 56
    taxation, 25
autopilots, 264
    in private aircraft, 100
    on hydrofoils, 68
AV8B aircraft, 241
avionics, in private aircraft, 100
axle loading, on freight trains, 45
azimuth, 72

B

B-52 aircraft, 240
B-58 bomber, 96
Babbage, Charles, 260, *260*
BAC 111 aircraft, 89
    use in general aviation, 101, *101*
BAC 146 aircraft, 95
Baekeland, Leon, 209
Baird, John L., 274
Bakelite, 209
ballast
    for railroads, 40
    on inland ships, 67
ballista, 220-1
balloons
    hot air, 88, *88*
    military applications, 238, *238*
barges, on inland waterways, 67
Barton, Otis, 76, *76*
bathyscaphes, 76
bathyspheres, 76, *76*, 77
batteries, 182
    for electric cars, 56-7
battering rams, *220*
Battersea Power Station (London), *179*
battle-cruisers, 246
Beebe, William, 76, *76*
behavioral studies, 16-17
Bell, Alexander Graham, *276*, 277
Bell and Hiller helicopters, 105
Bell Telephone Labs
    semiconductor research, 214
    speech recognition research, 152
Bell X-22A aircraft, 106
Bell XS-1 research aircraft, 96
"bends" (affecting divers), 80
Berliner, Emile, 156
beryllia, 203
beryllium fibers, 213
Big Bird satellites, 250
binary numbers, 261-2
biological warfare, 230-1, *230*
blind landing facilities, 108
blind people, ultrasonic aid, 165
Blue Riband, 64
boatbuilding, use of plywood, 212, *212*
bobsleds, 36
Boeing KC-97 Stratofreighters and
    KC-135 Stratotankers, 99
Boeing 707, 727, 737 and 747
    aircraft, 89, 92, *94*, 99
    noise reduction, 171
Boeing 747-200F, 99
Boeing 757 and 767 aircraft, 92, *93*
Boeing Stratocruiser aircraft, 89
Boeing YC-14 aircraft, 107, *107*
bola (South American weapon), 220
bomber aircraft, early types, 239
bombs, 229
boomerang, 220, *220*
boron fibers, *12*, 13, 213
bow and arrow, 220
Braille, 259
braking

hydrokinetic, 47-8
    on freight trains, 45
breakwaters, model testing, 135
breathing equipment for divers, 80, *80*
Bréguet, Louis, 104
bridges, 50-1
    pontoon, 67
bridle-chord bridges, 51
Bristol 170 aircraft, 98
Bristol Jupiter aero engine, 88
Britannia bridge (Menai Straits), 50
British Empire, contributing to growth
    of passenger liners, 64
bromine, extracted from sea, 83
bronchoscopes, *129*
bronze, discovery, 200, *200*
Browning, John M., 223
Brunel, Sir Marc, 53
Bryan, Michael, 167
"bubble cars," *55*, 56
built-in test equipment (BITE), 95
bulk cargoes, 62
bulk carriers, 63, *63*
bulk tankers, 31
bullets, 221, 222
buoys, 75
buses, 30, *30*
business aviation, 100

C

C-Bahn (Cabin taxi) system, 55, *55*
cablecars, 43
Cabtrack system, 55
Cahill, Thaddeus, 155
calculators, 214
    powered by photoelectricity, 133, *133*
Calder Hall nuclear power station, *193*
camber on roads, 32
cameras, 124-7
Canadair CL-215 aircraft, 102, *102*
canals
    barges, 66
    development, 23, *23*
    to inland ports, 75
    tunnels, 52
cancer, laser surgery, 138, *139*
canning operations, 266
cannon, 221
cantilever bridges, 50
    record spans, 51
Caravelle aircraft, 89, *89*, 92
    rear engine mounting, 92, *92*
carbon dioxide lasers, 137
    military applications, 140
carbon fibers, 213, *213*
    use in propeller blades, 90
cargo vessels, potential of
    submarines, 79
cargoes
    classification, 62
    handling at ports, 74
Carothers, Wallace Hume, 209
Cassegrain telescope, 121

cassette tape recorders, 156
    see also tape recording
catenaries, 43
cathode ray displays, linked to
    sonar, 165
Caus, Salomon de, 184
cavitation, 161-2, *161*
cellophane, 209
cellular biology, 14
celluloid, 209
cellulose, as natural polymer, 208-9, *208*
ceramics, industrial applications, 202-3
Cessna 172 aircraft, 100, *100*
Challenger aircraft, 101
Challenger, HMS, 83
change, rate of
    discerning trends, 18
    in human environment, 17
Channel Tunnel, 53
chariots, Egyptian, 22, *22*
chassis, automobile, 27
chauffeurs, 25
chemical warfare, 230-1, *230*
Chesapeake Bay Bridge-Tunnel, 53, *53*
chess games, computerized, 215
Chicago, USS, 219
Chinese
    development of harness, 22
    early inventions, 11
chlorine gas, 230
Christie, J. Walter, 243
chromatic aberration, 120
"Chunnel," 53
Cierva, Juan de la, 104
ciphers, 251
cities
    economic problems of
    maintenance, 54
    limitations imposed on transport
      systems, 54, 55
    new, construction of, 57, *57*
    see also urban travel
Citroën car company, 26
civil engineering
    materials science applications, 13
    see also bridges
            tunnels
cleaning, ultrasonic, 161
Clipper ships, 61
cloverleaf road junctions, 35
clutch (automobile), 25
coaches, *see* railroads: coaches
coal
    formation, 176
    reserves, 177
    used in power generation, 179, *179*
coal mining
    automated railroad, 41
    cost in human terms, 316
    early records, 176
    in 1790s, *176*
    new techniques, 177
coaxial cable, 278, *278*
cobalt, from sea bed, 83
Cockerell, Sir Christopher, 70, *70*
codes, 251
collisions

liquid and gas coolants, 192–3
nuclear weapons, 232–3, *232*
numbers, use of 254
numerals, arabic, 256
nylon, 209, *209*

O

oars, propulsion by, 60
oceanography
    satellite information, 295
    use of bathyspheres, *77*
Odeillo solar furnace, *9*, 186, *186*
offroad racing (ORR), 29
office buildings, 19
oil
    consumption rate, 177
    from offshore fields, 82
    North Sea production platform, *59*
    reduced supplies, encouraging
        quieter transport methods, 169
    used in power generation, 179
Olson, Harry F., *152*
Omega navigation system, 72, 112
onager, 221
one-way streets, 34
Onnes, Heike Kamerlingh, 216
optical fibers, 139, *139*, 204–5
    use in voice transmission, 278
Orbiter space shuttle, 304, *305*, 307
Orbiting Astronomical Observatory
    (OAO), 310, *310*
Orbiting Solar Observatories (OSO),
    310
Orlon, 209–10
oscilloscopes, *148*
Otto, Nikolaus, 28
oxide ceramics, 203

P

Pacific, mining consortia interests, 84
packaging systems, automation of,
    265–6
paddlewheels, 61, 66
Paget, Sir Richard, 153
Palomares (Spain), loss of hydrogen
    bomb, 76, 79
pantographs, 43
parabolic mirrors, in solar furnaces, 186
paraffin, use in fuel cells, 183
Pascal, Blaise, 260
passenger aircraft, 88, *89*
passenger liners, 64–5, *64*
pedestrian crossing signals, *34*, 35
perceived noise decibel (PNdB), 168
percussion caps, 222
perigee, 290
permeable membranes, 211
PETN (pentaerythritol tetranitrate), 229
petroleum

political effects, 22
    tankers, 62, *63*
pharmaceutical industry, 15
photocells, 132
    in light meters, 125
    in robot tools and toys, 270
    use in manufacturing industry,
        132–3, *133*
photochromic lenses, 205
photoconductivity, applied to
    photocopiers, 259
photocopiers, 259
photoelectricity, 117, 132–3
photography, 124–7
    application to printing, 258–9
    color, 124, 126
    from reflecting telescopes, 121
    high speed, 125
    reconnaissance, *251*
    use in astronomy, 127
photogravure, 259
photomultipliers, 132, *132*
photons, 117, 132
Piccard, Auguste, 76
pickups, sound reproduction
    equipment, 156–7
Picturephone, *279*
pilotage, 72
Pioneer space probes, 296
pipe-laying, using submersible craft, 78
pipelines, from offshore oil and gas
    fields, 82, *82*
piracy, elimination of, 61
pistols
    automatic, *225*
    self-loading, 223
pitch (sound), 149
Plainview, USS, *69*
planets
    orbits, 291
    probes, 296, *296*
plasma rockets, 289
plastics, 201
    color availability, 210, *210*
    disposal of waste, *314*
    for ophthalmic lenses, 119
    forming processes, 211
    misuse of resources, 17
    molecular structure, 208, *208*
plates, photographic, 124
platforms, at railroad stations, 44, *45*
Plato, on nature of light, 116
pleasure boats, 66
Plimsoll, Samuel, 62
Plimsoll line, 62
Pluto (planet), 127
plutonium in fast-breeder reactors, 193
plywood, 212, *212*
POE vertical platform (diving
    operations), 81
point-contact, *214*
polar exploration vehicles, 36
Polaris missiles, 236, *248*, *249*
    propellant, 288
polaroid cameras, 126, *126*
pollution
    by oil, *17*

    from automobiles, 27, *27*
    problems, 11
    see also noise; waste materials
polyethylene, 210
polymers, 208–11
    natural, 208–9
    synthetic, 209
polytetrafluoroethylene (PTFE), 210
polyvinylchloride (PVC), 210
    health hazard, *201*
Pompidou Center (Paris), *199*
pontoon bridges, 67
pop music, 155
Popovich, Pavel, 298
population growth, 314, *315*
population movement, during
    Industrial Revolution, 11
portal bridges, 51
ports, 74–5, *74*
    inland, 75, *75*
Poseidon missile, propellant, 288
Post Office, underground railway, 40–1
Post Office Tower (London), *15*
potential energy, 174, *174*
powder trucks, 31
power generation,
    see electrical power generation
powersats, 196–7, *196*
Pratt and Witney engines, 61
printing
    color, 259
    early process, 257
    from clay tablets, *254*
    from stored metal type, 258
    moveable type, *10*, 11, 254
private flying, 100, *100*
process-control, *265*, 266, *266*, *267*
propellants
    for space shuttle, 305
    rocket, 286
        solid, 288
propeller shafts (automobiles), 25
propellers
    aircraft, 90, *90*, *91*
    for marine ACVs, 70
    on ships, 61
        on inland waterways, 67
    problems for hydrofoils, 69
prosthetics, 268–9, *269*
    cybernetic applications, 271, *271*
public transport vehicles, 54–5
Pyke, Geoffrey, 212
pykrete, 212
pyramids, 10
Pythagoras
    on nature of light, 116
    study of music, 154, *154*

Q

quadraphonic sound recording, 158
quartz-halogen light bulbs, 128–9
Quebec rail bridge, 51
Queen Elizabeth, liner, 64, *65*

PICTURE CREDITS

100   Robert Estall
101(T)   Dr Alan Beaumont
101(C)   M. Hardy/*Daily Telegraph* Colour Library
101(B)   Cessna Aircraft Company
102   Canadair Ltd.
103(T)   Aeropix, New Zealand
103(C)   Shell Photo Service
103(B)   Zefa
104(T)   *Radio Times* Hulton Picture Library
104(B)   Novosti
105(L)   Sikorsky Aircraft
105(R)   Zefa
106(T)   Popperfoto
106(B)   Photri
107(T)   Hawker Aircraft Company
107(B)   U.S. Air Force photo
108   I.B.A.
109(T)   Alan Benstead © Aldus Books
109(BL)   Central Office of Information/Plessey Navaids
109(BR)   Internationales Bildarchiv
110(L)   Photri
111   Aerofilms
112(T)   *Radio Times* Hulton Picture Library
112(C)   Photri
112(B)   Internationales Bildarchiv
113(T)   Lockheed Aircraft Corporation
113(B)   Günter Heil/Zefa
114   Paul Brierley
116(T)   Photo J.-L. Charmet
116–117(B)   © Aldus Books
117(TL)   Aldus Archives
117(TR)   Photri
118(T)   Institute of Ophthalmology, London
118(B)   Sidney W. Woods © Aldus Books, after A. B. McNaught and R. Callender, *Illustrated Physiology*, E. & S. Livingstone Ltd., Edinburgh, 1964
119(TL)   Seminario Cescovlie of Treviso/Fiorentini, Venice
119(TR)   Popperfoto
119(B)   Syndication International
120(L)   Courtesy Director General, Monumenti, Musei e Gallerie Pontificie, Photographic Archives of the Vatican Museums and Galleries
121(L)   Rex Features
121(R)   Robert Estall
122   Carl Zeiss, Oberkochen, West Germany
123(L)   United Kingdom Atomic Energy Authority
123(TR)   Paul Brierley
123(BR)   Photo R. W. Newman and J. J. Hren, E. & I. Experimental Station, Florida
124(T)   Ernst Leitz Wetzlar GmbH
124(BL)   Science Museum, London
124(BR)   © Aldus Books
125   *Daily Telegraph* Colour Library
126–127(T)   Ernst Leitz Wetzlar GmbH
126(BL)   Polaroid (U.K.) Ltd.
126(BR)   Pat Morris/*Daily Telegraph* Colour Library
127(B)   United States Information Service, London
128(L)   *Radio Times* Hulton Picture Library
128(R)   Central Office of Information/Precision Grinding Ltd.
129(L)   Robert Estall
129(TR)   Photo Barry Richards © Aldus Books
129(BR)   Optec Reactors Ltd., London
130(T)   Howard Sochurek/John Hillelson Agency
130(B)   *Radio Times* Hulton Picture Library
131   Photo Robert P. Kirschner, The University of Michigan, Ann Arbor
132(L)   Reproduced by permission of the Trustees of the British Museum
132(R), 133(T)   Paul Brierley
133(B)   Kaleidoscope
134(L)   Science Museum, London
134(R)   Paul Brierley
135(T)   Photri
135(B)   © Aldus Books
136(L)   Hughes Aircraft Company
136(R)   © Aldus Books
137(T)   G.P.O.
137(B)   Paul Brierley

138(L)   International Research & Development Co. Ltd.
138(R)   B.B.C. copyright photo
139(L)   *World Medicine*
139(TR)   Paul Brierley
139(B), 140(T)   © Aldus Books
140(C)   Compix
140(B)   Photri
141(R)   Ferranti Ltd.
142(T)   Curtis Publishing Company
142(B)   A. Howarth/*Daily Telegraph* Colour Library
143(T)   Lawrence Livermore Laboratory
143(B)   Photri
144(T)   Aldus Archives
144(B)   © Aldus Books
145(T)   Compix
145(B)   Internationales Bildarchiv
146   Robert Harding Associates
148   Paul Brierley
149(T)   John Messenger © Aldus Books
149(B)   Photos Dr. Hans Jenny, Dornach
150(T)   Lincoln Center for the Performing Arts, New York
150(C)   J. Allan Cash
150(B)   Bell Telephone Laboratories, Murray Hill, New Jersey
151   Keystone
152(L)   Courtesy of RCA Laboratories
152(R)   British Crown Copyright. Central Office of Information
153   *Science Journal*
154(T)   Internationales Bildarchiv
154(B)   David Paramor Collection, Newmarket
155(L)   Keystone
155(R)   Spectrum Colour Library
156(L)   British Crown Copyright. Science Museum, London
156(R)   E.M.I. Ltd., London
157(T)   M. Goddard/*Daily Telegraph* Colour Library
157(B)   E.M.I. Records Ltd.
158(T)   Sony (U.K.) Ltd.
158(B)   Phil Stern/Globe Photos Inc., London
159(T)   Camera Press Ltd.
159(B)   Decca Record Company Ltd.
160(L)   Paul Brierley
160(R)   © Aldus Books
161(T)   Bavaria-Verlag
161(BL)   British Crown Copyright. National Physical Laboratory, Teddington
161(BR)   Brown & Goodman, *High Intensity Ultrasonics,* Iliffe Books Ltd., London
162(L)   Ultrasonics Ltd.
162(R)   Photo Ken Coton © Aldus Books, courtesy Upjohn Limited, Crawley, Sussex
163   Courtesy Faculty of Medicine, The University of Tokyo
164(T)   after D. G. Tucker, *Sonar in Fisheries,* Fishing News (Books) Ltd., London
164(B)   Wideroe's Flyveselskap A/S, Oslo
165(L)   U.S. Navy photo
165(R)   Courtesy Sikorsky Aircraft, Stratford, Connecticut
166(L)   Electroacoustics and Automation Laboratories, Centre de Recherche Physique, Marseilles
166(R)   Photo by permission of Dr. E. M. Bryan, University of Salford
167(T)   Paul Brierley
167(B)   Rex Features/SIPA Press
168–169(T)   Syndication International
168(B)   Popperfoto
169(B)   Paul Brierley
170(L)   Novosti
171(TL)   Photo J. Edward Bailey © 1965 Time Inc.
171(TR)   Courtesy Rolls-Royce Limited, Derby
171(B)   Spectrum Colour Library
172   Robert Harding Associates
174(L)   Picturepoint, London
174(R)   Robert Estall
175(L)   Gillian Newing © Aldus Books
175(R)   Behram Kapadia © Aldus Books

176(T)   Gulf Oil Corporation, Pittsburgh
176(B)   Walker Art Gallery, Liverpool
177(L)   American Gas Association
177(R)   Photo The Gas Council
178(T)   Allgemeine Elektrizitats-Gesellschaft
178(B)   British Crown Copyright. Science Museum, London
179   Robert Estall
180(T)   British Electrical Development Association
180(B)   British Crown Copyright. Central Office of Information
181(L)   J. G. Mason/*Daily Telegraph* Colour Library
181(R)   British Crown Copyright. Central Office of Information
182(L)   Ian Kestle © Aldus Books
182(R)   Aldus Archives
183(L)   Lent to the Science Museum, London, by the Smithsonian Institution, Washington, D.C.
183(R)   Central Office of Information/Chiloride Alcao Ltd.
184(T)(C)   Grumman Corporation
184(B)   Reproduced by permission of the Trustees of the British Museum
185   Novosti
186(L)   Central Office of Information/Department of Physics, Heriot Watt University, Edinburgh
187(T)   Zefa
187(BL)   Photri
187(BR)   *Daily Telegraph* Colour Library
188(T)   U.S. Department of Energy
188(B)   Dr. Alan Beaumont
189(T)   Harwell Design Studio/*Daily Telegraph* Colour Library
189(B)   Central Office of Information/Department of Mechanical Engineering, Edinburgh University
190(T)   Electricité de France, Paris
190(B)   © Aldus Books
191(T)   Ste Larderello, Pisa
191(B)   Morton Beebe & Associates/Photo Researchers Inc.
192(L)   Ian Kestle © Aldus Books
192(R)   U.S. Navy photo
193(T)   Ian Kestle © Aldus Books
193(B)   United Kingdom Atomic Energy Authority
194(T)   Homer Sykes/*Daily Telegraph* Colour Library
194(B)   Sidney W. Woods © Aldus Books
195(T)   Lawrence Livermore Laboratory
195(B)   Syndication International
196   Photri
197(L)   Hughes Aircraft Company
197(R)   British Crown Copyright. Central Office of Information
198   Eric Crichton/Bruce Coleman Ltd.
200(T)   *Radio Times* Hulton Picture Library
200(B)   Reproduced by permission of the Trustees of the British Museum
201   John Moss/Colorific!
202(T)   Bell Telephone Laboratories, Murray Hill, New Jersey
202(B)   © Aldus Books
203(T)   British Iron and Steel Federation
203(B)   United Kingdom Atomic Energy Authority
204(TL)   Pilkington Brothers Ltd.
204(BL)   Triplex Safety Glass Co. Ltd.
204–205(C)   *Daily Telegraph* Colour Library
205(R)   Maurice Broomfield, F.I.I.P., F.R.P.S.
206   Paul Brierley
207(TL)   Popperfoto
207(TR)   Courtesy of Kaiser Aluminium and Chemical Corporation
207(CB)   Ling-Temco-Vought
207(BR)   Sulzer Bros.
208(L)   Eldenbenz und Eglin, Basel
208(R)   Paul Brierley
209(L)   Spectrum Colour Library
209(R)   E. I. Du Pont de Nemours & Co., Inc.
210(T)   J. R. Geigy, S.A., Basel
210(B)   Shell Photo Service
211(L)   British Industrial Plastics Ltd. (a